Thermal Analysis Kinetics for Understanding Materials Behavior

Thermal Analysis Kinetics for Understanding Materials Behavior

Special Issue Editor
Sergey Vyazovkin

MDPI • Basel • Beijing • Wuhan • Barcelona • Belgrade • Manchester • Tokyo • Cluj • Tianjin

Special Issue Editor
Sergey Vyazovkin
University of Alabama at Birmingham
USA

Editorial Office
MDPI
St. Alban-Anlage 66
4052 Basel, Switzerland

This is a reprint of articles from the Special Issue published online in the open access journal *Molecules* (ISSN 1420-3049) (available at: https://www.mdpi.com/journal/molecules/special_issues/thermal_analysis_kinetics).

For citation purposes, cite each article independently as indicated on the article page online and as indicated below:

LastName, A.A.; LastName, B.B.; LastName, C.C. Article Title. *Journal Name* **Year**, *Article Number*, Page Range.

ISBN 978-3-03936-559-3 (Hbk)
ISBN 978-3-03936-560-9 (PDF)

© 2020 by the authors. Articles in this book are Open Access and distributed under the Creative Commons Attribution (CC BY) license, which allows users to download, copy and build upon published articles, as long as the author and publisher are properly credited, which ensures maximum dissemination and a wider impact of our publications.

The book as a whole is distributed by MDPI under the terms and conditions of the Creative Commons license CC BY-NC-ND.

Contents

About the Special Issue Editor vii

Preface to "Thermal Analysis Kinetics for Understanding Materials Behavior" ix

Tatsiana Liavitskaya and Sergey Vyazovkin
All You Need to Know about the Kinetics of Thermally Stimulated Reactions Occurring on Cooling
Reprinted from: *Molecules* **2019**, *24*, 1918, doi:10.3390/molecules24101918 1

Nicolas Sbirrazzuoli
Advanced Isoconversional Kinetic Analysis for the Elucidation of Complex Reaction Mechanisms: A New Method for the Identification of Rate-Limiting Steps
Reprinted from: *Molecules* **2019**, *24*, 1683, doi:10.3390/molecules24091683 17

Ina Keridou, Luis J. del Valle, Lutz Funk, Pau Turon, Lourdes Franco and Jordi Puiggalí
Non-Isothermal Crystallization Kinetics of Poly(4-Hydroxybutyrate) Biopolymer
Reprinted from: *Molecules* **2019**, *24*, 2840, doi:10.3390/molecules24152840 33

Yui Yamamoto and Nobuyoshi Koga
Thermal Decomposition of Maya Blue: Extraction of Indigo Thermal Decomposition Steps from a Multistep Heterogeneous Reaction Using a Kinetic Deconvolution Analysis
Reprinted from: *Molecules* **2019**, *24*, 2515, doi:10.3390/molecules24132515 57

Jiří Málek and Roman Svoboda
Kinetic Processes in Amorphous Materials Revealed by Thermal Analysis: Application to Glassy Selenium
Reprinted from: *Molecules* **2019**, *24*, 2725, doi:10.3390/molecules24152725 75

Nikita V. Muravyev, Alla N. Pivkina and Nobuyoshi Koga
Critical Appraisal of Kinetic Calculation Methods Applied to Overlapping Multistep Reactions
Reprinted from: *Molecules* **2019**, *24*, 2298, doi:10.3390/molecules24122298 91

Evangelia Tarani, George Z. Papageorgiou, Dimitrios N. Bikiaris and Konstantinos Chrissafis
Kinetics of Crystallization and Thermal Degradation of an Isotactic Polypropylene Matrix Reinforced with Graphene/Glass-Fiber Filler
Reprinted from: *Molecules* **2019**, *24*, 1984, doi:10.3390/molecules24101984 107

Zoi Terzopoulou, Evangelia Tarani, Nejib Kasmi, Lazaros Papadopoulos, Konstantinos Chrissafis, Dimitrios G. Papageorgiou, George Z. Papageorgiou and Dimitrios N. Bikiaris
Thermal Decomposition Kinetics and Mechanism of In-Situ Prepared Bio-Based Poly(propylene 2,5-furan dicarboxylate)/Graphene Nanocomposites
Reprinted from: *Molecules* **2019**, *24*, 1717, doi:10.3390/molecules24091717 125

Bertrand Roduit, Charles Albert Luyet, Marco Hartmann, Patrick Folly, Alexandre Sarbach, Alain Dejeaifve, Rowan Dobson, Nicolas Schroeter, Olivier Vorlet, Michal Dabros and Richard Baltensperger
Continuous Monitoring of Shelf Lives of Materials by Application of Data Loggers with Implemented Kinetic Parameters
Reprinted from: *Molecules* **2019**, *24*, 2217, doi:10.3390/molecules24122217 143

Walid M. Hikal and Brandon L. Weeks
Non-Isothermal Sublimation Kinetics of 2,4,6-Trinitrotoluene (TNT) Nanofilms
Reprinted from: *Molecules* **2019**, *24*, 1163, doi:10.3390/molecules24061163 . . . . **169**

Yong Joon Lee and Brandon L. Weeks
Investigation of Size-Dependent Sublimation Kinetics of 2,4,6-Trinitrotoluene (TNT) Micro-Islands Using In Situ Atomic Force Microscopy
Reprinted from: *Molecules* **2019**, *24*, 1895, doi:10.3390/molecules24101895 . . . . **177**

Pietro Bartocci, Roman Tschentscher, Ruth Elisabeth Stensrød, Marco Barbanera and Francesco Fantozzi
Kinetic Analysis of Digestate Slow Pyrolysis with the Application of the Master-Plots Method and Independent Parallel Reactions Scheme
Reprinted from: *Molecules* **2019**, *24*, 1657, doi:10.3390/molecules24091657 . . . . **187**

Musfirah Zulkurnain, V.M. Balasubramaniam and Farnaz Maleky
Effects of Lipid Solid Mass Fraction and Non-Lipid Solids on Crystallization Behaviors of Model Fats under High Pressure
Reprinted from: *Molecules* **2019**, *24*, 2853, doi:10.3390/molecules24152853 . . . . **203**

About the Special Issue Editor

Sergey Vyazovkin received his Ph.D. from Belorussian State University (1989). His research is concerned with the kinetics of thermally stimulated processes in condensed phase systems. He is a winner of Mettler-Toledo Award in thermal analysis and the James J. Christensen Award in calorimetry. Prof. Vyazovkin is Editor of *Thermochimica Acta* and a member of the editorial boards of *Macromolecular Rapid Communications, Macromolecular Chemistry and Physics, Molecules, Polymers*, and *Thermo*. He is Chair of the Kinetics Committee of the International Confederation for Thermal Analysis and Calorimetry. His research has been published in one book, five book chapters, and 200 peer-reviewed papers that have been cited over 18,000 times.

Preface to "Thermal Analysis Kinetics for Understanding Materials Behavior"

All materials respond to temperature. They also exist and function within limited timeframes. The science of kinetics combines time and temperature into a single framework that allows one to understand the thermal behavior of materials. This book provides a collection of papers dealing with the kinetic analysis of data obtained by differential scanning calorimetry (DSC) and thermogravimetric analysis (TGA). The papers are, collectively, very diverse in their content. Some provide general overviews of the kinetic approaches while other focus on advanced applications of the kinetic techniques to specific materials. However, they all adequately reflect the current state of affairs in the modern kinetics of thermally stimulated processes.

Sergey Vyazovkin
Special Issue Editor

Review

All You Need to Know about the Kinetics of Thermally Stimulated Reactions Occurring on Cooling

Tatsiana Liavitskaya and Sergey Vyazovkin *

Department of Chemistry, University of Alabama at Birmingham, 901 S. 14th Street, Birmingham, AL 35294, USA; tliavi@uab.edu
* Correspondence: vyazovkin@uab.edu

Received: 1 May 2019; Accepted: 17 May 2019; Published: 18 May 2019

Abstract: In this tutorial overview article the authors share their original experience in studying the kinetics of thermally stimulated reactions under the conditions of continuous cooling. It is stressed that the kinetics measured on heating is similar to that measured on cooling only for single-step reactions. For multi-step reactions the respective kinetics can differ dramatically. The application of an isoconversional method to thermogravimetry (TGA) or differential scanning calorimetry (DSC) data allows one to recognize multi-step kinetics in the form of the activation energy that varies with conversion. Authors' argument is supported by theoretical considerations as well as by experimental examples that include the reactions of thermal decomposition and crosslinking polymerization (curing). The observed differences in the kinetics measured on heating and cooling ultimately manifest themselves in the Arrhenius plots of the opposite curvatures, which means that the heating kinetics cannot be used to predict the kinetics on cooling. The article provides important background knowledge necessary for conducting successful kinetic studies on cooling. It includes a practical advice on optimizing the parameters of cooling experiments as well as on proper usage of kinetic methods for analysis of obtained data.

Keywords: activation energy; Arrhenius equation; cooling; crosslinking; decomposition; isoconversional method; model-free kinetics; rate constant

1. Introduction

This article summarizes our experience in studying the kinetics of thermally stimulated reactions taking place during continuous cooling. We have studied the kinetics of thermal decomposition and thermal polymerization by using two methods of thermal analysis: differential scanning calorimetry (DSC) and thermogravimetric analysis (TGA). Note, DSC and TGA are used broadly for kinetic studies of these types of reactions [1–3]. However, these studies are conducted routinely on heating, i.e., by continuously raising temperature. The use of a continuous heating program is primarily the matter of convenience. As long as a reaction is initiated thermally it is convenient to perform it by gradually adding heat. The energy gained is then converted into continuously intensified translational motion of molecules and vibrational motion of chemical bonds. At some point the bonds starts breaking producing reactive species that initiate the process. Ultimately, continuous heating drives the reaction to completion. The data obtained in a continuous heating run can then be used to gain some insights into the reaction mechanism as well as to develop a kinetic model suitable for research and industrial applications.

We have strived to accomplish the same tasks but using continuous cooling runs. Note that cooling segments have been employed in kinetic studies when using the techniques of temperature modulated DSC and TGA as well as of controlled rate thermal analysis. However, in these techniques the cooling segments are invariably combined with heating segments. When applied to thermally stimulated reactions these techniques do not employ an overall cooling program, i.e., the temperature at completion of a reaction is not lower than at its initiation. Rather, the temperature at completion is either higher (overall heating program) or the same (quasi-isothermal program) as the temperature at initiation. In our studies, the kinetics has been measured entirely under continuous cooling, i.e., when temperature decreases progressively throughout the reaction progress. Our research was motivated by both scholastic and pragmatic interest described in the following sections. When we initiated this research we had ventured in truly uncharted territory. Suffice to say that when we started we simply had troubles to detect the DSC signal on cooling for a reaction that had been well studied on heating. Over time, we have learned how to obtain reliable kinetic data, perform kinetic computations, and interpret the obtained results. This is the experience that we share in the present article. Its objective is to create a single reference that covers all basics needed for one to successfully conduct a kinetic study of thermally stimulated reactions occurring on continuous cooling. To accomplish this objective, we discuss scholastic and pragmatic motivation, the theory and praxis of measurements, kinetic computations, and representative examples.

2. Scholastic Motivation

As already stated, DSC and TGA are routinely used to measure the kinetics of thermally stimulated reaction on heating. However, obtaining the results exclusively on heating unavoidably gives rise to a limited picture of the process. Consider a mathematical function of two independent variables x and y, e.g., $f(x,y) = mx + ny$, with positive values of m and n. Simultaneously increasing both variables results in increasing $f(x,y)$ that makes it difficult to explore the individual effects of x and y because the effects are similar. On the other hand, when the variables are changed in the opposite ways, i.e., one increases and another decreases, they have distinctly different effects on $f(x,y)$ that affords the possibility of better understanding of the individual effects. Respectively, an analogy can be drawn with thermally stimulated reactions whose rate depends on conversion, α and temperature, T. Carrying out a reaction on continuous heating and cooling allows one to vary T in opposite directions and, thus, to learn more about the individual effects of both variables, i.e., to obtain a more complete kinetic picture of the process.

Although any reaction can be performed on cooling, the most interesting results should be expected for multi-step reactions. This is because for multi-step reactions the effective activation energy typically depends on both temperature and conversion [4,5]. In this situation, one can expect different results for heating and cooling. This can be illustrated by a simplistic algebraic example. Consider a hypothetical reaction that takes place in the temperature region from 300 to 400. To keep things as simple as possible we will use dimensionless temperature. Let us introduce the α vs T functions that change from 0 to 1 in that temperature range on heating:

$$\alpha^+ = 0.01T - 3 \tag{1}$$

and on cooling:

$$\alpha^- = -0.01T + 4 \tag{2}$$

Let us now introduce a dimensionless activation energy that depends on both α and T in some very simple form:

$$E = \alpha \cdot T \tag{3}$$

Now we can obtain temperature (Equations (4) and (5)) and conversion (Equations (6) and (7)) dependences for the activation energy on heating and cooling by rearranging and substituting Equations (1) and (2) into the activation energy expression, Equation (3). The results of these manipulations are shown below in Equations (4)–(7):

$$E^+(T) = 0.01T^2 - 3T \tag{4}$$
$$E^-(T) = -0.01T^2 + 4T \tag{5}$$
$$E^+(\alpha) = 100\alpha^2 + 300\alpha \tag{6}$$
$$E^-(\alpha) = -100\alpha^2 + 400 \tag{7}$$

These dependences are illustrated graphically in Figure 1A,B. Even though from the E vs α trends it seems that the values do not differ significantly, especially for $\alpha < 0.5$, the temperature dependences of the activation energy demonstrate opposite trends. This means that in the heating experiment the activation energy increases with increasing temperature, whereas on cooling it decreases with increasing temperature.

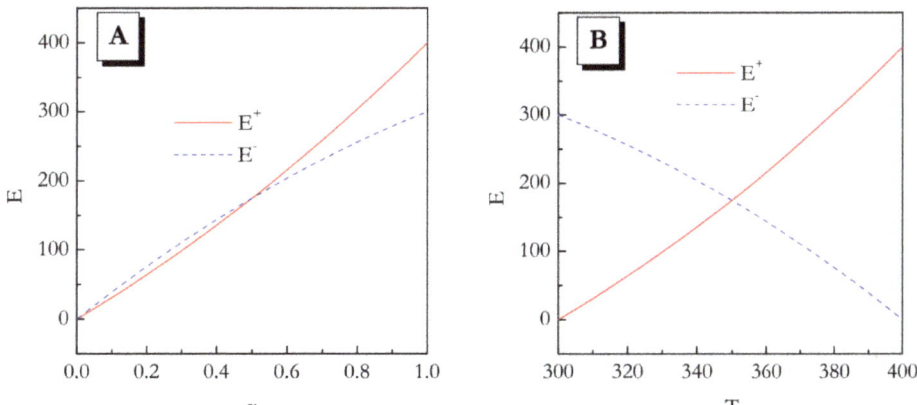

Figure 1. Schematic presentation of the E vs α (**A**) and E vs T (**B**) dependences for heating and cooling based on Equations (1)–(7).

Naturally, the activation energy represents the slope of the Arrhenius plot:

$$\ln k = \ln A - \frac{E}{RT} \tag{8}$$

where k is the rate constant, and A is the preexponential factor. The opposite directions of a change in E with increasing T means that the Arrhenius plots have to have the opposite curvatures. That is, if on heating the plot is concave up (E increase with T), on cooling it would be concave down (E decreases with T). Therefore, the heating and cooling kinetics cannot be reduced to each other. The major conclusion of this simplistic example is that for the multi-step reactions whose activation energy depends on both temperature and conversion, the kinetic parameters evaluated on heating may not suitable for predicting the kinetic behavior on cooling.

If a reaction is single-step, its activation energy is constant, i.e., independent of either α or T. Then, in line with the aforementioned arguments, the kinetics on heating and cooling should be the same. In this case, the kinetic parameters evaluated on heating would be suitable for predicting the kinetic behavior on cooling.

The considerations presented in this section allow one to generate the following central hypothesis for kinetic studies on continuous cooling. For a single-step reaction the kinetics on heating and cooling should be practically identical, whereas the respective kinetics should differ significantly for multi-step processes. Of course, these two types of processes are easy to identify by employing an isoconversional method to evaluate the activation energy as a function of conversion.

3. Pragmatic Motivation

The existence and importance of reactions occurring on cooling has been recognized in the literature. The issue has been discussed in connection with decomposition of amino acids [6], pyrolysis of hydrocarbons [7], cracking of heavy oils [8], pyrolysis of heterogeneous materials such as almond shells, municipal solid waste, lignin, and polyethylene [9]. It has also been brought up in regard to epoxy materials crosslinking [10,11] and vulcanization of rubber [12–15]. That is, there is no doubt that cooling is an integral part of many manufacturing processes and, as such, it affects the properties of the final product. Therefore, studying the kinetics of thermally stimulated reactions on cooling is not only of scholastic but also of practical interest.

In the absence of such studies, it is often assumed that the reaction rate is negligibly small during cooling or that kinetics determined on heating can be applied to the cooling conditions [10,16,17]. However, as demonstrated in the previous section, the similarity of the kinetics on heating and cooling should be expected only for single-step reactions. For multi-step reactions, the respective kinetics are likely to be different that means that one has to study the kinetic on cooling in order to understand the processes that involve cooling segments. Despite the obvious need, there had been no systematic studies of thermally stimulated reactions on cooling until our initiatory work [18]. A likely reason for this situation is that performing kinetic measurements on cooling is quite challenging. The issue is discussed in the next two sections.

4. Theory of Measurements on Cooling

The major problem with performing kinetic measurements on cooling is that the respective experiments have to satisfy two opposing conditions. First, cooling has to be initiated from the temperature, T^*, at which a reaction proceeds rapidly enough to follow its rate reliably. Second, this temperature has to be reached so that the reaction cannot proceed to any significant extent. The idea of satisfying both conditions experimentally is presented in Figure 2. First off, it is necessary to outrun the reaction while heating. It means that the temperature needs to be raised over a period of time that is markedly shorter than the characteristic reaction time, $\tau(T)$. One can define $\tau(T)$ via the reciprocal rate constant as [19]:

$$\tau(T) = k^{-1} = \left[A \exp\left(\frac{-E}{RT}\right)\right]^{-1} \qquad (9)$$

To make our estimates more realistic we take the values of A and E as 10^5 s^{-1} and 60 kJ mol^{-1} that would be characteristic of a process such as polymerization. If the reaction temperature is raised at constant rate β_H, the temperature, T^* will be reached at the following time:

$$t_H(T^*) = \frac{T^* - T_0}{\beta_H} \qquad (10)$$

where T_0 is the starting, e.g., ambient, temperature, and the subscript H denotes heating. Outrunning the reaction on heating means satisfying the condition $t_H(T^*) << \tau(T^*)$. For instance (Figure 2), if the reactant is heated rapidly, e.g., at 300 K min^{-1} to T^* = 410K (point A), the heating time $t_H(T^*)$ would be close to 0.4 min. The characteristic reaction time at this temperature is roughly 8 min. That is, the reaction would not have time to proceed to any significant extent before reaching T^*. Once T^* is reached, we start cooling the reactant slowly, e.g., at β_C = −0.5 K min^{-1}. The respective cooling time is then determined as:

$$t_C(T) = \frac{T - T^*}{\beta_C} \qquad (11)$$

Since $|\beta_C| << \beta_H$, the time scale of the cooling run will soon become similar to the time scale of the characteristic reaction time, which can be expressed as $t_C(T) \sim \tau(T)$ (temperature drops below that at point B in Figure 2). The conditions of the similar time scales would hold until decreasing temperature slows down the reaction so that characteristic reaction time becomes longer than the time scale of cooling (temperature drops below that at point C in Figure 2). Then, the reaction would cease to be measurable. The region between points B and C, would provide a temperature window, ΔT, within which the reaction kinetics can be measured during continuous cooling.

As follows from the above discussion, successful measurements of the reaction kinetics on cooling are contingent on an appropriate selection of the experimental parameters β_H, β_C, and T^*. Basic principles of selecting and optimizing these parameters are discussed in the next section.

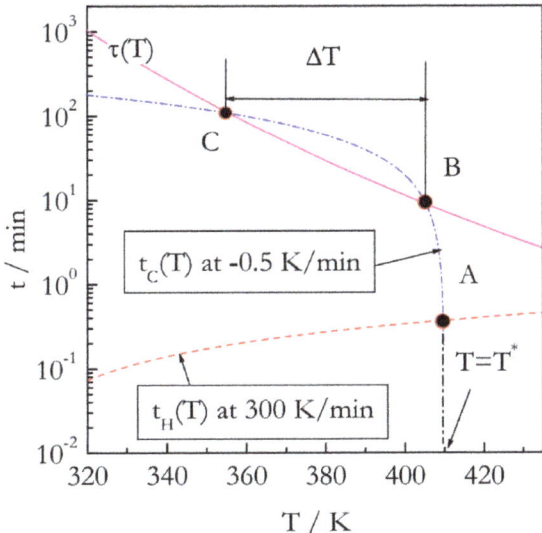

Figure 2. Temperature dependence of characteristic time, $\tau(T)$ (solid line) for a reaction with E = 60 kJ mol^{-1} and A = 10^5 s^{-1}. Turning temperature, T^* = 410 K. Dash line: heating time, dash-dot line: cooling time. Time is on log scale.

5. Praxis of Measurements on Cooling

As discussed above, the experiments on cooling have to start at elevated temperature and complete while the system is cooling down. In DSC or TGA this is done by rapidly heating a reagent to an elevated temperature, T^*, where the reaction rate is sufficiently fast. Once this temperature is reached the fast heating is switched to a slow cooling. For a limited number of reactions with a very long induction period such as decomposition of NiC$_2$O$_4$ [20] this procedure can be modified. The instrument can first be preheated to T^*. Then, the reagent can be introduced into the preheated furnace for a short isothermal hold to stabilize the temperature followed by a cooling segment. However, this works only when a reaction does not start immediately, i.e. has a significant induction period. If a reaction has no induction period, opening and closing the furnace will cause significant disturbance in the experimental data at the beginning of the reaction and these would not be suitable for kinetic analysis.

The majority of reactions do not have a long induction period, thus, a fast heating segment is necessary. In a regular DSC or TGA, the fastest available heating rate is about hundreds of degrees per minute. Such heating rates are entirely sufficient to minimize the reaction progress during heating. However, the use of excessively fast heating rates may introduce unnecessary disturbance into the measured signal. When the instrument switches from fast heating to slow cooling, it continues to heat up overshooting the set turning temperature. Overshooting is larger for faster heating rates in general and for the TGA measurements in particular due to slower response time of the latter. Generally, for the DSC experiments when using the heating rate of 100 °C min^{-1}, overshooting is not much larger than 5 °C, and the instrument catches up with the cooling program within ~10 s. As for the TGA measurements, overshooting is much larger and the adjustment time is longer. Thus, the disturbance in the measured signal results from the fact that when one temperature program is switched to another one, the instrument unavoidably needs some time to adjust to the new temperature program. This effect is especially strong when switching between heating and cooling. Also, the theory of the kinetic measurements on cooling requires switching from a fast heating rate to a slow cooling rate that additionally prolongs the adjustment time. Nevertheless, the overall disturbance in the instrumental signal can be diminished by properly choosing experimental parameters of the experiments.

The choice of the heating rate depends on the instrument used and nature of the process studied. For instance, for reactions with sigmoid or accelerating types of kinetics, the heating rate in DSC runs can be as slow as 30 °C min^{-1} [21]. For this type of kinetics the rate is slowest when the reaction starts. This means two things. First, such kinetics are easy to outrun on heating. Second, the signal disturbance caused by switching from heating to cooling does not have a significant effect on the data collected during the cooling segment. The longer the induction period is, the slower the heating rate can be. For instance, for decomposition of NiC_2O_4 a rather slow heating rate of 5 °C min^{-1} has been adequate for obtaining reliable data on cooling.

It is significantly more difficult to deal with reactions that follow decelerating kinetics. This type of kinetics demonstrates the fastest rate at the beginning of a reaction. These reactions are more difficult to outrun on heating. Also, the signal disturbance caused by switching from heating to cooling may have a rather strong effect on the data measured during cooling. Thus, to diminish the detrimental impact of the signal disturbance the heating rate in the respective DSC measurements should be at least 100 °C min^{-1} [21].

For TGA measurements, the situation is more complicated. As discussed above, when a fast heating is switched to a slow cooling, there is a stabilization period during which the instrument adjusts to the new temperature program. The length of the stabilization period is associated with the instrument response time or the signal time constant. The larger the signal time constant is, the longer it takes to stabilize the signal after switching the program. For instance, in the TGA and DSC instruments used for our measurements on cooling [18,20–23], the signal time constants were 14 and 1.7 s, respectively. Thus, TGA stabilizes much slower. Consequently, the use of TGA imposes additional limitations on the choice of the heating rates. Thus, even for the processes with decelerating type of kinetics, the heating rate should not exceed 20 °C min^{-1}. Because of the slow instrumental response time the fast heating program cannot be switched immediately to the slow cooling. Instead, TGA continues to heat up the system for some time, reaching temperatures significantly larger than T^*. This, in turn, results in non-negligible reaction progress attained before the slow cooling program is established. Obviously, the faster the heating rate is, the more temperature overshoots above T^*. At the same time, the use of excessively slow heating rate leads to a significant reaction progress during heating that, as a consequence, introduces larger error in the kinetic parameters estimated on cooling. In our experience the heating rates around 15–20 °C min^{-1} seem optimal.

The second experimental parameter that must be controlled is the turning temperature, T^*. It determines how fast the reaction will proceed when the cooling segment starts. The higher the turning temperature is, the faster the reaction is and, as a result, the measurements on cooling will require the use of faster heating and cooling rates. Faster heating rate is required to minimize the reaction progress during the initial heating. Higher turning temperature also requires the use of faster cooling rates. This is because higher T^* means that reaction rate is faster and that the process will be complete within a narrower temperature range during cooling. The use of narrow temperature ranges is generally undesirable because it leads to a greater uncertainty in evaluating the kinetic parameters. To expand the temperature range, one has to use faster cooling rates. As a rule of thumb, for autocatalytic (sigmoid) kinetics the turning temperature should be chosen as a DSC peak temperature of a heating run at 5 °C min^{-1}. For the decelerating kinetics, T^* should be selected ~10 °C higher. This is because on cooling the process decelerates due to decrease in temperature and decelerating reaction model. Faster deceleration leads to incomplete reaction on cooling and inability to evaluate kinetic parameters for higher extents of conversion.

The aforementioned recommendations about choosing the turning temperature are mostly relevant to the measurements made by DSC that has a shorter response time and, thus, better control over the temperature program. Due to slower response time of TGA, the respective runs on cooling should generally be performed to keep the reaction rate slower. Therefore, the turning temperatures in the TGA runs should be kept about 10°C lower than in the DSC runs. For instance, decomposition of $Li_2SO_4 \cdot H_2O$ on cooling has been studied by using both experimental techniques, DSC and TGA [18,20,21]. In the DSC runs, the turning temperatures have been selected to be 100 and 110 °C whereas in the TGA measurements T^* has been 95 °C [18,20]. As discussed before, the use of slower heating rates is required in the TGA measurements to minimize overheating. For example, for decomposition of $Li_2SO_4 \cdot H_2O$ the heating rate has been 15 °C min^{-1} [20]. At the same time, such a low heating rate will unavoidably lead to partial reaction progress during heating. Thus, lower T^* is chosen to reasonably minimize the effect of the partial reaction progress during heating on the overall kinetics on cooling. However, it should be kept in mind that the turning temperature cannot be lowered significantly. Low T^* can lead to incomplete reaction on cooling that would make it impossible to evaluate the kinetic parameters reliably within the whole range of conversions.

The last experimental parameter that needs to be optimized is the cooling rate. As mentioned before, the cooling rates need to be much slower than the heating ones. They are selected to be slower to ensure that the reaction reaches its completion during the cooling segment. For the DSC measurements, the cooling rates are typically chosen in the range of 0.1 to 2 °C min^{-1}. For the slower cooling rate, 0.1 °C min^{-1} the reaction typically happens within a couple of degrees whereas for the faster one it may cover 20–40 °C range. A wider temperature range covered on cooling allows for evaluating kinetic parameters with a smaller uncertainty.

In the measurements on cooling, the reaction continuously decelerates due to decrease in temperature. Thus, if the heating rate and the turning temperature are selected properly, the use of slower cooling rates typically leads to reaching 100% of conversion whereas for the faster ones it can be significantly smaller. It is not recommended to use the fast rates that lead to less than 90% conversion. This applies to both TGA and DSC measurements. However, as discussed earlier, the TGA should generally be performed under the conditions that maintain slower reaction rate. As a consequence, the fastest cooling rates in TGA should be slower than in DSC, typically not faster than 1 °C min^{-1}. This enables the reaction system to stay longer at higher temperature and, thus, to attain full or nearly full conversion.

To obtain adequate results for the kinetics on cooling, one should perform measurements with at least five different cooling rates. For the TGA experiments where cooling temperature region can be even narrower, more cooling rates may be required. It is noteworthy that the abovementioned recommendations on selecting the experimental parameters are not a set of strict rules. Rather, they are general directions on how to get a handle on controlling kinetic measurements on cooling. Each particular process may require a somewhat different experimental setup to achieve optimal condition of the measurements. It should also be remembered that all the experimental parameters are interconnected. If one parameter is changed, the other should also be readjusted. For instance, if the turning temperature is increased, the heating rate should be increased as well to minimize the reaction progress during heating. At the same time, the faster cooling rate can also be increased providing a wider temperature range for the cooling measurements.

Another important feature of the experiments on cooling is that they require knowing the final extent of the reactant conversion. In the heating measurements, the reaction normally attains complete conversion because the system is continuously heated promoting the reaction. On cooling, temperature decreases and so does the reaction rate. At some point, the temperature may become so low that process virtually ceases before reaching 100% conversion. Another reason why the reaction can stop is specific to polymerization reactions and associated with vitrification [22,23]. Our simulations have demonstrated that in the case of incomplete reactions one has to use the absolute values of conversion in order to retrieve correct kinetic parameters from the data [18]. For example, if at the fastest cooling rate the ultimate extent of conversion reached is 60%, for performing isoconversional calculations one has to use only the portion limited to 60% conversion from the data obtained at slower cooling rates.

For decomposition reactions or any processes that involve a mass change, determination of the final extent of conversion is relatively straightforward. For both DSC and TGA data, it is done by measuring the mass before and after the run and comparing their ratio to the one experimentally evaluated on heating. This is especially important when the ultimate mass loss cannot be evaluated theoretically from the stoichiometry as in degradation of polymers. As for the stoichiometric reactions such as dehydration of hydrates one can use a theoretically evaluated mass loss for evaluating the complete extent of conversion. In the TGA runs, the mass is monitored continuously. Thus, it is possible to accurately detect partial reaction progress during heating. Therefore, to account for the related reaction progress and more accurately evaluate the conversion, it is recommended not to use the mass at the turning temperature. It is more appropriate to use the mass at the beginning of detectable decomposition that may start on heating.

The kinetics of the processes that do not involve change in mass (e.g., polymerization) can be studied by DSC. In this case, determination of the extent of conversion relies only on measuring the reaction heat. This creates another challenge in the measurements on cooling. Switching the temperature programs at the turning temperature, causes a disturbance in the DSC signal. As a result, one cannot use regular DSC software to subtract the baseline because the baseline in the vicinity of the turning temperature is highly uncertain. This issue arises for any reaction. However, for the reactions that involve a change in the mass, the final extent of conversion can be established by weighing the sample before and after the experiment. Difficulty in subtracting the baseline can be overcome by using software such as MS Excel, and performing this procedure manually. By conducting multiple DSC runs with an empty pan, we have established that switching from fast heating to slow cooling causes a heat flow perturbation that decays in a manner similar to that defined by the Kohlrausch, Williams, and Watts (KWW) function [24,25]:

$$y = 1 - \exp\left[-\left(\frac{t}{\tau_{ef}}\right)^{\gamma}\right] \quad (12)$$

where γ is the stretch exponent, t is the time, and τ_{ef} is the effective relaxation time. This equation has been transformed to fit the baseline by taking y as $(HF_0 - HF_t)/(HF_0 - HF_f)$. This transformation gives rise to the following equation [18]:

$$HF_t = HF_f + \exp\left[-\left(\frac{T^* - T}{T_{ef}}\right)^{\gamma}\right](HF_f - HF_0) \quad (13)$$

where HF_t is the current heat flow, HF_0 and HF_f are the heat flow values at the beginning and end of the reaction, T^* is the turning temperature, and T_{ef} and γ are adjustable parameters. The parameters of the fit need to be chosen to satisfy the following criteria. First, the reaction should start as close as possible to the turning temperature and end where the DSC signal merges into the baseline value, i.e. when the process stops producing the latent heat. Second, the enthalpy of the reaction on cooling determined after subtracting the baseline should be consistent with the value determined in the heating runs. As mentioned before, for decomposition reactions, the thermal effect can be additionally verified by weighing the sample before and after the experiment. As for the processes that do not involve a mass change, it has been determined that the enthalpy of the reaction stays reasonably constant for one cooling rate and does not change much if the fitting parameters are somewhat varied. One also should rely on the value of the thermal effect obtained on heating as a reference value of the enthalpy for the measurements on cooling. It is noteworthy that even for the regular heating DSC measurements the choice of the baseline implies certain variability. Nevertheless, it is recommended to first practice to subtract a baseline in the cooling runs for the processes with a single-step kinetics where the activation energy is expected to be the same on heating and cooling and then move to more complicated reactions.

6. Kinetic Computations for Cooling Data

All our kinetic computations on the cooling data were performed based on the recommendations of the ICTAC Kinetics Committee [26]. The effective activation energy, E_α, was evaluated as a function of conversion with the help of an advanced isoconversional method [27] that is suitable for kinetic analysis of data obtained at any temperature program, T(t). The method has been used successfully for a large variety of thermally stimulated processes [1,2,4], including those that occur on cooling, i.e., crystallization [28–30], gelation [31–33], and solid-solid transition [34]. It affords evaluating E_α as a function of conversion α, by finding a minimum the function:

$$\Psi(E_\alpha) = \sum_{i=1}^{n} \sum_{j \neq i}^{n} \frac{J[E_\alpha, T_i(t_\alpha)]}{J[E_\alpha, T_j(t_\alpha)]} \qquad (14)$$

where:

$$J[E_\alpha, T_i(t_\alpha)] \equiv \int_{t_\alpha - \Delta\alpha}^{t_\alpha} \exp\left[\frac{-E_\alpha}{RT_i(t)}\right] dt \qquad (15)$$

and n is the number of the temperature programs. It is important to note that this method belongs to the class of flexible [4,35] integral methods, i.e., the methods in which the user can control the limits of integration. This is essential for isoconversional calculations on cooling data because some of the most popular methods (e.g., Ozawa, Flynn and Wall, Starink, etc) [26] are rigid integral methods. They are applicable exclusively to the heating data. The issue is discussed in detail elsewhere [4,30,35]. As an alterative to the rigid integral methods one can also use differential isoconversional methods.

In addition to the E_α vs α dependences we also evaluated the E_α vs T dependences. The latter is readily evaluated by replacing each value of α with the mean value of the temperatures, T_α related to this α at different cooling rates.

The preexponential factor, A_α as a function of conversion was determined by substituting the E_α values into the equation of the compensation effect [26]:

$$\ln A_\alpha = a + bE_\alpha \qquad (16)$$

The parameters a and b were estimated via fitting the pairs of E_i and $\ln A_i$ into Equation (16). The E_i and $\ln A_i$ values were found by substituting the reaction models, $f_i(\alpha)$, into the linear form of the basic rate equation [26]:

$$\ln\left(\frac{d\alpha}{dt}\right) - \ln[f_i(\alpha)] = \ln A_i - \frac{E_i}{RT} \qquad (17)$$

For each $f_i(\alpha)$, the E_i and $\ln A_i$ values were evaluated respectively from the slope and intercept of the linear plot of the left-hand side of Equation (17) vs the reciprocal temperature. As determined previously [18], four $f(\alpha)$ functions that represent the power law (P2, P3, P4) and Avrami-Erofeev (A2) models [26] are sufficient for obtaining accurate values of the preexponential factor. Here we must stress the need of using a differential method for evaluating the E_i and $\ln A_i$ values. Recall, that all our calculations are done on cooling data. However, most popular model-fitting methods (e.g., Coats-Redfern) [26] are rigid integral methods and, as stated earlier, inapplicable to cooling data.

Finally, we used the values of E_α and $\ln A_\alpha$ to determine the rate constant. It has been done by substituting the respective values into the Arrhenius Equation (8) so that for each given temperature T_α the rate constant is determined as [36]:

$$\ln k = \ln A_\alpha - \frac{E_\alpha}{RT_\alpha} \qquad (18)$$

7. Representative Examples

Having covered the practical aspects of the data analysis, we can now return to the central hypothesis of the kinetic studies on cooling. As already stated, for single-step reactions the kinetics on heating and cooling are expected to be nearly identical, whereas for multi-step reactions significantly different. The single-step reactions demonstrate activation energies that do not vary with conversion. Figure 3 shows isoconversional activation energy evaluated for the thermal decomposition of ammonium nitrate (AN) [20] and nonstoichiometric polymerization of diglycidyl ether of bisphenol A (DGEBA) epoxy and m-phenylenediamine (m-PDA) [22]. Both reactions are examples of a single-step process. As seen in Figure 3, for these processes the activation energy evaluated on heating agrees well with that determined on cooling. Similar results have been obtained for the preexponential factor [20,22]. The similarity of the Arrhenius parameters for heating and cooling translates naturally into similar Arrhenius plots (Figure 4). One can see that for the considered reactions of decomposition and polymerization the Arrhenius plots estimated by Equation (18) for heating and cooling experiments practically coincide with each other.

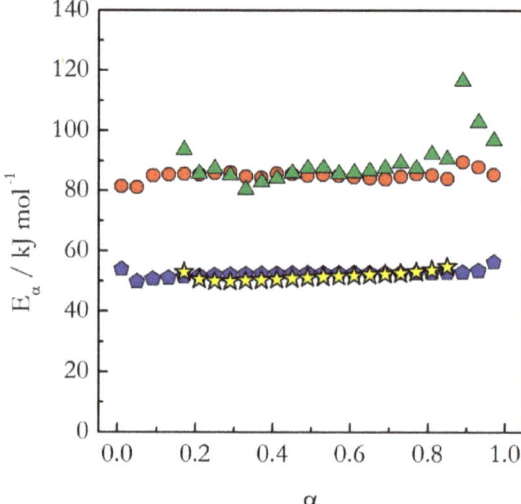

Figure 3. The activation energy as a function of conversion for the thermal decomposition of AN (heating (red circles) and cooling (green triangles)) and nonstoichiometric polymerization of DGEBA and m-PDA (heating (blue pentagons) and cooling (yellow stars)) evaluated on heating and cooling [20,22]. Partially adapted from Ref. [20] with permission from the PCCP Owner Societies.

This means that as long as the reaction is single-step, one should not expect any significant difference in the kinetics determined on heating and cooling. In other words, if a reaction exhibits constant activation energy one can use the Arrhenius parameters evaluated on heating to predict the kinetic behavior on cooling.

On the other hand, multi-step reactions typically exhibit the activation energy that varies with conversion. As illustrated above by using a purely algebraic example, one may expect significant differences in the kinetics measured on heating and cooling when the activation energy depends on both temperature and conversion. A representative example of such reaction is crosslinking polymerization of DGEBA with m-PDA. This type of reactions tends to manifest a transition from a kinetic to a diffusion regime so that their overall kinetics is described by rate equations that include the reaction and diffusion rate constants [4]. The respective rate equation proposed by Vyazovkin and Sbirrazzuoli yields the following form of the effective activation energy [37]:

$$E = \frac{E_D k(T) + E k_D(T, \alpha)}{k(T) + k_D(T, \alpha)} \quad (19)$$

where E_D and E are the activation energies of diffusion and reaction, $k(T)$ is the reaction rate constant, and $k_D(T,\alpha)$ is the diffusion rate constant. The latter depends not only on temperature but also on conversion because diffusion slows down with increasing the extent of polymerization. Therefore, the activation energy for crosslinking polymerization is generally a function of both conversion and temperature.

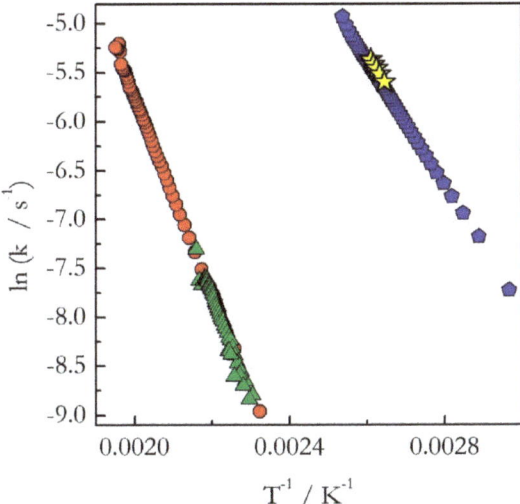

Figure 4. The rate constant as a function of reciprocal temperature for the thermal decomposition of AN (heating (red circles) and cooling (green triangles)) and nonstoichiometric polymerization of DGEBA and m-PDA (heating (blue pentagons) and cooling (yellow stars)) evaluated on heating and cooling. Adapted from Ref. [23] with permission from Elsevier.

We have studied the crosslinking polymerization of DGEBA with m-PDA on heating and on cooling from 140 °C [22,23]. The application of the advanced isoconversional method to the obtained data has yielded the dependences of E_α on α presented in Figure 5. The activation energy decreases with the reaction progress on both heating and cooling. However, the values do not match each other. A more dramatic difference is revealed when the activation energy is plotted against temperature as shown in Figure 6. The E_α on T dependences evaluated on heating and cooling exhibit the opposite trends. The activation energy decreases with increasing temperature on heating and increases with increasing temperature on cooling.

As explained earlier, the opposite trends in E with respect to temperature mean the opposite curvatures of the respective Arrhenius plots. The experimental Arrhenius plots determined by Equation (18) are shown in Figure 7.

Indeed, the plot obtained for heating is concave down (E decreases with T), whereas for cooling it is concave up (E increases with T). Obviously, the lines of the opposite curvatures cannot coincide. This means that the kinetics evaluated on heating cannot be used to predict the kinetic behavior during cooling. This example confirms clearly our argument that generally one should not assume that the kinetics measured on heating should be similar to that measured on cooling. This assumption holds true for single-step reactions, i.e., reactions for which the activation energy does not vary with conversion. If the activation energy is found to vary with conversion, one should avoid extrapolating heating kinetics to the cooling conditions. Instead, the cooling kinetics should be studied on their own to determine adequate kinetic parameters.

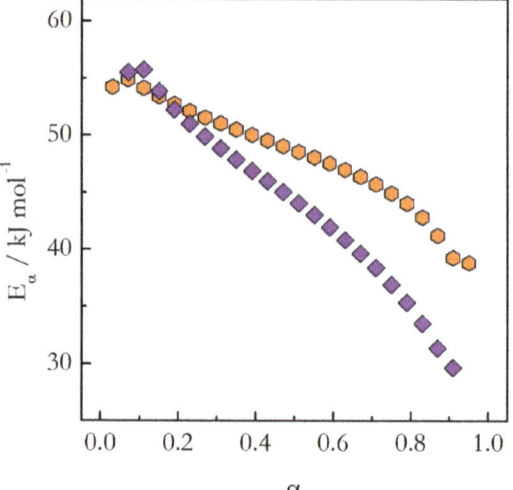

Figure 5. The activation energy as a function of conversion for the stoichiometric polymerization of DGEBA and m-PDA evaluated on heating (orange hexagons) and cooling (violet diamonds) [22,23]. Adapted from Ref. [22] with permission from Wiley.

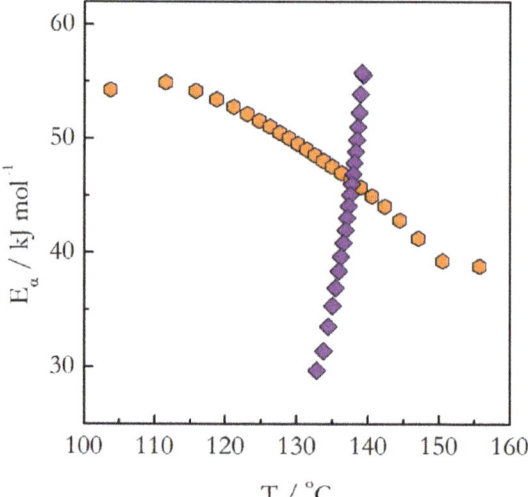

Figure 6. The isoconversional activation energy as a function of temperature for the stoichiometric polymerization of DGEBA and m-PDA evaluated on heating (orange hexagons) and cooling (violet diamonds) [22,23]. Adapted from Ref. [22] with permission from Wiley.

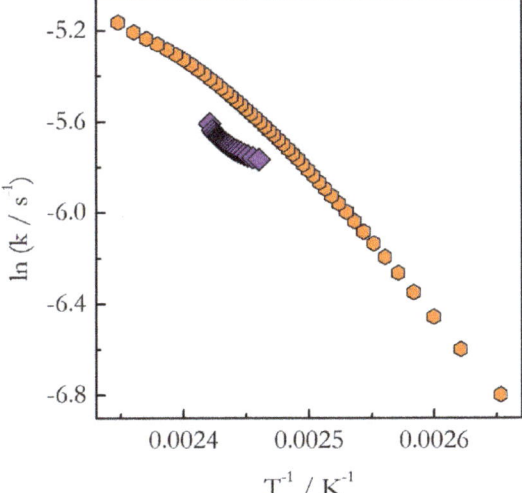

Figure 7. The rate constant as a function of reciprocal temperature for the stoichiometric polymerization of DGEBA and m-PDA evaluated on heating (orange hexagons) and cooling (violet diamonds) [23]. Adapted from Ref. [23] with permission from Elsevier.

8. Conclusions

In the real world, many thermally stimulated reactions occur during continuous cooling. However, the kinetics of thermally stimulated reactions are customarily studied under continuous heating conditions. The relevance of such studies to the cooling conditions rests upon the assumption that the kinetics on cooling are the same as on heating. Our studies demonstrate that this holds only for single-step reactions, whereas for multi-step reactions the difference between the respective kinetics can be very significant. This highlights the practical need in continuing systematic studies of thermally stimulated reactions occurring during continuous cooling. Conducting such studies presents a number of challenges not encountered in routine kinetic studies on heating. This overview article has shared our experience in addressing these challenges. Together with the other issues discussed, this paper is expected to provide sufficient background knowledge to help other workers to have a quick and successful start in their kinetic studies of thermally stimulated reactions that occur during continuous cooling.

Author Contributions: Conceptualization, S.V.; Methodology, T.L., S.V.; Validation, T.L.; Formal Analysis, T.L., S.V.; Investigation, T.L.; Data Curation, T.L.; Writing-Original Draft Preparation, T.L.; Writing-Review & Editing, S.V.; Visualization, T.L., S.V.; Supervision, S.V.; Project Administration, S.V.

Funding: This research received no external funding.

Conflicts of Interest: The authors declare no conflict of interest.

References

1. Vyazovkin, S.; Sbirrazzuoli, N. Isoconversional Kinetic Analysis of Thermally Stimulated Processes in Polymers. *Macromol. Rapid Commun.* **2006**, *27*, 1515–1532. [CrossRef]
2. Vyazovkin, S. Isoconversional kinetics of polymers: The decade past. *Macromol. Rapid Commun.* **2017**, *38*, 1600615. [CrossRef] [PubMed]
3. Galwey, A.K.; Brown, M.E. *Thermal Decomposition of Ionic Solids*; Elsevier: Amsterdam, The Netherlands, 1999.
4. Vyazovkin, S. *Isoconversional Kinetics of Thermally Stimulated Processes*; Springer: Heidelberg, Germany, 2015.
5. Vyazovkin, S. A time to search: Finding the meaning of variable activation energy. *Phys. Chem. Chem. Phys.* **2016**, *18*, 18643–18656. [CrossRef]

6. Ross, D.S. Cometary Impact and Amino Acid Survival—Chemical Kinetics and Thermochemistry. *J. Phys. Chem. A* **2006**, *110*, 6633–6637. [CrossRef]
7. Garcia, A.N.; Font, R.; Marcilla, A. Kinetic studies of the primary pyrolysis of municipal solid waste in a Pyroprobe 1000. *J. Anal. Appl. Pyrol.* **1992**, *23*, 99–119. [CrossRef]
8. Yang, J.; Wang, L.; Tian, C.; Xiao, J.; Yang, C. Improved method for kinetic parameters estimation of non-isothermal reaction: Application to residuum thermolysis. *Fuel Process. Technol.* **2012**, *104*, 37–42. [CrossRef]
9. Font, R.; Marcilla, A.; Garcia, A.N.; Caballero, J.A.; Conesa, J.A. Kinetic models for the thermal degradation of heterogeneous materials. *J. Anal. Appl. Pyrol.* **1995**, *32*, 29–39. [CrossRef]
10. Gross, T.S.; Jafari, H.; Tsukrov, I.; Bayraktar, H.; Goering, J. Curing cycle modification for RTM6 to reduce hydrostatic residual tensile stress in 3D woven composites. *J. Appl. Polym. Sci.* **2016**, *133*, 43373. [CrossRef]
11. Kim, H.S.; Lee, D.G. Avoidance of fabricational thermal residual stresses in co-cure bonded metal-composite hybrid structures. *J. Adhes. Sci. Technol.* **2006**, *20*, 959–979. [CrossRef]
12. Khouider, A.; Vergraud, J.M. Effect of temperature of motionless air on the cure of vulcanizates after removal from the mold. *J. Appl. Polym. Sci.* **1986**, *32*, 5301–5313. [CrossRef]
13. Abdul, M.; Vergnaud, J.M. Vulcanization progress in rubber sheets during cooling in motionless air after extraction from the mold. *Thermochim. Acta* **1984**, *76*, 161–170. [CrossRef]
14. Tadlaoui, S.; Azaar, K.; El Brouzi, A.; Granger, R.; Vergnaud, J.M. Increase in cure of thermosets after extraction out of the mould. *Eur. Polym. J.* **1993**, *29*, 585–591. [CrossRef]
15. Warley, R.L. Simulation of the effect of variation in the cooling cycle on the state of cure in a rubber component. *J. Elastom. Plast.* **2013**, *45*, 33–46. [CrossRef]
16. Li, N.; Li, Y.; Jelonnek, J.; Link, G.; Gao, J. A new process control method for microwave curing of carbon fibre reinforced composites in aerospace applications. *Composites Part B* **2017**, *122*, 61–70. [CrossRef]
17. Yebi, A.; Ayalew, B. *Model-Based Optimal Control of Layering Time for Layer-by-Layer UV Processing of Resin Infused Laminates*; American Control Conference: Boston, MA, USA, 2016.
18. Liavitskaya, T.; Vyazovkin, S. Discovering the kinetics of thermal decomposition during continuous cooling. *Phys. Chem. Chem. Phys.* **2016**, *18*, 32021–32030. [CrossRef]
19. Frank-Kamenetskii, D.A. *Diffusion and Heat Transfer in Chemical Kinetics*; Plenum: New York, NY, USA, 1969.
20. Liavitskaya, T.; Guigo, N.; Sbirrazzuoli, N.; Vyazovkin, S. Further insights into the kinetics of thermal decomposition during continuous cooling. *Phys. Chem. Chem. Phys.* **2017**, *19*, 18836–18844. [CrossRef] [PubMed]
21. Liavitskaya, T.; Vyazovkin, S. Delving into the kinetics of reversible thermal decomposition of solids measured on heating and cooling. *J. Phys. Chem. C* **2017**, *121*, 15392–15401. [CrossRef]
22. Liavitskaya, T.; Vyazovkin, S. Kinetics of thermal polymerization can be studied during continuous cooling. *Macromol. Rapid Commun.* **2018**, *39*, 1700624. [CrossRef]
23. Liavitskaya, T.; Vyazovkin, S. Is the kinetics of crosslinking polymerization the same on heating and cooling? *Polymer* **2019**, *161*, 8–15. [CrossRef]
24. Matsuoka, S. *Relaxation Phenomena in Polymers*; Oxford University Press: New York, NY, USA, 1992.
25. Debenedetti, P.G. *Metastable liquids. Concepts and Principles*; Princeton University Press: Princeton, NJ, USA, 1996.
26. Vyazovkin, S.; Burnham, A.K.; Criado, J.M.; Pérez-Maqueda, L.A.; Popescu, C.; Sbirrazzuoli, N. ICTAC kinetics committee recommendations for performing kinetic computations on thermal analysis data. *Thermochim. Acta* **2011**, *520*, 1–19. [CrossRef]
27. Vyazovkin, S. Modification of the integral isoconversional method to account for variation in the activation energy. *J. Comp. Chem.* **2001**, *22*, 178–183. [CrossRef]
28. Vyazovkin, S.; Sbirrazzuoli, N. Isoconversional Analysis of the Nonisothermal Crystallization of a Polymer Melt. *Macromol. Rapid Commun.* **2002**, *23*, 766–770. [CrossRef]
29. Vyazovkin, S.; Sbirrazzuoli, N. Isoconversional approach to evaluating the Hoffman-Lauritzen parameters (U^* and K_g) from the overall rates of nonisothermal crystallization. *Macromol. Rapid Commun.* **2004**, *25*, 733–738. [CrossRef]
30. Vyazovkin, S. Nonisothermal crystallization of polymers: Getting more out of kinetic analysis of differential scanning calorimetry data. *Polym. Cryst.* **2018**, *1*, e10003. [CrossRef]

31. Chen, K.; Vyazovkin, S. Temperature dependence of sol-gel conversion kinetics in gelatin-water system. *Macromol. Biosci.* **2009**, *9*, 383–392. [CrossRef]
32. Guigo, N.; Sbirrazzuoli, N.; Vyazovkin, S. Atypical gelation in gelatin solutions probed by ultra fast calorimetry. *Soft Matter* **2012**, *8*, 7116–7121. [CrossRef]
33. Espinosa-Dzib, A.; Vyazovkin, S. Gelation of Poly(Vinylidene Fluoride) Solutions in Native and Organically Modified Silica Nanopores. *Molecules* **2018**, *23*, 3025. [CrossRef] [PubMed]
34. Farasat, R.; Yancey, B.; Vyazovkin, S. High Temperature Solid-solid Transition in Ammonium Chloride Confined to Nanopores. *J. Phys. Chem. C* **2013**, *117*, 13713–13721. [CrossRef]
35. Vyazovkin, S. Modern Isoconversional Kinetics: From Misconceptions to Advances. In *Handbook of Thermal Analysis and Calorimetry: Recent Advances, Techniques and Applications*, 2nd ed.; Vyazovkin, S., Koga, N., Schick, C., Eds.; Elsevier: Amsterdam, The Netherlands, 2018; pp. 131–172.
36. Liavitskaya, T.; Birx, L.; Vyazovkin, S. Thermal Stability of Malonic Acid Dissolved in Poly(vinylpyrrolidone) and Other Polymeric Matrices. *Ind. Eng. Chem. Res.* **2018**, *57*, 5228–5233. [CrossRef]
37. Vyazovkin, S.; Sbirrazzuoli, N. Mechanism and kinetics of epoxy-amine cure studied by differential scanning calorimetry. *Macromolecules* **1996**, *29*, 1867–1873. [CrossRef]

© 2019 by the authors. Licensee MDPI, Basel, Switzerland. This article is an open access article distributed under the terms and conditions of the Creative Commons Attribution (CC BY) license (http://creativecommons.org/licenses/by/4.0/).

Article

Advanced Isoconversional Kinetic Analysis for the Elucidation of Complex Reaction Mechanisms: A New Method for the Identification of Rate-Limiting Steps

Nicolas Sbirrazzuoli

University Côte d'Azur, Institute of Chemistry of Nice, UMR CNRS 7272, 06100 Nice, France; Nicolas.SBIRRAZZUOLI@unice.fr or Nicolas.SBIRRAZZUOLI@univ-cotedazur.fr

Received: 9 April 2019; Accepted: 25 April 2019; Published: 30 April 2019

Abstract: Two complex cure mechanisms were simulated. Isoconversional kinetic analysis was applied to the resulting data. The study highlighted correlations between the reaction rate, activation energy dependency, rate constants for the chemically controlled part of the reaction and the diffusion-controlled part, activation energy and pre-exponential factors of the individual steps and change in rate-limiting steps. It was shown how some parameters computed using Friedman's method can help to identify change in the rate-limiting steps of the overall polymerization mechanism as measured by thermoanalytical techniques. It was concluded that the assumption of the validity of a single-step equation when restricted to a given α value holds for complex reactions. The method is not limited to chemical reactions, but can be applied to any complex chemical or physical transformation.

Keywords: kinetic analysis; isoconversional methods; polymerization mechanisms; curing; epoxy; DSC; thermoanalytical techniques

1. Introduction

The control of polymerization reactions of thermosetting material is of major importance to reach the optimal properties of the cured polymer or composite. These complex reactions often involve several chemical and diffusion steps that make the elucidation of the reaction mechanism very difficult. The use of classical analytical techniques is often limited by high change in the viscosity of the reaction medium during polymerization. Thus, Differential Scanning Calorimetry (DSC) and rheometry have been widely used to monitor complex cure kinetics that depend on both time, temperature and heating (cooling) rate. The final degree of crosslinking, which is correlated to the extent of conversion (α), is linked to the temperature program used for curing. In addition, there is a direct link between the final properties of the polymer and (i) the molecular weight, (ii) the extent of conversion as evaluated by DSC and (iii) the glass transition temperature (T_g). As an illustration of this, the Di Benedetto equation shows the link between the glass transition temperature and the extent of conversion [1–3].

In order to gain more insight into the elucidation of the reaction mechanism of complex reactions, the objective of the work is to show how isoconversional kinetic analysis can provide information on the rate-limiting steps involved in complex chemical reactions or physical transformations by an analysis of the activation energy dependency, called the E_α dependency. In the case of single-step processes, the activation energy computed with an isoconversional method leads to a constant value with the extent of conversion (α). Nevertheless, this case is not frequently observed. In the case of complex reactions or complex physical transformations, analysis of the E_α dependency and its variations indicate the presence of a complex mechanism and may give important insight into change in the rate-limiting steps. For this purpose, a complex chemical polymerization was selected and simulated using data obtained in a previous experimental work [4]. This complex cure is of peculiar interest as it involves several steps consisting of an autocatalytic step, a first-order reaction and a diffusion-controlled part at the end of the reaction. This mechanism was frequently observed in the crosslinking of various epoxy–amine systems and was selected because it represents a good example of reaction complexity [5]. The first stages of the reaction are often controlled by an autocatalytic process, followed by epoxy–amine addition and a diffusion-controlled part after gelation when the viscosity of the reaction medium reaches high values as a result of the high increase of the molecular weight due to crosslinking. The specific polymerization system used as an example was made of 1,3-phenylenediamine (m-PDA) and diglycidyl ether of bisphenol A (DGEBA), obtained from Sigma-Aldrich and used as received. DGEBA has a molecular weight of about 355 g mol^{-1}, a glass transition temperature T_g of about −20 °C (midpoint DSC) and an epoxy equivalent (determined by 1 H NMR) of about 175 g equiv^{-1}. The example proposed here to illustrate the method corresponds to a polymerization reaction which is well described by an autocatalytic model for the chemically controlled part of the reaction and by a diffusion model for the end of the reaction. Nevertheless, this procedure can be extended to other complex processes, such as parallel, competitive (consecutive) or other mechanisms.

2. Theoretical Part

Isoconversional methods are amongst the more reliable kinetic methods for the treatment of thermoanalytical data, see for example [5–7]. The Kinetics Committee of the International Confederation for Thermal Analysis and Calorimetry (ICTAC) has recommended the use of multiple temperature programs for the evaluation of reliable kinetic parameters [6]. The main advantages of isoconversional methods are that they afford an evaluation of the effective activation energy, E_α, without assuming any particular form of the reaction model, $f(\alpha)$ or $g(\alpha)$, and that a change in the E_α variation, called the E_α dependency, can generally be associated with a change in the reaction mechanism or in the rate-limiting step of the overall reaction rate, as measured with thermoanalytical techniques.

Polymerizations are frequently accompanied by a significant amount of heat released; thus, cure kinetics can be easily monitored by DSC. It is generally assumed that the heat flow measured by calorimetry is proportional to the process rate [5–7]. Thus, the extent of conversion at time t, α_t, is computed according to Equation (1), as follows:

$$\alpha_t = \frac{\int_{t_i}^{t}(dQ/dt)\,dt}{\int_{t_i}^{t_f}(dQ/dt)\,dt} = \frac{Q_t}{Q_{tot}} \tag{1}$$

where t_i represents the first integration bound of the DSC signal and t_f is the last integration bound selected when the reaction is finished. (dQ/dt) is the heat flow measured by DSC at time t, Q_{tot} is the total heat released (or absorbed) by the reaction and Q_t is the current heat change.

The general form of the basic rate equation is usually written as [5]:

$$\frac{d\alpha}{dt} = A\exp\left(-\frac{E}{RT}\right)f(\alpha) \tag{2}$$

where T is the temperature, $f(\alpha)$ is the differential form of the mathematical function that describes the reaction model that represents the reaction mechanism, E is the activation energy and A is the pre-exponential factor.

The advanced non-linear isoconversional method (NLN) [8–11] used in this study is presented in Equations (3) and (4) and was derived from Equation (2):

$$\Phi(E_\alpha) = \sum_{i=1}^{n} \sum_{j \neq i}^{n} \frac{J[E_\alpha, T_i(t_\alpha)]}{J[E_\alpha, T_j(t_\alpha)]} \quad (3)$$

$$J[E_\alpha, T(t_\alpha)] = \int_{t_\alpha - \Delta\alpha}^{t_\alpha} \exp\left[\frac{-E_\alpha}{RT(t)}\right] dt \quad (4)$$

where E_α is the effective activation energy. The E_α value is determined as the value that minimizes the function $\Phi(E_\alpha)$. This non-linear kinetic method (referred as NLN) allows one to handle a set of n experiments carried out under different arbitrary temperature programs $T_i(t)$ and uses a numerical integration of the integral with respect to the time. For each i-th temperature program, the time $t_{\alpha,i}$ and temperature $T_{\alpha,i}$ related to selected values of α are determined by an accurate interpolation using a Lagrangian algorithm [11,12]. Numerical integration is performed using trapezoidal rule. Several possibilities are proposed for the initial estimate E_0 of E_α in the non-linear procedure. The method developed by Sbirrazzuoli and implemented in his internally generated software can treat any kind of isothermal or non-isothermal data from DSC, calorimetry (C80, for example), Thermogravimetric Analysis (TGA), Dynamic Mechanical Analysis (DMA), or rheometry [9,11–16]. This software was used in this study to compute a value of E_α for each value of α between 0.02 and 0.98 with a step of 0.02. This advanced non-linear isoconversional method (NLN) was applied in this study.

Another isoconversional method can be derived by the linearization of Equation (2) and is known as Friedman's method [12,17]:

$$\ln\left(\frac{d\alpha}{dt}\right)_{\alpha,i} = \ln[A_\alpha f(\alpha)] - \frac{E_\alpha}{RT_{\alpha,i}} \quad (5)$$

Application of this method requires the knowledge of the reaction rate $(d\alpha/dt)_{\alpha,i}$ and of the temperature $T_{\alpha,i}$ corresponding to a given extent of conversion, for the i temperature programs used. The advantages of differential methods such as Friedman's method (referred as FR) are that they use no approximations and can be applied to any temperature program. As for NLN, the interpolation is made using a Lagrangian algorithm. This does not hold for usual integral methods, but is also the case for the non-linear advanced isoconversional method previously described. Nevertheless, simulations have shown that differential isoconversional methods can sometimes reveal numerical instability [12]; therefore, before using Friedman's method it was checked that the obtained results were consistent with those obtained with the NLN method.

Equation (5) shows that the intercept of the Friedman's plot led to the determination of the term $[A_\alpha f(\alpha)]$. This term represents the product between the pre-exponential factor A_α and the mathematical function $f(\alpha)$ that describes the reaction mechanism. Once E_α and $[A_\alpha f(\alpha)]$ have been evaluated it is possible to compute the reaction rate $(d\alpha/dt)$ for each value of α using Equation (6):

$$\left(\frac{d\alpha}{dt}\right)_\alpha = [A_\alpha f(\alpha)] \exp\left(-\frac{E_\alpha}{RT_\alpha}\right) \quad (6)$$

The terms $(d\alpha/dt)_\alpha$, $[A_\alpha f(\alpha)]$ and $\exp(-E_\alpha/RT)$ of Equation (6) were evaluated for each α. If the reaction rate increased and the term $\exp(-E_\alpha/RT)$ decreased (i.e., E_α increased), then it was concluded that an increase of the term $[A_\alpha f(\alpha)]$ compensated for the decrease of the exponential term. This corresponds to a change in the pre-exponential factor and/or to the reaction mechanism. Thus, the aim of this work is to show how the analysis of these variations in association with the E_α dependency can be used to identify changes in the reaction mechanism.

Figure 1 show the variation of the extent of conversion (α) with temperature (T) for three heating rates and the corresponding reaction rate ($d\alpha/dt$). Equation (2) is the equation of a single-step process that does not apply to complex polymerization, which is a multi-step process. When applying an isoconversional method, the computations are performed for a constant value of α. Thus, Equation (2) is transformed into Equation (6). In this case, the hypothesis of a single-step process is only applied for each constant α value used for the computation, which corresponds, for non-isothermal data, to a very narrow temperature range. Therefore, it can be assumed that the validity of a single-step equation for a given α value generally holds, even for complex reactions. In addition, the Arrhenius equation is only applied to a narrow temperature region related to this α value.

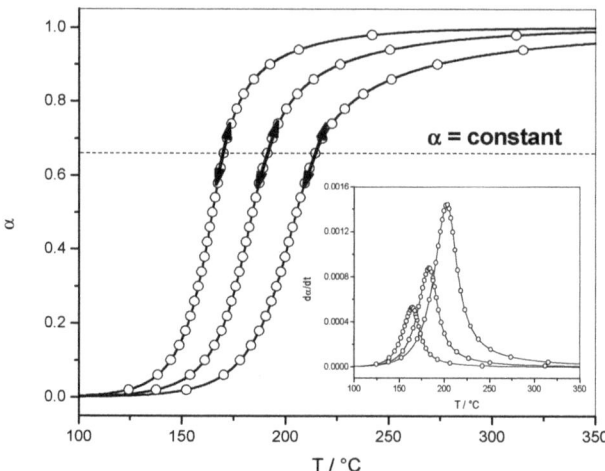

Figure 1. Non-isothermal data. Example of variation of the extent of conversion (α) with temperature (T). Inset: corresponding variation of the reaction rate ($d\alpha/dt$) with temperature. Isoconversional methods are based on the assumption of the hypothesis of a single-step process only for each α value and the Arrhenius equation applies to a narrow temperature region related to this α value.

Isoconversional methods require the performance of a series of experiments at different temperature programs and yield the values of effective activation energy E_α as a function of the extent of conversion α. A significant variation of E_α with α indicates that the process is kinetically complex and the E_α dependencies evaluated by an isoconversional method allow for meaningful mechanistic and kinetic analyses and for understanding multi-step processes, as well as for reliable kinetic predictions. Model-free isoconversional kinetic methods are a powerful tool to gain information on the reaction complexity through the E_α dependency determination. A change in the slope of the E_α dependency may generally be associated with a change in the rate-limiting step of the overall reaction mechanism. Isoconversional methods are based on the isoconversional principle that states that the reaction rate is only a function of temperature for a given constant value of the extent of conversion. Thus, the E_α dependency can be evaluated without any assumption of the reaction mechanism, as illustrated by Equation (7):

$$\left[\frac{d\ln(d\alpha/dt)}{dT^{-1}}\right]_\alpha = \left[\frac{d\ln k(T)}{dT^{-1}}\right]_\alpha + \left[\frac{d\ln f(\alpha)}{dT^{-1}}\right]_\alpha = -\frac{E_\alpha}{R} \qquad (7)$$

In this equation, the derivative of the term containing $f(\alpha)$ is zero because each computation is performed for a constant value of α (isoconversional methods).

3. Data Simulation

Two sets of simulated data were generated and analyzed for non-isothermal conditions using the heating rates of 1, 2 and 4 K·min^{-1} and for isothermal conditions at four temperatures of 120, 140, 160 and 180 °C. Usually, it is recommended to use three to five temperature programs [6]. In this work only three heating rates were used for the E_α dependency computations because using three or four heating rates led to the same values as the computation was performed using simulated data.

In the first set (set 1), a complex reaction involving an autocatalytic step, a first-order step and a diffusion process was simulated according to the following equations [18]:

$$k_D(T, \alpha) = D_0 \exp\left(-\frac{E_D}{RT} + K\alpha\right) \tag{8}$$

$$k_C(T, \alpha) = A_1 \exp\left(-\frac{E_1}{RT}\right) + A_2 \exp\left(-\frac{E_2}{RT}\right)\alpha^m \quad [19, 20] \tag{9}$$

$$k_{ef} = k_C^{-1} + k_D^{-1} \quad [18] \tag{10}$$

$$\frac{d\alpha}{dt} = k_{ef}(1 - \alpha)^n \tag{11}$$

$$\alpha_{i+1} = \alpha_i + \int_{t_i}^{t_{i+1}} \left(\frac{d\alpha}{dt}\right)_i dt \tag{12}$$

and using the following parameters [4]: $A_1 = 20739.00$ s^{-1}, $E_1 = 70.0$ kJ·mol^{-1}, $m = 1$, $A_2 = 499.00$ s^{-1}, $E_2 = 45.0$ kJ·mol^{-1}, $n = 1$, $D_0 = 1.43$ s^{-1}, $E_D = 4.4$ kJ·mol^{-1}, $K = -7.06$. Here k_D, k_C and k_{ef} respectively represent the specific rate constant for diffusion, the rate constant for the chemically controlled reaction and the effective rate constant. D_0 represents the pre-exponential factor of the diffusion-controlled reaction, A_1 the pre-exponential factor for the non-catalyzed reaction and A_2 the pre-exponential factor for the catalyzed reaction. K is a constant accounting for the effect of the chemical reaction on the change in diffusivity. E_D, E_1 and E_2 respectively represent the activation energy of the diffusion-controlled reaction, the activation energy of the non-catalyzed reaction and the activation energy of the catalyzed reaction. m and n are kinetic exponents [19,20].

A second set of data (set 2) involving a first-order step and a diffusion process was simulated according to the same procedure, wherein Equation (9) was replaced by Equation (13):

$$k_C(T) = A_1 \exp\left(\frac{-E_1}{RT}\right) \tag{13}$$

and using the following parameters: $A_1 = 1762.24$ s^{-1}, $E_1 = 50.0$ kJ·mol^{-1}, $n = 1$, $D_0 = 1.43$ s^{-1}, $E_D = 4.4$ kJ·mol^{-1}, $K = -7.06$.

4. Results

4.1. Autocatalytic Reaction with Diffusion-Controlled Part (Data Set 1)

4.1.1. Reaction Rate and Extent of Conversion for Non-Isothermal and Isothermal Conditions

Figure 2 shows the variation of the reaction rate and of the extent of conversion with temperature for three heating rates (non-isothermal data). The curves shifted to higher temperatures when the heating rate was increased. Figure 3 shows the variation of the reaction rate and of the extent of conversion with time for four temperatures (isothermal data). It can be observed that the maximum of the reaction rate was not obtained for $\alpha = 0$, as would be the case for a reaction order mechanism. Another characteristic feature of the autocatalytic model is that the isothermal α–t curves presented typical sigmoidal shapes, as shown in Figure 3.

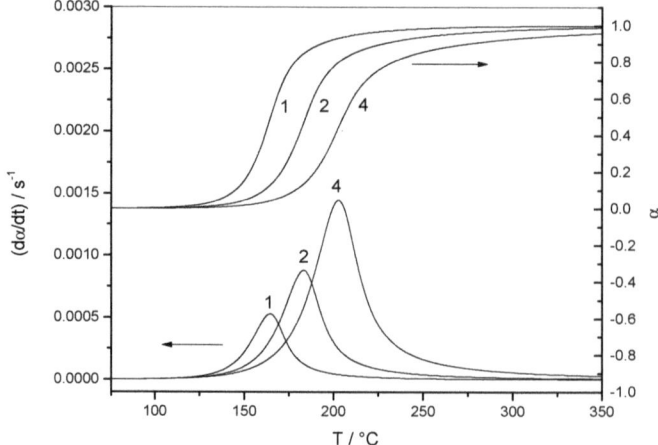

Figure 2. Non-isothermal data. Variation of the reaction rate (dα/dt) and the extent of conversion (α) with temperature (T) for data set 1. The heating rate of each experiment (in K min^{-1}) is indicated by each curve.

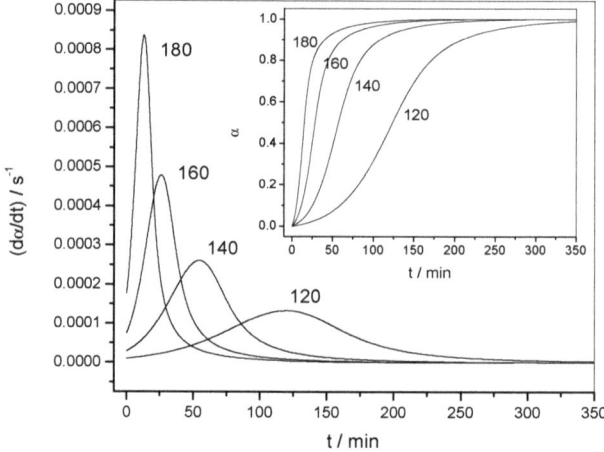

Figure 3. Isothermal data. Variation of the reaction rate (dα/dt) with time (t) for data set 1. Inset: Variation of the extent of conversion (α) with time (t). The temperature of each experiment (in °C) is indicated by each curve.

4.1.2. Dependence of the Effective Activation Energy and of the Pre-Exponential Factor

The dependence of the effective activation energy (E_α) with the extent of conversion (α) is presented in Figure 4 (left axis). Note that FR and NLN methods gave similar results in this case. The complexity of the mechanism is perfectly reflected by the important E_α dependence observed. The first value obtained was $E_\alpha = 63.4$ kJ·mol^{-1} for $\alpha = 0.02$ with the NLN method. This value is close to the value of $E_1 = 70.0$ kJ·mol^{-1} used in the simulation for the activation energy of the uncatalyzed reaction. For $\alpha = 0.001$ this value would be $E_\alpha = 69.0$ kJ·mol^{-1} which is very close to E_1. The lowest activation energy value was $E_\alpha = 5.1$ kJ·mol^{-1} for $\alpha = 0.98$ (NLN method), which is very close to the activation energy of diffusion ($E_D = 4.4$ kJ·mol^{-1}).

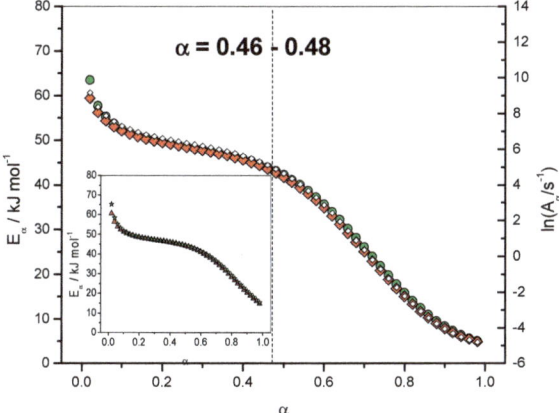

Figure 4. Dependence of the effective activation energy (E_α) and pre-exponential factor ($\ln A_\alpha$) with the extent of conversion (α). Non-isothermal data. Open lozenges: pre-exponential factor, red solid lozenges: E_α computed with FR method, green circles: E_α computed with NLN method. Inset: isothermal data. Solid triangles: E_α computed with FR method, solid stars: E_α computed with NLN method.

Figure 4 (right axis) give the results obtained for the dependence of the logarithm of the pre-exponential factor ($\ln A_\alpha$) with the extent of conversion (α). This dependence was computed using the compensation parameters method described by Sbirrazzuoli [11]. For this computation the model-fitting method of Achar–Brindley–Sharp was used in the interval $0.10 < \alpha < 0.40$ to evaluate the relationship between E and $\ln A$. A good correlation ($r^2 = 0.99903$) between these two parameters was obtained using the models F1 (Mampel, first order), A2 (Avrami–Erofeev), D3 (three-dimensional diffusion), R3 (contracting sphere) and D2 (two-dimensional diffusion) of ref. [6]. For $\alpha = 0.02$, $\ln(A_\alpha/s^{-1})$ was found to be 9.16, which is very close to the value used in the simulation, i.e., $\ln(A_1/s^{-1}) = 9.94$. The lowest value for $\ln(A_\alpha/s^{-1})$ was -4.82, which can be associated with $\ln(D_0/s^{-1}) = 0.36$. The dependence of the effective activation energy (E_α) with temperature (T) is presented in Figure 5 in association with the corresponding activation energies of the different steps, i.e., E_1, E_2 and E_D.

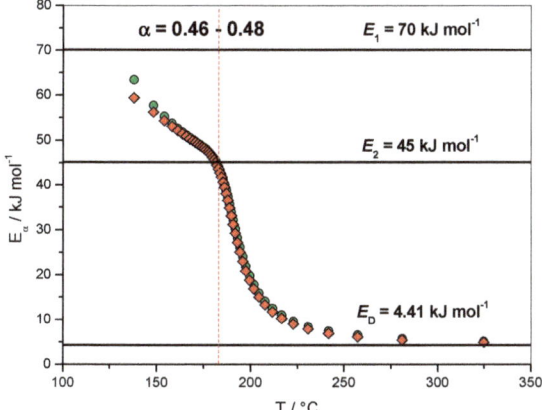

Figure 5. Non-isothermal data. Dependence of the effective activation energy (E_α) with temperature. Red solid lozenges: E_α computed with FR method, green open circles: E_α computed with NLN method.

4.1.3. Variation of the Reaction Rate with the Extent of Conversion

The terms $[A_\alpha f(\alpha)]$ and $\exp[-E_\alpha/(RT_\alpha)]$ can be evaluated using Equation (6), then a value of the reaction rate can be computed for each extent of conversion as the product of $[A_\alpha f(\alpha)]$ by $\exp[-E_\alpha/(RT_\alpha)]$. Figure 6 shows the relation between the dependence of the effective activation energy (E_α) with the extent of conversion (α) and the reaction rate ($d\alpha/dt$). The maximum of the reaction rate ($d\alpha/dt$) was located in the range of $0.46 < \alpha < 0.48$. Figure 6 shows that this corresponds to a change in the slope of the E_α dependence.

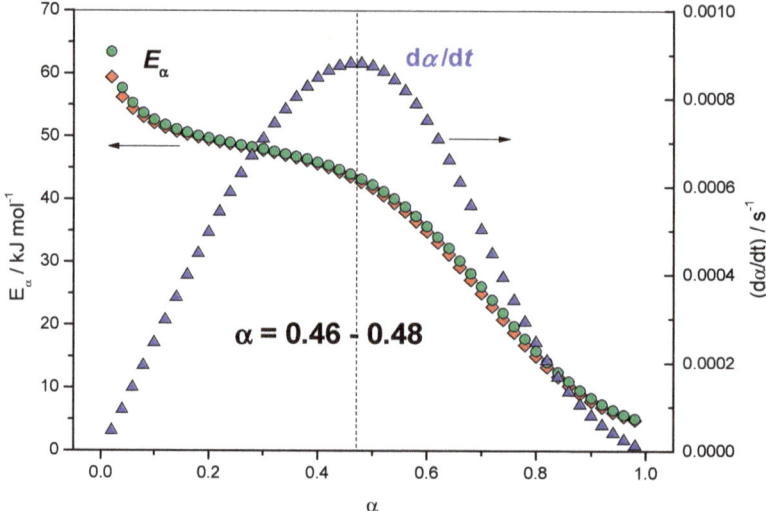

Figure 6. Dependence of the effective activation energy (E_α) with the extent of conversion (α). Red lozenges: E_α computed with FR method, green circles: E_α computed with NLN method, blue triangles: ($d\alpha/dt$) computed as the product of $[A_\alpha f(\alpha)]$ by $\exp[-E_\alpha/(RT_\alpha)]$ (Equation (6)).

The comparison of the terms $[A_\alpha f(\alpha)]$, $\exp[-E_\alpha/(RT_\alpha)]$ and ($d\alpha/dt$) can be used to identify rate-limiting steps in the overall reaction rate. The principle of this method is explained below and illustrated by Figure 7. From $\alpha = 0.02$ to 0.46, ($d\alpha/dt$) increases and the term $[A_\alpha f(\alpha)]$ decreases, so the increase of ($d\alpha/dt$) is attributed to the increase of the term $\exp[-E_\alpha/(RT_\alpha)]$, i.e., a decrease of E_α. Thus, the increase of the rate is mainly due to a favorable energetic term. When $\alpha \geq 0.48$, ($d\alpha/dt$) decreases, the term $\exp[-E_\alpha/(RT_\alpha)]$ still increases and the term $[A_\alpha f(\alpha)]$ still decreases. This indicates that the term $[A_\alpha f(\alpha)]$ dominates and corresponds to a change in the rate-limiting step at this stage of the reaction. The decrease of the rate is attributed to a change in the mechanism, which may originate from a change in $f(\alpha)$ or from an entropic change (A_α). This entropic change may be due to a change in configuration or a decrease in the efficiency of collisions. Thus, it is identified as a change in the rate-limiting step in the overall reaction rate for $\alpha = 0.46$–0.48. Table 1 gives some values of the various terms of Figure 7 used to identify a change in the rate-limiting step for $0.46 < \alpha < 0.48$.

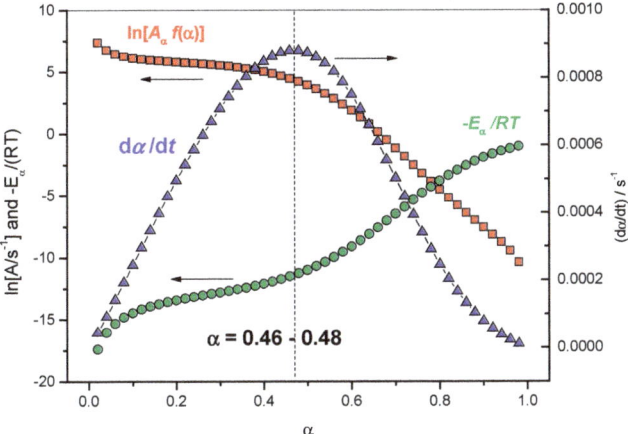

Figure 7. Red squares: variation of ln $[A_\alpha f(\alpha)]$ with α, green circles: variation of $-E_\alpha/(RT_\alpha)$ with α, blue triangles: variation of $(d\alpha/dt)$ computed as the product of $[A_\alpha f(\alpha)]$ by $\exp[-E_\alpha/(RT_\alpha)]$ with α.

Table 1. Some values of the various terms used to identify a change in the rate-limiting step.

$T_\alpha/°C$	E_α (FR)/kJ·mol^{-1}	α	ln (A_α/s^{-1})	$[A_\alpha f(\alpha)]/s^{-1}$	$\exp(-E_\alpha/(RT_\alpha)$	$(d\alpha/dt)/s^{-1}$
181.56	45.05	0.42	5.47	130.25	6.632×10^{-6}	8.638×10^{-4}
182.34	44.39	0.44	5.30	108.30	8.076×10^{-6}	8.746×10^{-4}
183.11	43.63	0.46	5.11	87.43	1.007×10^{-5}	8.802×10^{-4}
183.89	42.76	0.48	4.89	68.30	1.289×10^{-5}	8.801×10^{-4}
184.66	41.78	0.50	4.63	51.45	1.699×10^{-5}	8.740×10^{-4}
185.45	40.68	0.52	4.35	37.26	2.313×10^{-5}	8.618×10^{-4}

Once the change in the rate-limiting step was identified for $\alpha = 0.46$, an analysis of Figure 4 showed that $\ln(A_\alpha/s^{-1}) = 5.11$ and $E_\alpha = 44.0$ kJ·mol^{-1} for $\alpha = 0.46$, which are close to the values used in the simulation for the catalyzed reaction, i.e., $\ln(A_2/s^{-1}) = 6.21$ and $E_2 = 45.0$ kJ·mol^{-1}. The change in the curvature of the E_α dependence occurred exactly in this region of $\alpha = 0.46$–0.48, as seen in Figures 4–6. This result shows that, in addition to giving information on the rate-limiting steps, the E_α and $\ln(A_\alpha)$ dependencies can be used as estimate values of kinetic parameters to be used in a non-linear fitting procedure [9].

4.1.4. Variation of the Rate Coefficients with Extent of Conversion

Figure 8 shows the variation of the rate coefficients k_D, k_C and k_{ef} with the extent of conversion (α) k_D always decreased with α, while k_C always increased. Initial values of k_D were high, while they were low for k_C. This is the opposite at the end where the values of k_D were lower than the values of k_C. This is in perfect agreement with a chemical control at the beginning of the reaction ($k_C \ll k_D$) followed by a diffusion control at the end ($k_D \ll k_C$). The variation of k_{ef} is more complex, as reflected by Equation (10), showing an increasing trend at the beginning and a decreasing tendency at the end. The previously reported value of $\alpha = 0.46$–0.48 ($T \sim 183$ °C), attributed to a change in the rate-limiting step, corresponded to the point at which k_C started to become higher than k_{ef} and $k_D/k_C < 10$ (9.6). Generally, it is estimated that a factor of 100 is the minimum required to neglect one reaction to another. This value was obtained for $\alpha = 0.24$ ($T \sim 173$ °C), $k_D/k_C \approx 100$ (102.7).

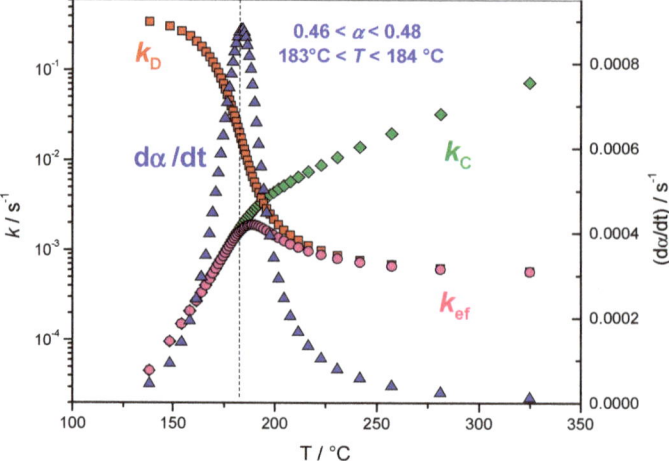

Figure 8. Blue triangles: variation of (dα/dt) computed as the product of [$A_\alpha f(\alpha)$] by exp[$-E_\alpha/(RT_\alpha)$] with temperature, red squares: variation of k_D with temperature, green lozenges: variation of k_C with temperature, magenta circles: variation of k_{ef} with temperature.

4.1.5. Variation of the Overall Rate Coefficient $k(T)$ and of the Effective Rate Coefficient $k_{ef}(T)$ with Reciprocal Temperature

The variations of ln $k(T)$ and ln $k_{ef}(T)$ with reciprocal temperature are presented in Figure 9. The values of α = 0.46–0.48 corresponded to the temperature at which the two rate constants were very close. The closest values were reached for α = 0.54.

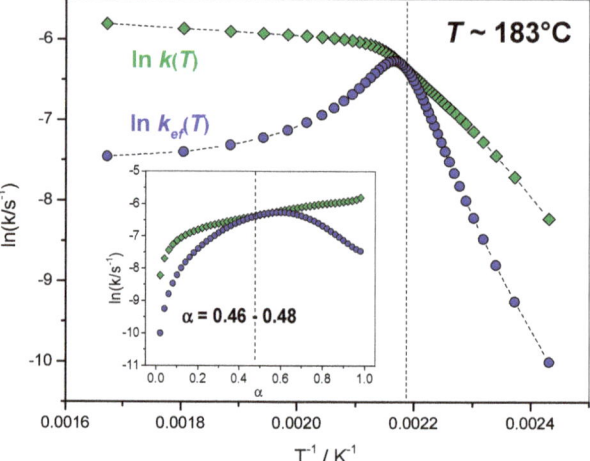

Figure 9. Green lozenges: variation of ln $k(T)$ computed as ln $k(T) = \ln(A_\alpha) - E_\alpha/(RT_\alpha)$ as a function of reciprocal temperature, blue circles: variation of ln $k_{ef}(T)$ computed using Equation (10) as a function of reciprocal temperature.

4.1.6. Fit of the E_α Dependence with the Sourour and Kamal and Diffusion Models

The isoconversional principle (Equation (7)) was applied to the autocatalytic equation of Sourour and Kamal (Equation (9)). This model is used to describe the initial stages of the reaction when it is chemically controlled [4,5,18]:

$$E_\alpha = \frac{k_1(T)E_1 + k_2(T)E_2\,\alpha^m}{k_1(T) + k_2(T)\,\alpha^m} \tag{14}$$

$$E_\alpha = \frac{(A_1/A_2)\exp(-E_1/RT)E_1 + \exp(-E_2/RT)E_2\,\alpha^m}{(A_1/A_2)\exp(-E_1/RT) + \exp(-E_2/RT)\,\alpha^m} \tag{15}$$

The isoconversional principle (Equation (7)) was also applied to the diffusion model (Equation (8)) used to describe the end of the reaction [4,5,18]:

$$E_\alpha = \frac{k(T)E_D + k_D(T,\alpha)E_2}{k(T) + k_D(T,\alpha)} \tag{16}$$

$$E_\alpha = \frac{(A_2/D_0)\exp(-E_2/RT)E_D + \exp(-E_D/RT + K\alpha)E_2}{(A_2/D_0)\exp(-E_2/RT) + \exp(-E_D/RT + K\alpha)} \tag{17}$$

Note that, according to Equation (15), at the lowest extent of conversion ($\alpha \to 0$) E_α tends toward the activation energy of the uncatalyzed reaction ($E_\alpha \to E_1$), in agreement to what was reported in the analysis of Figure 4.

The results of the fit of Equations (15) and (17) are given in Table 2. Some differences were observed between the simulated values and the values resulting from the non-linear fit. A higher discrepancy was obtained for A_1/A_2 (5918.03). This value should be 41.56. Nevertheless, a good agreement was found between the simulated and experimental data resulting from the non-linear fit for the other parameters. The result of this fit is presented in Figure 10.

Table 2. Parameters obtained by fitting Equations (15) and (17).

$2 < \alpha < 46\%$	A_1/A_2	$E_1/\text{kJ·mol}^{-1}$	$E_2/\text{kJ·mol}^{-1}$	m	MSSD [a]
Autocatalytic	5918.03	79.8	38.2	0.97	0.4435
$48 < \alpha < 98\%$	A/D_0	$E_2/\text{kJ·mol}^{-1}$	$E_D/\text{kJ·mol}^{-1}$	K	MSSD [a]
Diffusion	362.67	48.9	4.9	−7.84	0.0023

[a] Mean of the sum of squared deviations $MSSD = (1/n)\sum_{i=1}^{n}(E_{calc} - E_{ref})^2 / E_{ref}$.

Table 3. Parameters obtained by fitting Equations (18) and (19).

	A_1/s^{-1}	A_2/s^{-1}	$E_1/\text{kJ·mol}^{-1}$	$E_2/\text{kJ·mol}^{-1}$	m	MSSD [a]
$2 < \alpha < 46\%$	20756.17	498.52	67.6	42.6	1.2	0.6011
$2 < \alpha < 24\%$	20258.31	510.84	76.1	46.7	1.3	0.0345
	A_2/s^{-1}	D_0/s^{-1}	$E_2/\text{kJ·mol}^{-1}$	$E_D/\text{kJ·mol}^{-1}$	K	MSSD [a]
$48 < \alpha < 98\%$	498.9	1.43	48.8	4.9	−7.87	0.0024

[a] Mean of the sum of squared deviations $MSSD = (1/n)\sum_{i=1}^{n}(E_{calc} - E_{ref})^2 / E_{ref}$.

Equations (15) and (17) can be expanded to allow computations of the pre-exponential factors, as proposed by Sbirrazzuoli in [4]. The resulting Equations (18) and (19) can be fitted to estimate A_1, A_2 and D_0. The results are given in Table 3.

$$E_\alpha = \frac{A_1 \exp(-E_1/RT)E_1 + A_2 \exp(-E_2/RT)E_2\alpha^m}{A_1 \exp(-E_1/RT) + A_2 \exp(-E_2/RT)\alpha^m} \tag{18}$$

$$E_\alpha = \frac{A_2 \exp(-E_2/RT)E_D + D_0 \exp(-E_D/RT + K\alpha)E_2}{A_2 \exp(-E_2/RT) + D_0 \exp(-E_D/RT + K\alpha)} \tag{19}$$

It can be seen that the parameters are in very good agreement with the reference values and the values of A_1, A_2 are very close the reference values in this case. For the initial part of the reaction, the restriction of the fit to the interval $0.02 < \alpha < 0.24$ resulted in an improved accuracy (lower MSSD). This confirms the previous statement that when $k_D/k_C \approx 100$ it is possible to neglect the diffusion reaction, while for higher temperatures (or values of α) it is not completely negligible. Although the fit was better for parameters of Table 3 in comparison with those of Table 2, it is impossible to see the difference on the graph (inset of Figure 10). For the autocatalytic model of Sourour and Kamal, the accuracy of the fit was improved by addition of more flexibility when moving from Equation (15) to Equation (18). Nevertheless, this increased the possibilities of reaching local minima, resulting in several sets of parameters leading to an accurate fit of the data. The use of parameters estimated by the advanced isoconversional method, as initial values of the non-linear fit, greatly facilitated the achievement of meaningful parameters and not only fitting parameters, especially for the initial part of the reaction. However, the existence of local minima is a problem that cannot be underestimated and that is difficult to avoid when fitting complex mechanisms which involve many kinetic parameters to be determined using non-linear fits. This is less problematic for the end of the reaction, i.e., for the fit of Equation (8). The use of a genetic algorithm could be an efficient method to avoid this kind of problem [21].

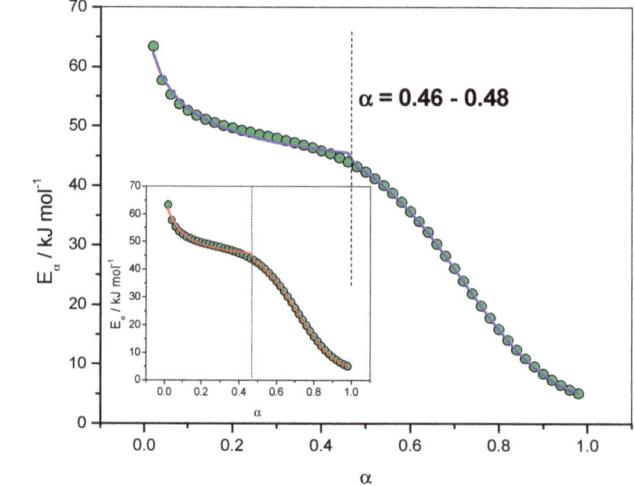

Figure 10. Circles: E_α dependency obtained with NLN method, line: fit using the autocatalytic model for $0.02 < \alpha < 0.46$ and the diffusion model for $0.48 < \alpha < 0.98$ with the parameters of Table 2. Inset: same plot for the data of Table 3.

4.2. First-Order Reaction with Diffusion-Controlled Part (Data Set 2)

Figure 11 shows the variation of the reaction rate and of the extent of conversion with temperature for three heating rates. In comparison with what was observed for data set 1, which include an autocatalytic step (Figure 2), the shift to higher temperatures upon increasing the heating rate was much lower in this case. This shows the difference between the reaction order and the autocatalytic mechanism for non-isothermal data. Figure 12 shows the variation of the reaction rate and of the extent of conversion with time for four temperatures (isothermal data). As expected, the maximum of the reaction rate was obtained for $\alpha = 0$ in this case and the isothermal α–t curves presented the characteristic shapes of a reaction order model.

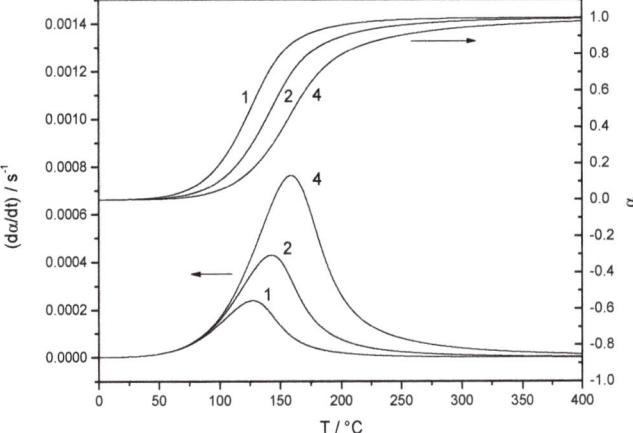

Figure 11. Non-isothermal data. Variation of the reaction rate (dα/dt) and extent of conversion (α) with temperature (T) for data set 2. The heating rate of each experiment (in K min^{-1}) is indicated by each curve.

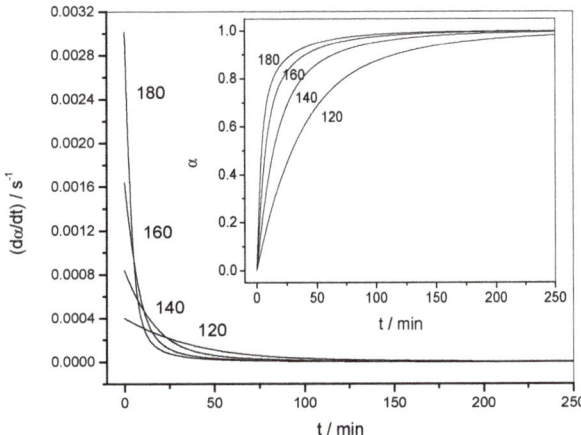

Figure 12. Isothermal data. Variation of the reaction rate (dα/dt) with time (t) for data set 2. Inset: Variation of the extent of conversion (α) with time (t). The temperature of each experiment (in °C) is indicated by each curve.

The dependence of the effective activation energy (E_α) with extent of conversion (α) is presented in Figure 13 (left axis) and the dependence with temperature is presented in Figure 14. Note that FR and NLN methods gave similar results in this case. The complexity of the mechanism is perfectly reflected by the important E_α dependence observed. The first value obtained was around 50 kJ·mol^{-1}, which is in perfect agreement with the value of the activation energy of the reaction order reaction E_1 (50.0 kJ·mol^{-1}). Figure 13 (right axis) gives the results obtained for the dependence of the logarithm of the pre-exponential factor (lnA_α/s^{-1}) with the extent of conversion (α). ln(A_α/s^{-1}) was found to be 6.61 and E_α = 50.2 kJ·mol^{-1} for α = 0.02, which are close to the values used in the simulation, i.e., ln(A_1/s^{-1}) = 7.47 and E_1 = 50.0 kJ·mol^{-1}. For α = 0.98, ln(A_α/s^{-1}) was found to be −5.65 and E_α = 5.5 kJ·mol^{-1}, which are also close to the values used in the simulation, i.e., ln(D_0/s^{-1}) = 0.36 and E_D = 4.4 kJ·mol^{-1}.

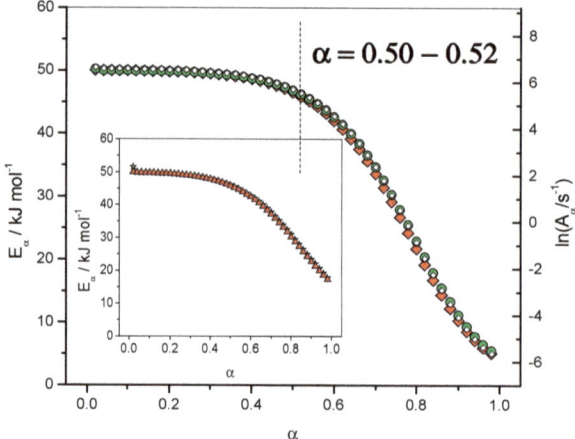

Figure 13. Dependence of the effective activation energy (E_α) and pre-exponential factor ($\ln A_\alpha$) with the extent of conversion (α). Non-isothermal data. Open lozenges: pre-exponential factor computed using compensation parameters method, red lozenges: E_α computed with FR method, green circles: E_α computed with NLN method. Inset: isothermal data. Red triangles: E_α computed with FR method, green stars: E_α computed with NLN method.

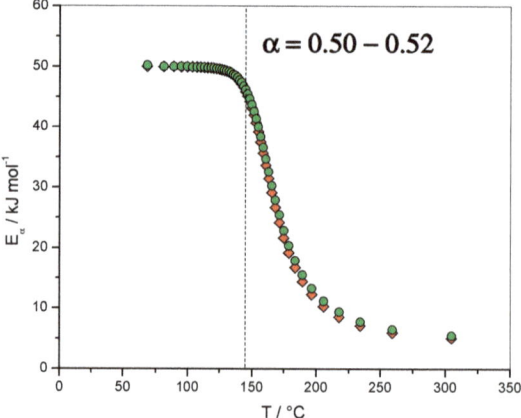

Figure 14. Non-isothermal data. Dependence of the effective activation energy (E_α) with temperature (T). Red lozenges: E_α computed with FR method, green circles: E_α computed with NLN method.

Application of the compensation parameters method [11] in the interval $0.05 < \alpha < 0.25$ (Achar-Brindley–Sharp's differential method, models F1, A3, A2, D3, $r^2 = 0.999999$) permit the identification of the F1 model (Mampel, first order) and the kinetic parameters of the reaction order reaction. The values found were $\ln(A/s^{-1}) = 7.45$ and $E = 49.9$ kJ·mol^{-1}, while the kinetic parameters used in the simulation were $\ln(A_1/s^{-1}) = 7.47$ and $E_1 = 50.0$ kJ·mol^{-1}.

In the case of two reactions with similar activation energies, the identification of the contribution of each individual reaction using the E_α dependency could be more difficult. Nevertheless, it is highly probable that these reactions would have different pre-exponential factors. In this case, we proposed to identify the complexity of the overall process by analyzing the A_α dependency computed with the compensation method.

5. Conclusions

A kinetic model has been proposed that simulates complex polymerizations with high accuracy. The first stages of the reaction were described by an autocatalytic process, followed by epoxy–amine addition and a diffusion-controlled model at the end of the reaction. Isoconversional methods—based on the assumption of the hypothesis of a single-step process only for each α value and the application of the Arrhenius equation to a very narrow temperature region related to this α value—give important insights into the change in the rate-limiting steps by analysis of the E_α dependency and its variations. In addition to this, the comparisons of the terms $(d\alpha/dt)$, $[A_\alpha f(\alpha)]$ and $\exp[-E_\alpha/(RT_\alpha)]$ evaluated by Friedman's method can help to identify the change in the rate-limiting steps of the overall mechanism as measured by thermoanalytical techniques. It was also concluded that the assumption of the validity of a single-step equation, when restricted to a given α value, holds for complex reactions.

Funding: This research received no external funding.

Conflicts of Interest: The author declares no conflict of interest.

References

1. Pascault, J.P.; Sautereau, H.; Verdu, J.; Williams, R.J.J. *Thermosetting polymers*; Marcel Dekker Inc.: New York, NY, USA, 2002.
2. Hale, A.; Macosko, C.W.; Bair, H.E. Glass Transition Temperature as a Function of Conversion in Thermosetting Polymers. *Macromolecules* **1991**, *24*, 2610–2621. [CrossRef]
3. Pascault, J.P.; Williams, R.J.J. Glass Transition Temperature Versus Conversion Relationships for Thermosetting Polymers. *J. Polym. Sci. Part B Polym. Phys.* **1990**, *28*, 85–95. [CrossRef]
4. Alzina, C.; Sbirrazzuoli, N.; Mija, A. Hybrid Nanocomposites: Advanced Nonlinear Method for Calculating Key Kinetic Parameters of Complex Cure Kinetics. *J. Phys. Chem. B* **2010**, *114*, 12480–12487. [CrossRef] [PubMed]
5. Vyazovkin, S.; Sbirrazzuoli, N. Isoconversional kinetic analysis of thermally stimulated processes in polymers. *Macromol. Rapid Comm.* **2006**, *27*, 1515–1532. [CrossRef]
6. Vyazovkin, S.; Burnham, A.K.; Criado, J.M.; Pérez-Maqueda, L.A.; Popescu, C.; Sbirrazzuoli, N. ICTAC kinetics committee recommendations for performing kinetic computations on thermal analysis data. *Thermochim. Acta* **2011**, *520*, 1–19. [CrossRef]
7. Vyazovkin, S. *Isoconversional Kinetics of Thermally Stimulated Processes*; Springer: Berlin, Germany, 2015.
8. Vyazovkin, S. Evaluation of activation energy of thermally stimulated solid-state reactions under arbitrary variation of temperature. *J. Comput. Chem.* **1997**, *18*, 393–402. [CrossRef]
9. Sbirrazzuoli, N.; Vincent, L.; Vyazovkin, S. Comparison of several computational procedures for evaluating the kinetics of thermally stimulated condensed phase reactions. *Chemometr. Intell. Lab* **2000**, *54*, 53–60. [CrossRef]
10. Vyazovkin, S. Modification of the integral isoconversional method to account for variation in the activation energy. *J. Comput. Chem.* **2001**, *22*, 178–183. [CrossRef]
11. Sbirrazzuoli, N. Determination of pre-exponential factors and of the mathematical functions $f(\alpha)$ or $G(\alpha)$ that describe the reaction mechanism in a model-free way. *Thermochim. Acta* **2013**, *564*, 59–69. [CrossRef]
12. Sbirrazzuoli, N. Is the Friedman method applicable to transformations with temperature dependent reaction heat? *Macromol. Chem. Phys.* **2007**, *208*, 1592–1597. [CrossRef]
13. Sbirrazzuoli, N.; Vincent, L.; Vyazovkin, S. Electronic solution to the problem of a kinetic standard for DSC measurements. *Chemometr. Intell. Lab* **2000**, *52*, 23–32. [CrossRef]
14. Sbirrazzuoli, N.; Brunel, D.; Elegant, L. Different kinetic equations analysis. *J. Therm. Anal.* **1992**, *38*, 1509–1524. [CrossRef]
15. Sbirrazzuoli, N.; Girault, Y.; Elegant, L. Simulations for evaluation of kinetic methods in differential scanning calorimetry. Part 3—peak maximum evolution methods and isoconversional methods. *Thermochim. Acta* **1997**, *293*, 25–37. [CrossRef]
16. Falco, G.; Guigo, N.; Vincent, L.; Sbirrazzuoli, N. FA polymerization disruption by protic polar solvent. *Polymers* **2018**, *10*, 529. [CrossRef] [PubMed]

17. Friedman, H.L. Kinetics of thermal degradation of char-forming plastics from thermogravimetry. Application to a phenolic plastic. *J. Polym. Sci. Part C* **1964**, *6*, 183–195. [CrossRef]
18. Vyazovkin, S.; Sbirrazzuoli, N. Mechanism and kinetics of epoxy-amine cure studied by differential scanning calorimetry. *Macromolecules* **1996**, *29*, 1867–1873. [CrossRef]
19. Ryan, M.E.; Dutta, A. Kinetics of epoxy cure: A rapid technique for kinetic parameter estimation. *Polymer* **1979**, *20*, 203–206. [CrossRef]
20. Sourour, S.; Kamal, M.R. Differential scanning calorimetry of epoxy cure: Isothermal cure kinetics. *Thermochim. Acta* **1976**, *14*, 41–59. [CrossRef]
21. Yamini, G.; Shakeri, A.; Vafayan, M.; Zohuriaan-Mehr, M.J.; Kabiri, K.; Zolghadr, M. Cure kinetics of modified lignosulfonate/epoxy blends. *Thermochim. Acta* **2019**, *675*, 18–28. [CrossRef]

© 2019 by the author. Licensee MDPI, Basel, Switzerland. This article is an open access article distributed under the terms and conditions of the Creative Commons Attribution (CC BY) license (http://creativecommons.org/licenses/by/4.0/).

Article

Non-Isothermal Crystallization Kinetics of Poly(4-Hydroxybutyrate) Biopolymer

Ina Keridou [1], Luis J. del Valle [1,2], Lutz Funk [3], Pau Turon [3], Lourdes Franco [1,2,*] and Jordi Puiggalí [1,2,*]

1. Departament d'Enginyeria Química, Universitat Politècnica de Catalunya, Escola d'Enginyeria de Barcelona Est-EEBE, c/Eduard Maristany 10-14, 08019 Barcelona, Spain
2. Center for Research in Nano-Engineering, Universitat Politècnica de Catalunya, Campus Sud, Edifici C', c/Pasqual i Vila s/n, E-08028 Barcelona, Spain
3. B. Braun Surgical, S.A. Carretera de Terrassa 121, 08191 Rubí (Barcelona), Spain
* Correspondence: Lourdes.Franco@upc.edu (L.F.); Jordi.Puiggali@upc.edu (J.P.); Tel.: +34-93-401-1870 (L.F.); +34-93-401-5649 (J.P.)

Academic Editor: Sergey Vyazovkin
Received: 16 June 2019; Accepted: 2 August 2019; Published: 5 August 2019

Abstract: The non-isothermal crystallization of the biodegradable poly(4-hydroxybutyrate) (P4HB) has been studied by means of differential scanning calorimetry (DSC) and polarizing optical microscopy (POM). In the first case, Avrami, Ozawa, Mo, Cazé, and Friedman methodologies were applied. The isoconversional approach developed by Vyazovkin allowed also the determination of a secondary nucleation parameter of 2.10×10^5 K^2 and estimating a temperature close to 10 °C for the maximum crystal growth rate. Similar values (i.e., 2.22×10^5 K^2 and 9 °C) were evaluated from non-isothermal Avrami parameters. All experimental data corresponded to a limited region where the polymer crystallized according to a single regime. Negative and ringed spherulites were always obtained from the non-isothermal crystallization of P4HB from the melt. The texture of spherulites was dependent on the crystallization temperature, and specifically, the interring spacing decreased with the decrease of the crystallization temperature (T_c). Synchrotron data indicated that the thickness of the constitutive lamellae varied with the cooling rate, being deduced as a lamellar insertion mechanism that became more relevant when the cooling rate increased. POM non-isothermal measurements were also consistent with a single crystallization regime and provided direct measurements of the crystallization growth rate (G). Analysis of the POM data gave a secondary nucleation constant and a bell-shaped G-T_c dependence that was in relative agreement with DSC analysis. All non-isothermal data were finally compared with information derived from previous isothermal analyses.

Keywords: poly(4-hydroxybutyric acid); bioabsorbable sutures; crystallization kinetics; non-isothermal crystallization; isoconversional methods; spherulites; synchrotron radiation

1. Introduction

Poly(4-hydroxybutyrate) (P4HB) is a biodegradable, linear, and aliphatic polyester with interesting applications in the biomedical field. This is mainly associated with its use as a wound closure material [1]. This hydroxyalkanoate (HA) derivative can easily be produced from microorganisms (e.g., *Escherichia coli K12*) by recombinant fermentation under a deficit of nutrients or other stress limitations [2–4]. Basically, the polymer is employed by microorganisms (as other polyhydroxyalkanoates (PHAs), such as poly(3-hydroxybutyrate) (P3HB) and polyhydroxyvalerate (PHV), as an energy storage form produced by the carbon assimilation from glucose or starch sources. Biosynthesis is the only practical pathway to produce P4HB, since samples with a reasonable molecular weight cannot be obtained from conventional chemical synthesis (e.g., ring-opening polymerization of butyrolactone) [5,6].

P4HB has exceptional mechanical properties which allow the preparation of strong fibers with good retention in vivo [7]. The use of P4HB for soft tissue ligation was approved by the food and drug administration (FDA) in 2007, being commercialized as a long-term absorbable monofilament suture under the trademark of MonoMax (B. Braun Surgical, S. A.). In fact, P4HB is the only PHA allowed for regulatory agencies (e.g., USFDA and EU) to be used in clinical applications [8,9], such as reconstructive surgery, materials for cardiovascular applications, such as vascular grafts and stents [10], and devices for repair of hernias [11], ligaments and tendons, are other well-known applications of P4HB [1].

Probably the most remarkable feature of P4HB is not only its high ductility and flexibility (i.e., the elongation at break can reach a value of 1000%) but also the great interest for the preparation of copolymers with other HAs. Thus, the highly brittle and crystalline P3HB can easily be modified by copolymerization with different molar fractions of the 4HB monomer to obtain a group of materials with suitable mechanical strength and properties, including degradation rate [12].

On the other hand, P4HB (-[$O(CH_2)_3CO$]$_n$-) has different properties than the similar linear polyesters with a lower (i.e., polyglycolide, -[OCH_2CO]$_n$-, PGA) and higher (i.e., poly(ε-caprolactone), -[$O(CH_2)_5CO$]$_n$-, PCL) number of carbon atoms in their chemical repeat unit. Surprisingly, a continuous evolution of properties with the length of the repeat unit is not observed even while maintaining its parity, that is avoiding great changes of the crystalline structure. For example, typical elongation at break values are <3%, 1000%, and 80%, while elastic modulus values are 6900 MPa, 70 Mpa, and 400 MPa for PGA, P4HB, and PCL, respectively [10]. Differences in the degradation rate are clear (e.g., fast and slow for PGA and P4HB, respectively), and it is of interest to explore the use of combinations of such materials to tune the final degradability of the material [13].

The peculiar properties of P4HB have attracted great attention for the development of new materials for applications in the biomedical field. However, research concerning physical characterization is surprisingly relatively scarce, especially considering that the above-indicated differences with related polyesters. To the best of our knowledge works concerning physical characterization of P4HB only include the study of its crystalline structure by both X-ray and electron diffraction techniques [14–16], morphologic studies [16], evaluation of hydrolytic and enzymatic degradation mechanisms [16,17], and determination of basic thermal and mechanical properties [1].

Thermal properties of P4HB are strongly dependent on the preparation conditions. Samples are crystalline and have shown a preferential melting peak at 72 °C after annealing at an appropriate temperature, under stress conditions or from solution crystallization. By contrast, a decrease in the melting temperature at 58 °C is characteristic when samples are crystallized after melting. Control of the crystallization process seems fundamental considering the relatively low melting point that becomes close to room temperature and the derived applications where crystallinity plays a significant role (i.e., degradation rate and even elastic behavior). Isothermal crystallization studies of P4HB have recently been carried out considering both calorimetric (DSC) and optical microscopy (OM) experimental data [18]. The obtained data covered a very narrow range of crystallization temperatures (i.e., 24–38 °C and 37–49 °C for DSC and OM observations, respectively) due to experimental limitations (e.g., high primary nucleation and slow crystallization rate at low and high temperatures, respectively). Results indicated a crystallization process from the melt state defined by an averaged Avrami exponent of 2.56. Crystallization occurred according to a single regime characterized by a secondary nucleation constant of 1.69×10^5 K^2 (from DSC data) and 1.58×10^5 K^2 (from OM data).

Nowadays, efforts are focused on understanding the non-isothermal crystallization behavior of semi crystalline polymers since this is more appropriate to describe the usual processing conditions and can even be useful in the description of the crystallization process for a wider temperature range. Different methodologies have been developed to carry out the evaluation of non-isothermal crystallization from DSC and OM experimental data. Results are in general controversial when different methods are compared, especially when interpretation and theoretical approximations are not clear. The goals of the present paper are a) the specific study of non-isothermal crystallization of P4HB, a polymer with increasing applied interest and with a crystallization process scarcely studied; b) a comparison of the non-isothermal process with isothermal crystallization data recently evaluated, and c) the evaluation of the more significant highlights given by the different theories derived from DSC and OM data.

2. Results and Discussion

2.1. Limitations of the Avrami Analysis of Non-Isothermal Crystallization of P4HB

Figure 1 shows the dynamic DSC exotherms obtained by cooling melted P4HB samples at different rates. Logically, peaks moved progressively to lower temperatures as the cooling rate increased. Calorimetric data allowed the determination of the relative degree of crystallinity at any temperature, $\chi(T)$, for all cooling rates by the expression

$$\chi(T) = \frac{\int_{T_0}^{T}(dH_c/dT)dT}{\int_{T_0}^{T_\infty}(dH_c/dT)dT} \quad (1)$$

where dH_c is the enthalpy of crystallization released within an infinitesimal temperature range dT, T_0 denotes the initial crystallization temperature, and T_∞ is the temperature required to complete the crystallization process. Thus, the denominator corresponds to the overall enthalpy of crystallization for specific heating/cooling conditions. Note that this relative crystallinity is obviously higher than the real extent of crystallization, which is limited by the slow dynamics of polymeric molecular chains.

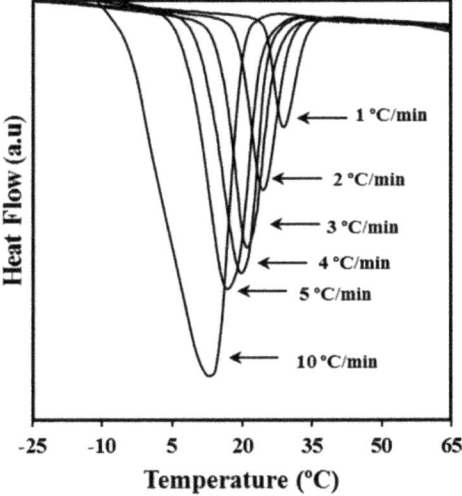

Figure 1. Dynamic differential scanning calorimetry (DSC) curves obtained at the indicated cooling rates for poly(4-hydroxybutyrate) (P4HB) crystallization from the melt state.

The time dependence of the degree of crystallinity (Figure 2a) can be derived considering the relationship

$$(t - t_0) = (T - T_0)/\varphi \quad (2)$$

where T_0 is the temperature when crystallization begins ($t = t_0$) and φ is the cooling rate.

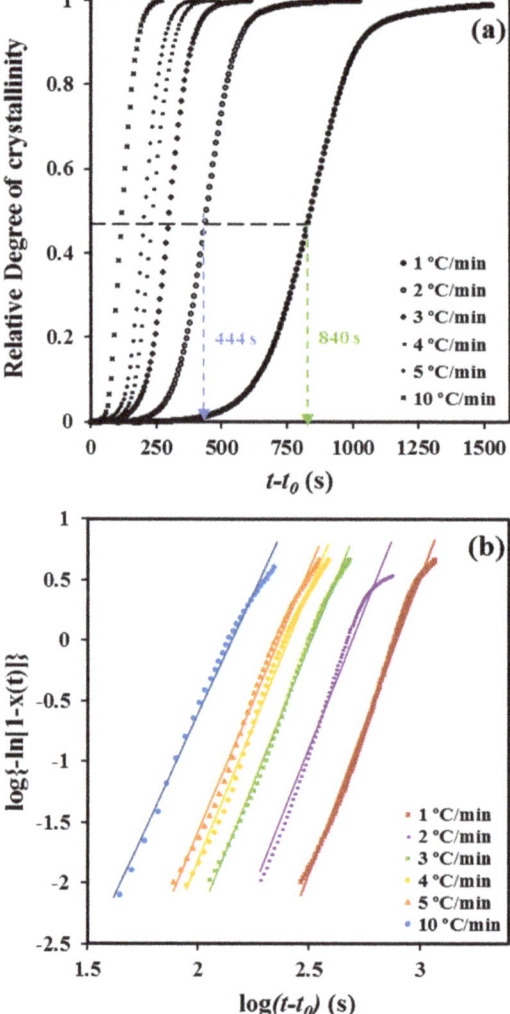

Figure 2. Time evolution of relative crystallinity (**a**) and Avrami plots, (**b**) at the indicated cooling rates for the non-isothermal crystallization of P4HB.

The typical Avrami analysis can be applied to these non-isothermal experiments based on Equation (3).

$$1 - \chi(t - t_0) = \exp\left[-Z(t - t_0)^n\right], \quad (3)$$

where Z is the temperature-dependent rate constant and n the Avrami exponent.

This exponent has a physical sense for isothermal crystallization of semi crystalline polymers despite being initially postulated for the study of the phase transformation of metals [19,20]. It has been established that the exponent varies according to the dimensionality of the crystal growth and the type of nucleation [21]. Namely, a time-dependent thermal nucleation (i.e., homogeneous nucleation and sporadic heterogeneous nucleation) can be differentiated from athermal nucleation (i.e., instantaneous heterogeneous nucleation) after evaluating the crystal dimensionality. Unfortunately, the direct application of the Avrami equation to the evaluation of non-isothermal crystallization merely corresponds to a mathematical fitting. This allows the evaluation of the variation of crystallinity with crystallization time, but parameters lose their physical meaning since, for example, values of the exponent become usually higher than four.

Plots of log $\{-\ln[1 - \chi(t - t_0)]\}$ versus log $(t - t_0)$ showed a good linearity (Figure 2b) before the start of the secondary crystallization process associated with the impingement of spherulitic crystals (i.e., the decrease of the dimensionality of the crystal growth). In fact, the linearity is observed, in our case, up to a relative crystallinity of 0.92.

Table 1 summarizes the main kinetic parameters deduced from the Avrami analysis, including the overall crystallization rate, k, calculated as $Z^{1/n}$. This rate has units of s^{-1} and consequently, can be used to compare data from crystallizations having different Avrami exponents. Note that the usually employed Z parameter is not useful since it has units of s^{-n} (i.e., it is dependent on the change of nucleation mechanism and crystal growth dimensionality). Logically, the crystallization became faster as the cooling rate increased and specifically a change from 1.30×10^3 to 4.40×10^3 was detected when the rate was increased from 1 °C/min to 5 °C/min. The general observed trend was the decrease of the Avrami exponent (being 5.0 the average value) when the cooling rates were increasing. This is clearly higher than the postulated value for a maximum crystal dimensionality and a homogeneous (or even a sporadic heterogeneous) nucleation, making the deduction of the crystallization mechanism as above indicated impossible. Nevertheless, the observed decrease suggests that the dimensionality decreased as crystallization was conducted faster. Reported data for the non-isothermal crystallization of the related PCL polyester also gave a high Avrami exponent (i.e., n between 3 and 4), although it was interpreted as a three-dimensional spherulitic growth with homogeneous nucleation [22].

Table 1. Non-isothermal crystallization kinetic parameters deduced from differential scanning calorimetry (DSC) experiments for poly(4-hydroxybutyrate) (P4HB).

Φ (°C/min)	n	Z (s^{-n})	$k \times 10^3$ (s^{-1})	$\tau_{1/2}$ (s)	$(1/\tau_{1/2}) \times 10^3$ (s^{-1})	$(Z/\ln 2)^{1/n} \times 10^3$ (s^{-1})
1	5.67	4.42×10^{-17}	1.30	840	1.19	1.39
2	5.18	1.47×10^{-14}	1.95	444	2.25	2.28
3	4.92	4.83×10^{-13}	3.13	300	3.33	3.37
4	4.24	6.89×10^{-11}	4.00	231	4.33	4.36
5	4.21	1.22×10^{-10}	4.42	207	4.83	4.82
10	3.98	2.54×10^{-9}	6.92	120	8.33	7.59

Table 1 also shows a satisfying agreement between the reciprocal crystallization half-times $(1/\tau_{1/2})$ that were directly determined from the experimental data and those that were deduced from the Avrami parameters (i.e., $1/\tau_{1/2} = (Z/\ln 2)^{1/n}$). The deduced parameters are at least appropriate to simulate the non-isothermal crystallization process.

2.2. Alternatives to the Avrami Analysis for the Non-Isothermal Crystallization of P4HB

Ozawa [23] proposed a modified Avrami equation that directly considers the effect of the cooling rate. The approach assumes that a non-isothermal process is the result of an infinite number of small isothermal steps. The Ozawa equation was formulated by applying the mathematical derivation of Evans [24] to the Avrami equation and considering a constant cooling rate.

$$1 - \chi(T) = \exp(-R(T)/\varphi^m) \qquad (4)$$

where $R(T)$ is a cooling function that depends on the temperature of the process and m is the so-named Ozawa exponent. The difference of the exponent with the above indicated Avrami exponent is not clear, and generally, it has been interpreted in the same way [25].

The exponent can be deduced from the plot of $\log\{-\ln[1 - \chi(T)]\}$ versus log φ for conversions determined at the same temperature and different cooling rates. The main limitation of the method is that the linearity is observed for a restricted range of cooling rates, as can be observed in Figure 3. In fact, the absence of linearity becomes more evident as the process becomes faster (i.e., the temperature is lower). In other words, the analysis is highly sensitive to the variation between primary and secondary crystallization processes. It was noted in the previous section that secondary crystallization only becomes significant when the degree of crystallinity becomes very high. Thus, plots performed at high crystallization temperatures are linear over a wide range of cooling rates due to the slow crystallization and the difficulty of entering into the secondary crystallization region.

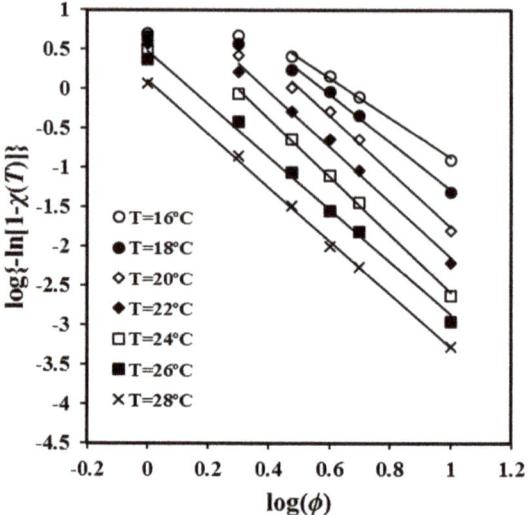

Figure 3. Plots of $\log\{-\ln[1 - \chi(T)]\}$ versus $\log\varphi$ for non-isothermal crystallizations of P4HB at the indicated temperatures.

The slopes of the different straight segments shown in Figure 3 are close to −3.5, which suggests athermal nucleation and three-dimensional spherulitic growth. The value of the Avrami exponent is in clear contradiction with the reported values from isothermal studies, a feature that is congruent with the previous discussion and that points out the limitation of the application of the Avrami analysis to non-isothermal crystallization studies. The values of the exponent clearly decreased at low crystallization temperatures when cooling rates were low as a consequence of the increasing secondary crystallization. Thus, exponents of 1.44, 1.87, and 2.35 were determined in the cooling rate region between 2 and 3 °C/min for temperatures of 16 °C, 18 °C, and 20 °C, respectively. Low values of 1.14 and 1.87 were estimated in the cooling rate region between 1 and 2 °C/min for temperatures of 22 °C and 24 °C, respectively.

Optical micrographs (Figures 4 and 5) taken during the non-isothermal crystallization clearly show the development of banded spherulites with a negative birefringence and increasing nucleation as a consequence of the temperature decrease. For instance, the dashed circles that are observed indicate the apparition of new nuclei and spherulites during the crystallization that was performed at a representative cooling rate of 0.5 °C/min. Obviously, these non-isothermal experiments cannot demonstrate that the crystallization takes place according to an athermal process (i.e., the apparition of new nuclei during crystallization at a given temperature), but this feature was corroborated in previous isothermal crystallization studies. The spherulitic texture was variable (e.g., the width of bands shown in Figure 5 continuously decreased during the crystallization that began at 45 °C and finished at 29 °C), since changes were expected between regions crystallized at high and low temperatures.

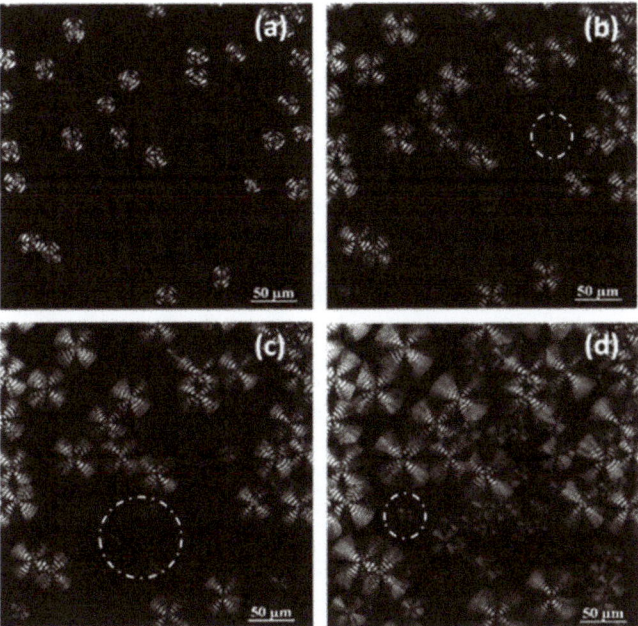

Figure 4. Optical micrographs of P4HB spherulites formed at the indicated non-isothermal crystallization times by cooling the sample at a rate of 0.5 °C/min from 45 °C, where some spherulites were isothermally grown (60 min) from the melt state. Micrographs were taken at temperatures of 44 °C (**a**), 38 °C (**b**), 36 °C (**c**), and 33 °C (**d**).

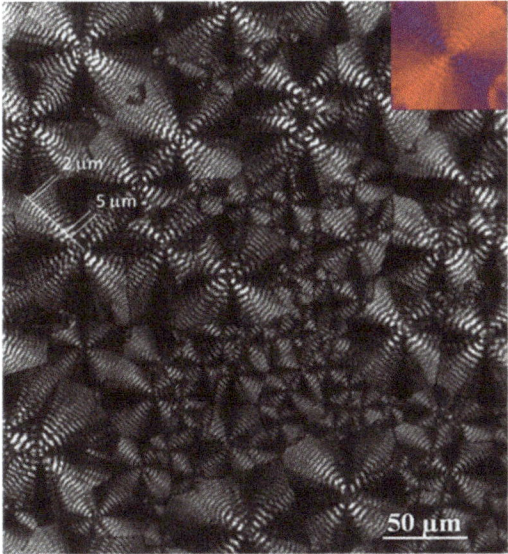

Figure 5. Optical micrograph at the end of the non-isothermal crystallization (0.5 °C/min) of P4HB. The decrease in the interring spacing during crystallization is evident. Inset shows a micrograph taken with a first-order red tint plate to determine the birefringence sign.

Liu et al. [26] postulated an alternative calorimetric analysis based on the combination of typical Avrami and Ozawa treatments. Equation (5) (also known as the Mo equation) was derived.

$$\log \varphi = \log F(T) - a \log (t - t_0) \tag{5}$$

where $F(T)$ is a new kinetic function, defined as $[\chi(T)/Z(T)]^{1/m}$, and a is the ratio between apparent Avrami and Ozawa exponents (n/m).

The main problem of the model is the non-clear physical sense of $F(T)$, which was defined as the cooling rate that must be chosen at a unit crystallization time to reach a certain crystallinity. $F(T)$ increases with increasing crystallinity, indicating that a higher cooling rate is required. The main interest of Liu analysis concerns the evaluation/quantification of how the modifications of a system (e.g., incorporation of additives) may be reflected in a more difficult crystallization (i.e., higher $F(T)$ values for a given crystallinity).

A plot of $\log \varphi$ versus $\log (t - t_0)$ yields a series of straight lines (Figure 6) that suggest the validity of the Mo equation for the P4HB system. The intercept and slope of these lines can be used to estimate the kinematic parameters. The values of $F(T)$ (Table 2) increased with crystallinity, indicating that the motion of molecular chains became slower, making the formation of crystals more difficult.

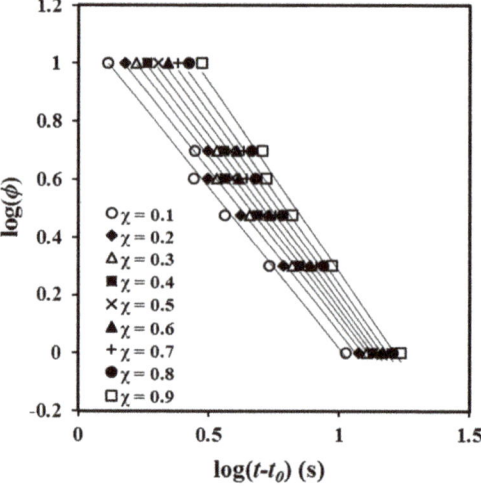

Figure 6. Plots of $\log \varphi$ versus $\log(t - t_0)$ at the indicated crystallinities for non-isothermal crystallization of P4HB.

Table 2. Values of kinetic parameters at a given crystallinity estimated from the combined Liu model for non-isothermal crystallization of P4HB. [1]

$\chi(T)$	a'	$F(T)$	r^2
0.1	−1.11	13.41	0.988
0.2	−1.12	15.70	0.990
0.3	−1.14	17.62	0.991
0.4	−1.16	19.74	0.991
0.5	−1.19	22.14	0.991
0.6	−1.22	24.85	0.990
0.7	−1.24	28.03	0.990
0.8	−1.28	32.30	0.989
0.9	−1.31	38.65	0.989

[1] r: correlation factor.

The second piece of information derived from the Liu and Mo analysis concerns the n/m ratio, which theoretically should be equal to 1 if equivalence of exponents is assumed. Different non-isothermal studies revealed as presumably a good equivalence between both exponents [25,27–29].

Table 2 shows that the values of a were almost constant and close to 1 (i.e., between 1.11 and 1.31). Nevertheless, the a values slightly increased with crystallinity (i.e., an increased dissimilarity between Avrami and Ozawa exponents was observed when crystallinity increased). Specifically, Avrami exponent became regularly higher than the Ozawa exponent. Note that Figure 3 demonstrates that Ozawa exponent changed and dramatically decreased at high crystallinity as a consequence of the great contribution of secondary crystallization. Previous studies performed with the related PCL polyester indicated a ratio higher than 1 and specifically an increase from 1.41 to 1.65 for conversions varying from 0.2 and 0.8 [22].

Cazé has also developed a methodology able to render an average value of the Avrami exponent for all the crystallization process [30]. The method hypothesizes that crystallization exotherms follow a Gaussian curve and considers three temperature inflection points: the onset temperature, the peak temperature, and the end crystallization temperature. The approach assumes that these three temperatures vary linearly with the cooling rate. A theoretical peak temperature T'_p and a new constant a' can be estimated, assuming the following equation:

$$\ln[1 - \ln(1 - \chi(T))] = a'(T - T'_p) \tag{6}$$

Plots of $\ln[1 - \ln(1 - \chi(T))]$ versus T at different cooling rates (Figure 7a) are linear and allow the indicated parameters (i.e., a' and T'_p) to be calculated (Table 3). It is worth noting that Equation (6) is confined to the primary crystallization regime. The range of crystallinities starts at 2% to ensure precision and cover data in such a way that the correlation coefficient is greater than 0.99. Table 3 also shows a favorable agreement between the peak temperature that was estimated assuming the Cazé model and that directly determined from the experimental DSC data.

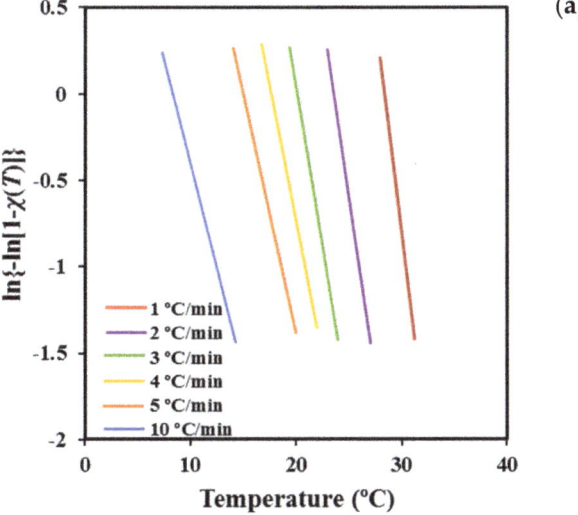

Figure 7. Cont.

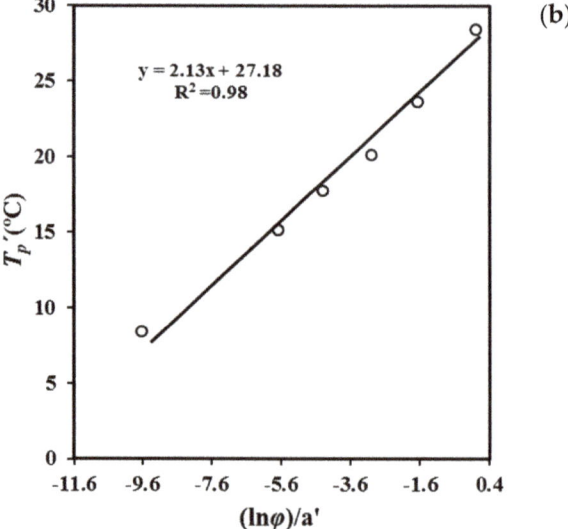

Figure 7. Plots of $\ln\{-\ln[1 - \chi(T)]\}$ against temperature for non-isothermal crystallization of P4HB with cooling rates as a parameter (**a**). Determination of the Avrami exponent using linear plots of T_p' against $(\ln \varphi)/a'$ (**b**).

Table 3. Characteristic crystallization parameters obtained for P4HB by using the methodology developed by Cazé et al. [30].

φ (°C/min)	a'	T_p' (°C)	$T_p{}^a$ (°C)
1	−0.49	28.43	28.92
2	−0.41	23.63	24.18
3	−0.37	20.15	20.86
4	−0.32	17.72	19.68
5	−0.28	15.14	16.87
10	−0.24	−9.61	13.27

[a] Temperature determined for the exothermic peak observed in the cooling scans.

The deduced peak temperatures can then be related to the cooling rate (Figure 7b) by the expression

$$T'_p = (m/a') \ln \varphi - b'/a' \qquad (7)$$

where b' is a new constant.

The plot of T'_p versus $\ln \varphi\, a'$ (Figure 7b) gives straight lines with a slope equal to the estimated Ozawa exponent m. The values obtained for P4HB are close to 2.13, which has physical meaning and appears very close to those deduced from the isothermal analysis [18] (i.e., exponents varied between 2.35 and 2.62, with 2.56 being the average value). Note that the derived value of m corresponds to an Avrami exponent of 2.57 to 2.87 if the n/m ratio deduced from the Liu model is applied. The obtained results support the suitability of the Cazé methodology, as previously reported in the non-isothermal study of different polymers [31].

2.3. Isoconversional Methods. Activation Energy

Evaluation of the activation energy of a non-isothermal crystallization from the melt was performed by using the isoconversional method of Friedman [32]. This considers that the energy barrier of crystallization can vary during the non-isothermal melt crystallization according to Equation (8).

$$d\chi(T)/dt = A \exp(-\Delta E/RT) f[\chi(T)] \qquad (8)$$

where A is a preexponential factor, and $f[\chi(T)]$ is the crystallization model. This method assumes that the activation energy is only constant at a given extent of conversion and for the narrow temperature region associated with this conversion.

In fact, the temperature-dependent activation energy is a consequence of the non-Arrhenius behavior expected for the crystallization process. In this sense, it should be indicated that a mistake is derived when other simpler isoconversional methods, such as Kissinger [33], Kissinger–Akahira–Sunose [34], Ozawa [35], and Flynn and Wall [36], are applied. These methods were also problematic, as indicated by Vyazovkin for crystallization processes that are defined by cooling rate values [25,37].

Crystallization experiments performed at different cooling rates allow obtaining values for ln $[d\chi(T)/dt]$ at different temperatures and crystallization degrees. For a given conversion, the slopes of the linear plots of ln $[d\chi(T)/dt]$ versus $1/T$ (Figure 8) determines ΔE.

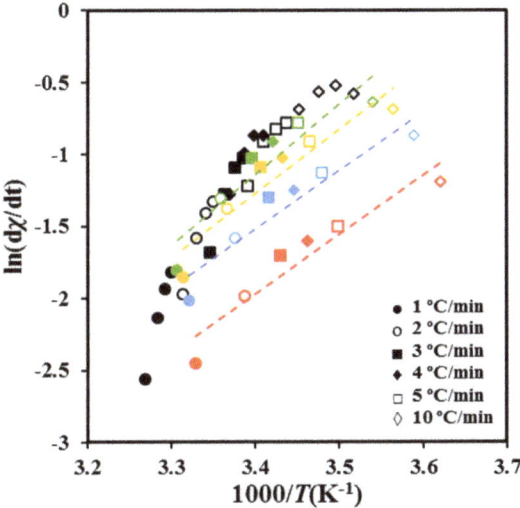

Figure 8. Plots of $\ln(d\chi/dt)_\chi$ versus $1/T$ for non-isothermal crystallization of P4HB at the indicated cooling rates. Data corresponding to different relative degrees of crystallinity are represented using symbols and lines of different colors (i.e., green, yellow, blue, and red correspond to 0.5, 0.6, 0.7, and 0.8 values, respectively).

As shown in Figure 9a, the deduced values of the activation energy are negative as expected in the temperature range from the melting point down to the temperature of the maximum crystallization rate. The energy sign indicates that crystallization rates increase with decreasing crystallization temperatures. It should be pointed out that the effective activation energy varies between −45 kJ/mol and −98 kJ/mol covering a wide range of energies. Published results concerning PCL showed a similar variation with a change from −49 kJ mol^{-1} to −110 kJ mol^{-1} with χt the ranging from 10 to 90% [38].

Finally, the activation energy can be correlated (Figure 9b) with the crystallization temperature by considering the average temperature associated with a given degree of crystallinity (Figure 9a). Note that the estimation of the activation energy is based on the application of Arrhenius equation within small temperature regions associated with given values of the degree of crystallinity. The plot clearly shows that the activation energy was negative at temperatures higher than that associated to the maximum crystallization rate and tended to zero when temperature decreased, as extensively discussed by Vyazovkin and Dranca [39].

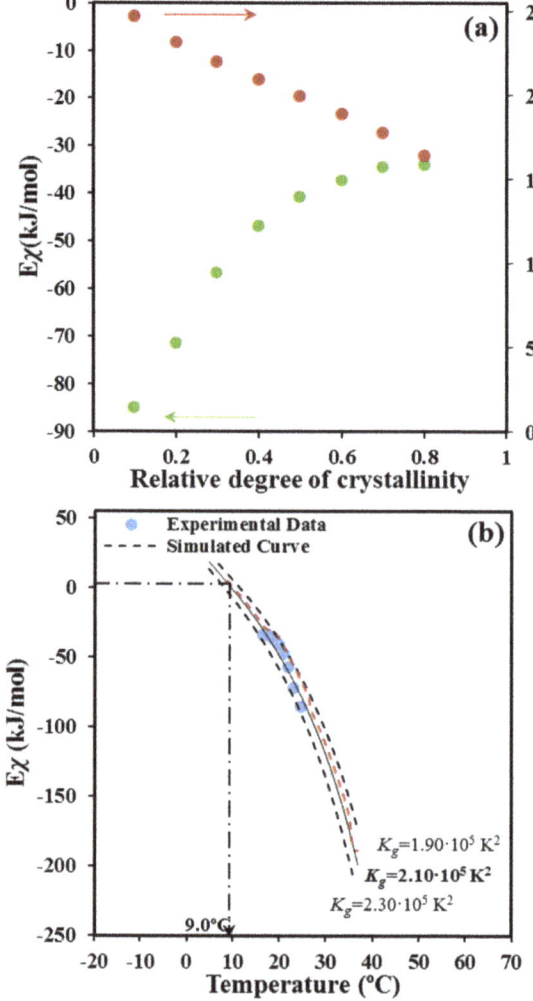

Figure 9. (a) Dependence of the activation energy of crystallization (●) and the average crystallization temperature (●) on crystallinity. (b) Experimental E_X vs. T data for non-isothermal crystallization of P4HB. The solid black line corresponds to the data calculated by Equation (11) and the optimized crystallization parameters. Arrow indicates the expected temperature for the maximum crystallization rate (i.e., the effective activation energy becomes equal to zero).

2.4. Secondary Nucleation Constant from Non-Isothermal Crystallization

DSC calorimetric data from non-isothermal crystallization experiments can also be employed to known crystal growth parameters as, for example, the secondary nucleation constant.

The Lauritzen–Hoffman model [40] is usually accepted to determine the crystal growth rate, G. According to this theory this rate is defined by two terms: a) The transport activation energy, U^*, which expresses the difficulty of crystallizing segments to move across the liquid–crystal interface, and b) the secondary nucleation constant, K_g, which evaluates the formation of new particles in the presence of an established population of previously formed particles. An increase of K_g indicates a greater difficulty for the surface of a growing lamellar crystal to act as an effective nucleus.

The Lauritzen–Hoffman equation is defined by

$$G = G_0 \exp[-U^*/(R(T_c - T_\infty))] \times \exp[-K_g/(T_c (\Delta T) f)] \quad (9)$$

where G_0 is a constant preexponential factor, T_∞ is the temperature below which molecular motion ceases, T_c is the selected crystallization temperature, R is the gas constant, ΔT is the degree of supercooling measured as the difference between the equilibrium melting temperature (T_m^0) and T_c (i.e., $\Delta T = T_m^0 - T_c$), and f is a correction factor accounting for the variation in the bulk melting enthalpy per unit volume with temperature ($f = 2T_c/(T_m^0 + T_c)$).

The temperature dependence of G follows a bell-shaped curve due to the two exponential terms of Equation (9). In general, crystallizations from the melt takes place at relatively low degrees of supercooling (right region of the curve) where the influence of the transport term is not highly relevant. In this case, it is usual to perform calculations with standard U^* and T_∞ values as those reported by Suzuki and Kovacs [41] (i.e., $U^* = 1500$ cal/mol and $T_\infty = T_g - 30$ K).

Hoffman and Lauritzen parameters can also be derived from non-isothermal crystallizations by using an isoconversional approach developed by Vyazovkin et al. [42] that has been satisfactorily tested for different polymers, such as poly(ethylene terephthalate) [42], poly(butylene naphthalate) [43], and poly(ethylene naphthalate) [39].

This isoconversional method is based on an explicit dependence of the activation energy on temperature (Equation (8)) that was derived assuming an equivalence of the temperature coefficients of the growth rate and the heat flow [44] (Equation (9)).

$$\Delta E_\chi = [U^* T^2/(T - T_\infty)^2] + [K_g R((T_m^0)^2 - T^2 - T_m^0 T)/((T_m^0 - T)^2 T)] \quad (10)$$

$$d(\ln \varphi)/T = d(\ln G)/T \quad (11)$$

The experimental temperature dependence of the activation energy that was determined in the previous section can be related to the theoretical one calculated from the right side of Equation (8). U^* and K_g parameters are selected to get the best fit between theoretical and experimental data, which is the process simplified when standard U^* values can be employed. T_m^0 and T_g were taken equal to 79.9 and −45.4 °C, as previously evaluated from the calorimetric analysis of P4HB [18].

Figure 9b shows that a reasonable fit between experimental and predicted values is attained with the set of parameters: $U^* = 1500$ cal/mol, $T_\infty = T_g - 30$ K and $K_g = 2.10 \times 10^5$ K^2. For the sake of completeness simulated curves for K_g values of 2.30×10^5 K^2 and 1.90×10^5 K^2 are also plotted (dashed lines), illustrating the impact caused by small changes in the nucleation parameter. In the same way, the non-significant influence caused by a change in U^* is also shown by the red dashed line calculated for $U^* = 1800$ cal/mol, $T_\infty = T_g - 30$ K, and $K_g = 2.10 \times 10^5$ K^2. Interestingly, the deduced parameters from the non-isothermal DSC data became very close to those evaluated from isothermal DSC experiments and direct OM measurements on the spherulitic growth. In this case values of the selected set of parameters became $U^* = 1500$ cal/mol, $T_\infty = T_g - 30$ K, and $K_g = 1.69 \times 10^5$ K^2 (DSC) and 1.58×10^5 K^2 (OM), which appear in acceptable agreement with those determined by the isoconversional methodology.

Figure 9b also shows that the activation energy becomes zero at a temperature of 9.0 °C. This zero of energy is associated with the maximum crystallization rate and therefore, should correspond to the maximum of the bell-shaped $G - T$ curve. Note that at higher temperatures (i.e., the region dominated by the secondary nucleation) the activation energy becomes negative and progressively increases with decreasing the temperature. This feature means that the crystallization rate is enhanced with decreasing temperatures as discussed at length by Vyazovkin and Dranca [39]. At lower temperatures than those corresponding to the maximum rate, the activation energy becomes positive, indicating that G decreases when crystallization temperature decreases. Results of the non-isothermal study show an impressive agreement with the maximum growth rate determined from isothermal measurements from both optical microscopy (i.e., 15.0 °C) and even the calorimetric data (i.e., 14.0 °C) [18].

Overall, crystallization rates determined from DSC data and applying the Avrami analysis (Table 1) can be employed to determine the Lauritzen and Hoffman parameters, considering a proportionality between k and G values. In this case, Equation (12) can be applied.

$$k = k_0 \exp[-U^*/(R(T_c - T_\infty))] \times \exp[-K_g/(T_c(\Delta T)f)]. \tag{12}$$

In addition, the temperature associated with each k value was taken as a rough approximation of the peak temperature determined for the DSC runs performed at the corresponding cooling rates.

The plot of $\ln k + U^*/(R(T_c - T_\infty))$ versus $1/[T_c(\Delta T)f)]$ gave a straight line with a slope (i.e., the K_g value) of 2.22×10^5 K^2. It is very interesting to point out the great agreement with the secondary nucleation constant determined from the evaluation of activation energies. The similarity between both analyses can also be observed in Figure 10, where simulated bell-shaped curves from both sets of data are plotted. The advantages of the isoconversional method are clear since additional information concerning energies are obtained and furthermore approximations related to temperature, associated with each cooling rate, avoided.

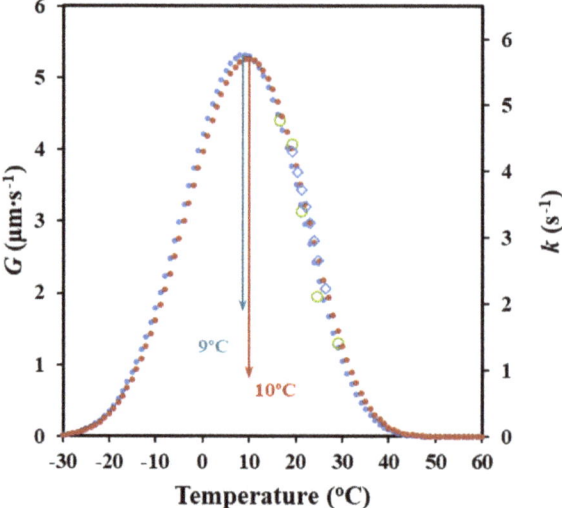

Figure 10. Comparison between bell-shaped curves of G/k temperature dependence obtained from DSC data and the two non-isothermal methodologies (from Avrami analysis (●) and activation energy (●) with their respective experimental results (o) (◊). An arbitrary value of G_0 has been selected to fit the maximum rate of both curves.

2.5. Non-Isothermal Crystallization Studies by Optical Microscopy

Spherulitic growth rates for non-isothermal crystallizations can also be determined by optical microscopy [45–47]. In this case the evolution of the spherulite radius (R) with temperature (T) is followed for a constant cooling rate (dT/dt) [45,46]. Specifically, the growth rate is given by Equation (13). It is necessary to select an appropriate cooling rate to get the maximum information concerning the $G - T_c$ curve. Furthermore, a set of cooling rates can be employed if necessary to extend the curve to higher or lower temperatures.

$$G = dR/dt = (dR/dT) \times (dT/dt) \tag{13}$$

Evolution of the radius versus crystallization temperature during each selected cooling run allows obtaining a plot which is then fitted to polynomial equations. The selected equation corresponds to the lower order that renders a good regression coefficient (r). Growing rates (dR/dT) are then calculated at each crystallization temperature from the first derivative function of the polynomial equation. Experimental problems lie in the choice of the cooling rate required to maximize the crystallization temperature range where radii can be well measured, making necessary, in some cases, the use of various rates.

Figure 11a shows the evolution of the crystal growth rate of P4HB spherulites with crystallization temperature for a cooling rate of 0.5 °C/min. This rate allowed to cover a wide range of experimental data that are comparable with those available from non-isothermal DSC experiments. This range was also clearly higher than that defined by both DSC and POM isothermal crystallizations [18]. Therefore, kinetic analysis was carried out considering only the cooling rate of 0.5 °C/min. A second-order equation ($R = 0.1091\ T^2 - 10.238\ T + 239.89$) gave a correlation coefficient of $r^2 = 0.996$, which was slightly better than those calculated for higher-order equations.

Equation (7) can be applied to obtain crystallization parameters by considering the above deduced $G - T_c$ data. Thus, the representation of $\ln G + U^*/(R (T_c - T_\infty))$ versus $1/[T_c\ (\Delta T)\ f)]$ gave a straight line with an intercept at the origin at $\ln G_0$ and a negative slope equal to K_g (Figure 11b). A single value of K_g was observed in agreement with the above reported DSC data. This single value is a clear indication of a process that took place according to a single crystallization regime. The Lauritzen–Hoffman theory postulated the possibility of three regimes according to the type of nucleation on the crystal surface, being in some cases even related to different morphologies (e.g., axialites, banded/ringed spherulites, and non-ringed ones). Regime II is usually associated with ringed spherulites as those observed for P4HB in the considered temperature range. This regime II obeys to a nucleation rate on the crystal surface that is comparable or even greater than the lateral growth rate.

$U^* = 1500$ cal/mol and $T_\infty = T_g - 30$ K values gave a straight line with $r^2 = 0.995$ and a K_g parameter of 1.25×10^5 K^2. Note that this constant is in complete agreement with that determined by the isoconversional methodology and even the Avrami analysis. Nevertheless, the higher discrepancy was derived from this methodology. It seems that the method has a great advantage due to its simplicity, but some cautions must be taken into account concerning the precision of the derived results. It should be pointed out that OM analyses are independent of the nucleation rate and that some discrepancies with DSC measurements can also be justified.

It is also interesting to note that K_g values determined from isothermal and non-isothermal crystallizations using DSC or POM data are close, with $1.7 \pm 0.5 \times 10^5$ K^2 being the average value. In the same way, temperatures corresponding to the maximum of the bell-shaped curves that express the dependence of crystal growth rate or the overall crystallization rate on the crystallization temperature showed minimum deviations (i.e., 13 ± 5 °C).

Figure 11a compares the plots of the experimental and simulated G values versus crystallization temperature. The simulated curve was obtained by applying Equation (9) and the deduced Lauritzen–Hoffman parameters. This curve had a typical bell shape and showed a maximum at 19 °C, which was relatively higher than the temperature deduced from the isoconversional methodology and the DSC non-isothermal data.

Figure 11c compares the bell-shaped curves obtained from all the performed crystallization studies. For the sake of simplicity, arbitrary units have been employed for the ordinate axis due to the different represented rates (i.e., G and k). A relatively good agreement was in general observed between the isothermal and the non-isothermal studies and between DSC and POM techniques. In general, non-isothermal methods has advantages derived from their higher simplicity, the larger number of available experimental data and finally, the closer fit to realistic processing conditions. Crystallization under isothermal and non-isothermal conditions is obviously different and consequently, slight differences, such as those detected for the secondary nucleation constant and the temperature associated with the maximum crystallization rate, can be expected.

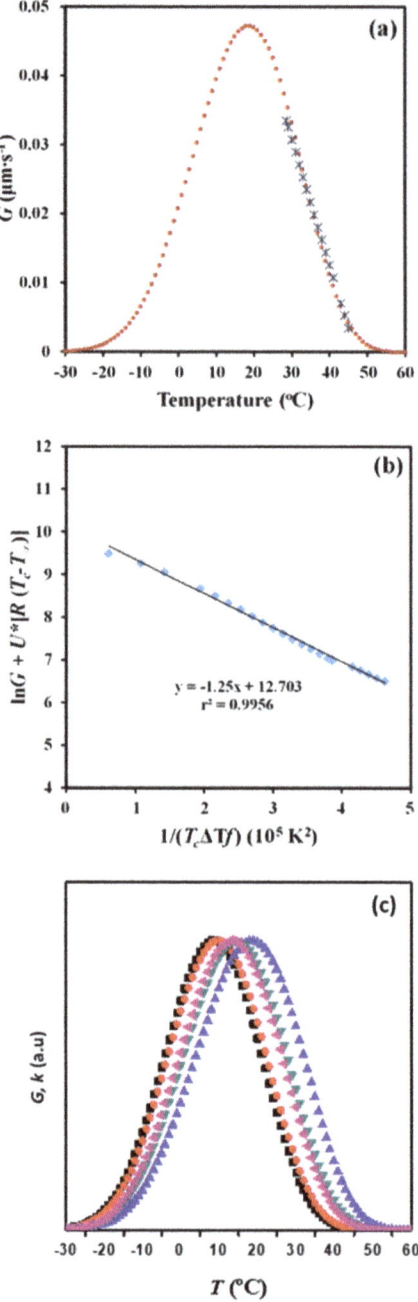

Figure 11. (**a**) Experimental and simulated dependence of the growth rate on crystallization temperature. (**b**) Plot of ln G + U^*/R $(T_c - T_\infty)$ versus $1/T_c$ $(\Delta T) f$ to determine the K_g secondary nucleation parameter of P4HB. (**c**) Comparison between bell-shaped curves derived from non-isothermal DSC data ((●) from isoconversional and (■) from Avrami analyses), non-isothermal OM data (▲) and isothermal DSC data (◄) and OM (▼) data.

2.6. Synchrotron Data on Non-Isothermal Crystallization of P4HB

Cooling rate has obviously an influence on the final morphology and even on the crystallinity of P4HB despite its rapid crystallization. Thus, melting enthalpy decreased from 39.6 kJ/mol to 35.2 kJ/mol when the cooling rate increased from 1 °C/min to 5 °C/min, while spherulitic size and texture changed (Figure 12). The average crystallization temperature decreased with the increase of the cooling rate (Figure 1) and consequently primary nucleation increased, leading to a decrease in the spherulite size (Figure 12). Spacing between rings was also temperature dependent as discussed earlier (Figure 5).

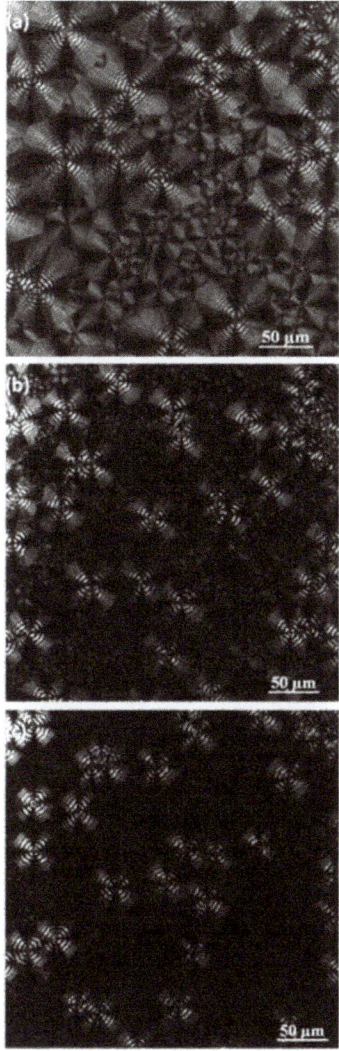

Figure 12. Optical micrographs taken at the end of the non-isothermal crystallizations of P4HB performed at cooling rates of 0.5 °C/min (**a**), 3 °C/min (**b**) and 5 °C/min (**c**).

Morphology of constitutive lamellae depends not only on the crystallization rate, as indicated for the twisting period (i.e., interring spacing), but also on the lamellar thickness. Small-angle X-ray scattering patterns (SAXS) taken in real-time during the cooling process allowed to follow the evolution of lamellar thickness during crystallization and also allowed us to compare morphological parameters for different cooling rates through the use of the normalized correlation function:

$$\gamma(r) = \int_0^\infty q^2 I(q) \cos(qr) dq / \int_0^\infty q^2 I(q) dq \quad (14)$$

where $I(q)$ is the intensity of the SAXS peak at each value of the scattering vector ($q = [4\pi/\lambda] \sin \theta = 2\pi/d$, θ and d being the Bragg angle and the Bragg spacing, respectively).

Long period, L_γ, amorphous layer thickness, l_a, and crystalline lamellar thickness, l_c, can be determined by the normalized one-dimensional correlation function [48] and applying Vonk's model [49] and Porod's law to perform extrapolations to low and high q values.

Figure 13a shows the evolution of the SAXS peak during the non-isothermal crystallization from the melt at a representative cooling rate of 7 °C/min. This scattering peak appeared at the same temperature as the wide-angle X-ray reflections, which are presumable for a crystallization, where the supramolecular structure (lamellae) is developed at the same time that the molecular arrangement took place. Wide-angle X-ray diffraction (WAXD) profiles show two main peaks at 0.406 nm and 0.388 nm that correspond to the (110) and (200) reflections of the orthorhombic structure ($a = 0.775$ nm, $b = 0.477$ nm, and c (fiber axis) $= 1.199$ nm) reported for P4HB [15]. Deconvolution of WAXD profiles allowed for the determination of the temperature evolution of crystallinity during cooling runs performed at three different representative rates (Figure 14). Results clearly indicated that the increase of the cooling rate led to a decrease of the temperature at which crystallization started and as well as of the crystallinity (i.e., crystallinities around 65%, 60%, and 46% were determined for 3 °C/min, 7 °C/min, and 10 °C/min, respectively). Logically, the crystallization rate diminished when temperatures approached −20 °C (i.e., at some degrees above the glass transition temperature of −45.4 °C). It is also clear that samples are mainly crystallized at a lower average temperature when the cooling rate increased, and consequently, some influence on the derived lamellar morphology should be expected.

Figure 13b compares the correlation functions corresponding to the end of crystallizations performed at 10 °C/min and 3 °C/min. Slight but significant changes can be detected, with peaks being clearly defined for the crystallization performed at the lower rate. In this case, the contrast between the amorphous and crystalline regions is increased. Lamellae became narrower (8.70 nm versus 9.70 nm for L_γ) for the crystallization performed at the lower rate, mainly as a consequence of the decrease in l_c (i.e., 7.08 nm versus 7.61 nm), although a slight decrease in l_a (i.e., 1.62 nm versus 2.09 nm) was also detected. Note that the average crystallization temperature is superior for samples crystallized at the lower cooling rate, consequently with an expected greater lamellar thickness. The opposite results that were attained can be explained due to a lamellar reinsertion mechanism which took place. This process is the consequence of the formation of thinner lamellar crystals between the loosely stacked primary lamellae and appears to be more significant when the crystallization process is slower. Figure 13c compares the correlation function obtained at the beginning, an intermediate stage, and the end of the crystallization process performed at a rate of 3 °C/min. It is clear that lamellar spacings decreased (i.e., 10.10/9.90/8.70 for L_γ, 2.09/1.98/1.62 for l_a and 8.01/7.92/7.08 for l_c) when the temperature did. The indicated evolution (also observed for the other rates) points out a lamellar insertion mechanism and a molecular rearrangement in the amorphous layer. The bilayer model demonstrated an improvement in the molecular arrangement in the crystalline domains as visualized by the increase in the contrast of the electronic density of amorphous and crystalline layers, or even by the increase in the crystallinity within the lamellar stacks ($X_c^{SAXS} = l_c/L_\gamma$) which varied from 0.79 to 0.81.

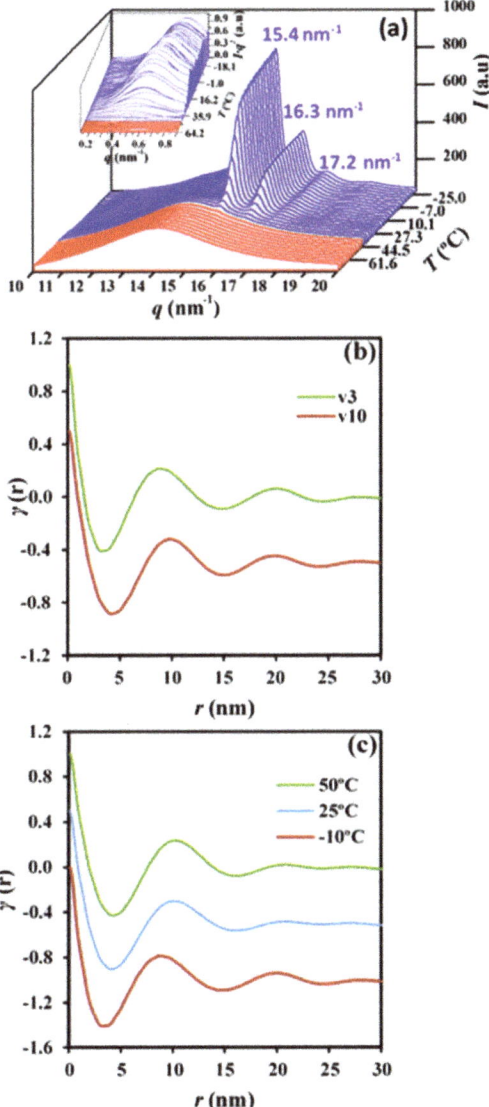

Figure 13. (**a**) Three-dimensional wide-angle X-ray diffraction (WAXD) and small-angle X-ray scattering patterns (SAXS) profiles of P4HB during cooling at a rate of 7 °C/min. (**b**) Correlation functions obtained after non-isothermal crystallizations of P4HB from the melt state performed at cooling rates of 3 and 10 °C/min. (**c**) SAXS correlation functions calculated at the indicated temperatures during the cooling run (3 °C/min) of a melted P4HB sample.

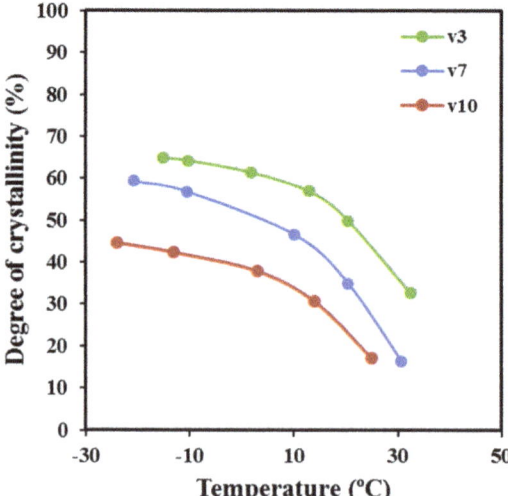

Figure 14. Evolution of the absolute degree of crystallinity determined from WAXD profiles with temperature for the indicated cooling rates.

3. Materials and Methods

3.1. Materials

Commercially available sutures of P4HB (Monomax™, violet sample, USP 1) were kindly supplied by B. Braun Surgical S.A. The weight and number average of molecular weights of Monomax™ samples were 215,000 and 68,000 g/mol, as determined by size exclusion chromatography (GPC).

3.2. Measurements

Molecular weight was estimated at room temperature by size exclusion chromatography (GPC) using a liquid chromatograph (model LC-8A, Shimadzu, Tokyo, Japan) equipped with an Empower computer program (Waters, Milford, MA, USA). A PL HFIP gel column (Polymer Lab, Agilent Technologies Deutschland GmbH, Böblingen, Germany) and a refractive index detector (Shimadzu RID-10A) were employed. The polymer was dissolved and eluted in 1,1,1,3,3,3-hexafluoroisopropanol containing CF_3COONa (0.05 M) at a flow rate of 0.5 mL/min (injected volume 100 µL, sample concentration 2.0 mg/mL). The number and weight average of molecular weights were estimated using polymethyl methacrylate standards.

Calorimetric data were obtained by differential scanning calorimetry with a TA Instruments Q100 series with T_{zero} technology and equipped with a refrigerated cooling system (RCS). Experiments were conducted under a flow of dry nitrogen with a sample weight of approximately 5 mg, and calibration was performed with indium. T_{zero} calibration required two experiments: The first was performed without samples while sapphire disks were used in the second. Non-isothermal crystallization studies were performed by cooling previously molten samples (5 min at 100 °C) at rates varying from 5 to 1 °C/min.

The spherulitic growth rate was determined by optical microscopy using a Zeiss Axioskop 40 Pol light polarizing microscope equipped with a Linkam temperature control system configured by a THMS 600 heating and freezing stage connected to an LNP 94 liquid nitrogen cooling system. Spherulites were grown from homogeneous thin films prepared from the melt. Small sections of these films were pressed or smeared between two cover slides and inserted into the hot stage, producing samples with thicknesses close to 10 µm in all cases. Samples were kept at approximately 100 °C for 5 min to eliminate sample history effects. Then the sample was quickly cooled to 45 °C and led to isothermally grown for 60 min to generate enough nuclei for measurements and avoid the induction step. Subsequently, the radius of growing spherulites was monitored during crystallization with micrographs taken with a Zeiss AxiosCam MRC5 digital camera at appropriate time intervals. A first-order red tint plate was employed to determine the sign of spherulite birefringence under crossed polarizers.

Wide-angle X-ray diffraction (WAXD) and small-angle X-ray scattering patterns (SAXS) data were obtained at the NCD beamline (BL11) of the ALBA synchrotron facility (Cerdanyola del Vallès, Barcelona, Spain), by using a wavelength of 0.100 nm. A WAXD LX255-HS detector from Rayonix and an ImXPAD S1400 photon-counting detector were employed. Polymer samples were confined between Kapton films. WAXD and SAXS diffraction patterns were calibrated with Cr_2O_3 and silver behenate (AgBh), respectively. WAXD peaks were deconvoluted with the PeakFit v4 program by Jandel Scientific Software. The correlation function and corresponding parameters were calculated with the CORFUNC program for Fiber Diffraction/Non-Crystalline Diffraction provided by the Collaborative Computational Project 13.

4. Conclusions

Non-isothermal crystallization studies of P4HB were performed by calorimetric experiments and allowed the determination of crystallization parameters by considering both the classical Avrami analysis and isoconversional methods. In the first case, the determined Avrami parameters had no physical sense but gave an estimation of the variation of the overall crystallization rate with crystallization temperature, when average crystallization temperatures were assumed for experiments performed at different cooling rates. Additionally, this approximation made feasible the evaluation of the secondary nucleation constant. The isoconversional analysis provided information concerning the activation energy and allowed the estimation of both the temperature associated with the maximum crystallization rate and the secondary nucleation constant. Both methodologies were in remarkable agreement with the derived secondary nucleation constants (i.e., 2.10×10^5 K^2 and 2.22×10^5 K^2), as well as the temperatures associated with the maximum growth rate/overall crystallization rate (i.e., 9 and 10 °C). Optical microscopy data allowed the estimation of crystal growth rates at different crystallization temperatures. Results were independent of primary nucleation and in relatively good agreement with calorimetric analyses. Discrepancies could be associated with a lower precision of the method or to the effect caused by thermal nucleation. The obtained results are also comparable with previously reported data from isothermal crystallization. In this case, DSC and OM analysis were in a satisfactory agreement suggesting a scarce influence of thermal nucleation. The cooling rate had a great effect on the morphology and texture of spherulites, as well as the twisting and thickness of the constitutive lamellar crystals. WAXD synchrotron experiments allowed for the determination of final crystallinities and demonstrated a lower crystallization when the cooling rate increased. More interestingly, SAXS data indicated a lamellar insertion mechanism during crystallization that led to a decrease of the crystalline layer thickness. This process was more significant when the cooling rate decreased. The crystallinity of lamellar stacks slightly increased during the cooling run due to the similar thickness evolution of amorphous and crystalline layers.

Author Contributions: I.K. performed experiments; L.J.d.V. was involved in synchrotron experiments; L.F. (Lourdes Franco). was involved in thermal characterization studies and analysis of data; P.T. provided technical and financial support; L.F. (Lutz Funk) provided commercial samples and technical advice; and J.P. supervised project direction.

Funding: The authors are in debt to support from MINECO and FEDER (RTI2018-101827-B-I00) and the Generalitat de Catalunya (2017SGR373). I.K. also acknowledges the financial support from B. Braun Surgical S.A. Diffraction experiments were performed at the NCD-SWEET beamline at ALBA Synchrotron with the collaboration of ALBA staff.

Conflicts of Interest: The authors declare no conflict of interest.

References

1. Williams, S.F.; Rizk, S.; Martin, D.P. Poly-4-hydroxybutyrate (P4HB): A new generation of resorbable medical devices for tissue repair and regeneration. *Biomed. Tech.* **2013**, *58*, 439–452. [CrossRef] [PubMed]
2. Ackermann, J.U.; Müller, S.; Lösche, A.; Bley, T.; Babel, W. Methylobacterium rhodesianum cells tend to double the DNA content under growth limitations and accumulate PHB. *J. Biotechnol.* **1995**, *39*, 9–20. [CrossRef]
3. Huisman, G.W.; Skraly, F. Biological systems for manufacture of polyhydroxyalkanoate polymers containing 4-hydroxyacids. U.S. Patent 6,316,262, 13 November 2001.
4. Dennis, D.E.; Valentin, H.E. Methods of making polyhydroxyalkanoates comprising 4-hydroxybutyrate monomer units. U.S. Patent 6,117,658, 12 September 2000.
5. Hori, Y.; Yamaguchi, A.; Hagiwara, T. Chemical synthesis of high molecular weight poly(3-hydroxybutyrate-co-4-hydroxybutyrate). *Polymer (Guildf)* **1995**, *36*, 4703–4705. [CrossRef]
6. Houk, K.N.; Jabbari, A.; Hall, H.K.; Aleman, C. Why δ-valerolactone polymerizes and gamma-butyrolactone does not. *J. Org. Chem.* **2008**, *73*, 2674–2678. [CrossRef] [PubMed]
7. Odermatt, E.K.; Funk, L.; Bargon, R.; Martin, D.P.; Rizk, S.; Williams, S.F. MonoMax suture: A new long-term absorbable monofilament suture made from poly-4-hydroxybutyrate. *Int. J. Polym. Sci.* **2012**, *2012*, 12. [CrossRef]
8. Zhou, X.-Y.; Yuan, X.-X.; Shi, Z.-Y.; Meng, D.-C.; Jiang, W.-J.; Wu, L.-P.; Chen, J.-C.; Chen, G.-Q. Hyperproduction of poly(4-hydroxybutyrate) from glucose by recombinant Escherichia coli. *Microb. Cell Fact.* **2012**, *11*, 54. [CrossRef] [PubMed]
9. Chen, G.Q.; Wu, Q. The application of polyhydroxyalkanoates as tissue engineering materials. *Biomaterials* **2005**, *26*, 6565–6578. [CrossRef] [PubMed]
10. Martin, D.P.; Williams, S.F. Medical applications of poly-4-hydroxybutyrate: A strong flexible absorbable biomaterial. *Biochem. Eng. J.* **2003**, *16*, 97–105. [CrossRef]
11. Plymale, M.A.; Davenport, D.L.; Dugan, A.; Zachem, A.; Roth, J.S. Ventral hernia repair with poly-4-hydroxybutyrate mesh. *Surg. Endosc.* **2018**, *32*, 1689–1694. [CrossRef] [PubMed]
12. Rao, U.; Sridhar, R.; Sehgal, P.K. Biosynthesis and biocompatibility of poly(3-hydroxybutyrate-co-4-hydroxybutyrate) produced by Cupriavidus necator from spent palm oil. *Biochem. Eng. J.* **2010**, *49*, 13–20. [CrossRef]
13. Generali, M.; Kehl, D.; Capulli, A.K.; Parker, K.K.; Hoerstrup, S.P.; Weber, B. Comparative analysis of poly-glycolic acid-based hybrid polymer starter matrices for in vitro tissue engineering. *Colloids Surf. B* **2017**, *158*, 203–212. [CrossRef] [PubMed]
14. Mitomo, H.; Kobayashi, S.; Morishita, N.; Doi, Y. Structural changes and properties of P (3HB-co-4HB). *Polym. Prepr.* **1995**, *44*, 3156.
15. Su, F.; Iwata, T.; Sudesh, K.; Doi, Y. Electron and X-ray diffraction study on poly(4-hydroxybutyrate). *Polymer (Guildf)* **2001**, *42*, 8915–8918. [CrossRef]
16. Su, F.; Iwata, T.; Tanaka, F.; Doi, Y. Crystal structure and enzymatic degradation of poly(4-hydroxybutyrate). *Macromolecules* **2003**, *36*, 6401–6409. [CrossRef]
17. Koyama, N.; Doi, Y. Effects of solid-state structures on the enzymatic degradability of bacterial poly(hydroxyalkanoic acids). *Macromolecules* **1997**, *30*, 826–832. [CrossRef]
18. Keridou, I.; del Valle, L.; Lutz, F.; Turon, P.; Yousef, I.; Franco, L.; Puiggalí, J. Isothermal crystallization kinetics of poly(4-hydroxybutyrate) biopolymer. *Materials* **2019**.
19. Avrami, M. Kinetics of phase change. I: General theory. *J. Chem. Phys.* **1939**, *7*, 1103–1112. [CrossRef]
20. Avrami, M. Kinetics of phase change. II Transformation-time relations for random distribution of nuclei. *J. Chem. Phys.* **1940**, *8*, 212–224. [CrossRef]
21. Mandelkern, L. Crystallization of Polymers. In *Kinetics and Mechanisms*, 2nd ed.; Cambridge University Press: Cambridge, UK, 2004; Volume 2.
22. Limwanich, W.; Phetsuk, S.; Meepowpan, P.; Kungwan, N. Kinetics Studies of Non-Isothermal Melt Crystallization of Poly (ε-caprolactone) and Poly (L-lactide). *Chiang Mai, J. Sci.* **2016**, *43*, 329–338.
23. Ozawa, T. Kinetics of non-isothermal crystallization. *Polymer (Guildf)* **1971**, *12*, 150–158. [CrossRef]

24. Evans, U.R. The laws of expanding circles and spheres in relation to the lateral growth of surface films and the grain-size of metals. *Trans. Faraday Soc.* **1945**, *41*, 365. [CrossRef]
25. Vyazovkin, S. Nonisothermal crystallization of polymers: Getting more out of kinetic analysis of differential scanning calorimetry data. *Polym. Cryst.* **2018**, *1*, e10003. [CrossRef]
26. Liu, T.; Mo, Z.; Wang, S.; Zhang, H. Non-isothermal melt and cold crystallization kinetics of poly(aryl ether ether ketone ketone). *Polym. Eng. Sci.* **1997**, *37*, 568–575. [CrossRef]
27. Márquez, Y.; Franco, L.; Turon, P.; Martínez, J.C.; Puiggalí, J. Study of non-isothermal crystallization of polydioxanone and analysis of morphological changes occurring during heating and cooling processes. *Polymers (Basel)* **2016**, *8*, 351. [CrossRef] [PubMed]
28. Tarani, E.; Wurm, A.; Schick, C.; Bikiaris, D.N.; Chrissafis, K.; Vourlias, G. Effect of graphene nanoplatelets diameter on non-isothermal crystallization kinetics and melting behavior of high density polyethylene nanocomposites. *Thermochim. Acta* **2016**, *643*, 94–103. [CrossRef]
29. Jape, S.P.; Deshpande, V.D. Nonisothermal crystallization kinetics of nylon 66/LCP blends. *Thermochim. Acta* **2017**, *655*, 1–12. [CrossRef]
30. Cazé, C.; Devaux, E.; Crespy, A.; Cavrot, J.P. A new method to determine the Avrami exponent by d.s.c. studies of non-isothermal crystallization from the molten state. *Polymer (Guildf)* **1997**, *38*, 497–502. [CrossRef]
31. Márquez, Y.; Franco, L.; Turon, P.; Puiggalí, J. Isothermal and non-isothermal crystallization kinetics of a polyglycolide copolymer having a tricomponent middle soft segment. *Thermochim. Acta* **2014**, *585*, 71–80. [CrossRef]
32. Friedman, H.L. Kinetics of thermal degradation of char-forming plastics from thermogravimetry. Application to a phenolic plastic. *J. Polym. Sci. Part C* **1964**, *6*, 183–195. [CrossRef]
33. Kissinger, H.E. Reaction Kinetics in Differential Thermal Analysis. *Anal. Chem.* **1957**, *29*, 1702–1706. [CrossRef]
34. Akahira, T.; Sunose, T. Joint Convention of Four Electrical Institutes. *Sci. Technol.* **1971**, *16*, 22–31.
35. Takeo, O. A New Method of Analyzing Thermogravimetric Data. *Bull. Chem. Soc. Jpn.* **1965**, *38*, 1881–1886.
36. Flynn, J.; Wall, L. General trement of the therogravimetry of polymers. *J. Res. Natl. Bur. Stand. (1934).* **1966**, *70A*, 487–523. [CrossRef]
37. Vyazovkin, S. Is the Kissinger equation applicable to the processes that occur on cooling? *Macromol. Rapid Commun.* **2002**, *23*, 771–775. [CrossRef]
38. Pires, L.S.O.; Fernandes, M.H.F.V.; de Oliveira, J.M.M. Crystallization kinetics of PCL and PCL–glass composites for additive manufacturing. *J. Therm. Anal. Calorim.* **2018**, *134*, 2115–2125. [CrossRef]
39. Vyazovkin, S.; Dranca, I. Isoconversional analysis of combined melt and glass crystallization data. *Macromol. Chem. Phys.* **2006**, *207*, 20–25. [CrossRef]
40. Lauritzen, J.I.; Hoffman, J.D. Extension of theory of growth of chain-folded polymer crystals to large undercoolings. *J. Appl. Phys.* **1973**, *44*, 4340–4352. [CrossRef]
41. Suzuki, T.; Kovacs, A.J. Temperature Dependence of Spherulitic Growth Rate of Isotactic Polystyrene. *Crit. Comp. Kinet. Theory Surf. Nucleation* **1970**, *1*, 82–100.
42. Vyazovkin, S.; Sbirrazzuoli, N. Isoconversional approach to evaluating the Hoffman-Lauritzen parameters (U* and Kg) from the overall rates of nonisothermal crystallization. *Macromol. Rapid Commun.* **2004**, *25*, 733–738. [CrossRef]
43. Achilias, D.S.; Papageorgiou, G.Z.; Karayannidis, G.P. Evaluation of the isoconversional approach to estimating the Hoffman-Lauritzen parameters from the overall rates of non-isothermal crystallization of polymers. *Macromol. Chem. Phys.* **2005**, *206*, 1511–1519. [CrossRef]
44. Toda, A.; Oda, T.; Hikosaka, M.; Saruyama, Y. A new method of analysing transformation kinetics with temperature modulated differential scanning calorimetry: Application to polymer crystal growth. *Polymer (Guildf)* **1997**, *38*, 231–233. [CrossRef]
45. Chen, M.; Chung, C.T. Analysis of crystallization kinetics of poly(ether ether ketone) by a nonisothermal method. *J. Polym. Sci. Part B Polym. Phys.* **1998**, *36*, 2393–2399. [CrossRef]
46. Di Lorenzo, M.L.; Cimmino, S.; Silvestre, C. Nonisothermal crystallization of isotactic polypropylene blended with poly(α-pinene). 2. Growth rates. *Macromolecules* **2000**, *33*, 3828–3832. [CrossRef]

47. Lorenzo, D.M.L. Determination of spherulite growth rates of poly(L-lactic acid) using combined isothermal and non-isothermal procedures. *Polymer (Guildf)* **2001**, *42*, 9441–9446. [CrossRef]
48. Vonk, C.G.; Kortleve, G. X-ray small-angle scattering of bulk polyethylene. *Kolloid Z Z Polym.* **1967**, *220*, 19–24. [CrossRef]
49. Vonk, C.G. A general computer program for the processing of small-angle X-ray scattering data. *J. Appl. Crystallogr.* **1975**, *8*, 340–341. [CrossRef]

Sample Availability: Not available.

© 2019 by the authors. Licensee MDPI, Basel, Switzerland. This article is an open access article distributed under the terms and conditions of the Creative Commons Attribution (CC BY) license (http://creativecommons.org/licenses/by/4.0/).

Article

Thermal Decomposition of Maya Blue: Extraction of Indigo Thermal Decomposition Steps from a Multistep Heterogeneous Reaction Using a Kinetic Deconvolution Analysis

Yui Yamamoto and Nobuyoshi Koga *

Chemistry Laboratory, Department of Science Education, Graduate School of Education, Hiroshima University, 1-1-1 Kagamiyama, Higashi-Hiroshima 739-8524, Japan
* Correspondence: nkoga@hiroshima-u.ac.jp; Tel.: +81-82-424-7092

Academic Editor: Sergey Vyazovkin
Received: 2 June 2019; Accepted: 6 July 2019; Published: 9 July 2019

Abstract: Examining the kinetics of solids' thermal decomposition with multiple overlapping steps is of growing interest in many fields, including materials science and engineering. Despite the difficulty of describing the kinetics for complex reaction processes constrained by physico-geometrical features, the kinetic deconvolution analysis (KDA) based on a cumulative kinetic equation is one practical method of obtaining the fundamental information needed to interpret detailed kinetic features. This article reports the application of KDA to thermal decomposition of clay minerals and indigo–clay mineral hybrid compounds, known as Maya blue, from ancient Mayan civilization. Maya blue samples were prepared by heating solid mixtures of indigo and clay minerals (palygorskite and sepiolite), followed by purification. The multistep thermal decomposition processes of the clay minerals and Maya blue samples were analyzed kinetically in a stepwise manner through preliminary kinetic analyses based on a conventional isoconversional method and mathematical peak deconvolution to finally attain the KDA. By comparing the results of KDA for the thermal decomposition processes of the clay minerals and the Maya blue samples, information about the thermal decomposition steps of the indigo incorporated into the Maya blue samples was extracted. The thermal stability of Maya blue samples was interpreted through the kinetic characterization of the extracted indigo decomposition steps.

Keywords: Maya blue; indigo; palygorskite; sepiolite; thermal decomposition; kinetic deconvolution analysis

1. Introduction

Thermal decomposition of inorganic solids is a complex heterogeneous reaction that is regulated by chemical kinetics and physio-geometrical constraints [1–3]. In addition, consecutive or concurrent reaction steps that originated from both chemical [4–9] and physio-geometrical reaction mechanisms can occur [10–16], wherein the individual reaction steps may be kinetically dependent on one another. For a multistep process in a homogeneous system, the kinetic behavior can be formalized using concentrations of reactant and intermediate and each reaction step can be verified according to probability considerations. By contrast, in the heterogeneous system, the rigorous formalization of the kinetic equation for multistep reactions is not so easy because of the physico-geometrical constraints of each reaction step and these complex interactions [17]. When each reaction step that exhibits Arrhenius-type temperature dependence can be approximated to be kinetically independent from the other steps, kinetic deconvolution analysis (KDA) can be applied to find an empirical solution for the kinetic description of multistep solid-state reactions [18,19], as follows:

$$\frac{d\alpha}{dt} = \sum_{i=1}^{N} c_i A_i \exp\left(-\frac{E_{a,i}}{RT}\right) f_i(\alpha_i) \text{ with } \sum_{i=1}^{N} c_i = 1 \text{ and } \sum_{i=1}^{N} c_i \alpha_i = \alpha \quad (1)$$

where α, c, A, E_a, and R are the fractional reaction, contribution, Arrhenius pre-exponential factor, apparent activation energy, and the gas constant, respectively. The subscript i denotes a reaction step out of a total N steps. The function $f(\alpha)$ describes the physico-geometrical reaction mechanism as formalized by considering the rate-limiting step of the reaction and the reaction geometry [1–3]. Despite the empirical nature of the kinetic analysis using KDA, the results provide necessary information to gain further insights into consecutive or concurrent kinetic features [20–22], as well as practically useful information about the multistep process, including the contribution (c_i) and apparent kinetic parameters (A_i, $E_{a,i}$, and $f_i(\alpha_i)$) of each reaction step i. Using the results of KDA, the overall reaction process, under a specific heating condition, can be reproduced or simulated. By comparing the results of KDA among a series of samples and under different reaction conditions, characteristics of multistep kinetic behavior can be correlated to different components of a composite sample [23–26] and specific reaction conditions [27,28]. KDA is also used to extract kinetic information about a selected reaction step from the overall process [29,30].

A multistep heterogeneous thermal decomposition can be observed for inorganic–organic hybrid materials. One example of such a material is Maya blue (MB), a well-known pigment used in the Mayan civilization. MB is a hybrid compound of a microporous clay mineral and indigo [31–33]. Palygorskite and sepiolite are the typical clay minerals used in the preparation of MB. These fibrous clay minerals exhibit external and internal nanochannels, which are typically filled with zeolitic water [34–41]. MB is produced by the replacement of the zeolitic water with indigo molecules [31,34–36]. Notably, MB exhibits high stabilities against thermal treatment, light exposure, and acid and/or base attacks. Consequently, the structural characteristics of MB have been intensively studied using spectroscopic techniques that include Fourier transform infrared (FT-IR) and Raman spectroscopies [31,34,35,41–44]. The thermal behavior of MB has also been studied using thermoanalytical techniques [34,36,42]. The formation of strong hydrogen bonds between the structural water of the clay mineral and the carbonyl and amino functional groups of the indigo molecules was reported as a possible reason for the high stabilities of MB [37]. MB with palygorskite as the clay mineral is expected to be more stable compared to MB with uses sepiolite, because of the higher number of hydrogen bonds available to form between the indigo and the substrate mineral [35]. It was also reported that the indigo molecules incorporated into the clay mineral substrate transform into dehydroindigo during the heating process used to prepare MB [37,43,45]. The thermal decomposition of MB begins just above room temperature and indicates partially overlapping multistep processes upon further heating, which may be composed of dehydration steps of zeolitic, coordinating, and structural waters; decomposition of hydroxides in the substrate mineral; and sublimation/decomposition of indigo molecules [34,36,42]. Indigo molecules that have been incorporated into the clay mineral are known to have a higher thermal stability, as confirmed by thermoanalytical curves, compared to pure indigo crystals [34].

Kinetic characterization of the thermal decomposition of MB is a promising approach to evaluating its thermal stability. However, its complex multistep thermal behavior interferes with a successful and straight-forward kinetic analysis. Application of KDA to the thermal decomposition of MB is one possible empirical method of separating the overlapping reaction steps and kinetically characterizing each one to determine the reaction steps that relate specifically to the thermal decomposition of indigo molecules. In the present study, MBs based on palygorskite (P-MB) and sepiolite (S-MB) were prepared by heating solid mixtures of indigo and clay minerals. The thermal decomposition processes of the purified clay mineral substrate samples and the MB samples were analyzed kinetically using KDA, after the necessary preliminary kinetic approaches. By comparing the kinetic results for the thermal decomposition of MB and its clay mineral substrates, the thermal decomposition steps of the indigo molecules were extracted. Using the kinetic information for the thermal decomposition steps of the indigo molecules, the kinetic stabilities of the indigo molecules incorporated in different clay mineral matrices were compared to one another and those of pure indigo crystals.

2. Results and Discussion

2.1. Sample Preparation and Characterization

The characterization details of the purchased palygorskite and sepiolite samples are described in Section S1(1) in the Supplementary Materials. Since a $CaCO_3$ impurity was found in both clay mineral samples, they were purified using HCl(aq) before use. Figure 1 represents the scanning electron microscope (SEM) images of the clay minerals after treatment with HCl(aq). The palygorskite sample is an agglomerate of needle-like crystal with a length of approximately 2–5 μm (Figure 1a). Agglomerates of columnar crystals, with a length of approximately 1–2 μm, are characteristic of the sepiolite sample (Figure 1b).

Figure 1. SEM images of the clay minerals after treatment with HCl(aq): (**a**) Palygorskite and (**b**) sepiolite.

The coloration changes in the indigo–clay mineral mixture before and after heating are described in Section S1(2) in the Supplementary Materials. Figure 2 compares thermogravimetry (TG)–derivative thermogravimetry (DTG)–differential thermal analysis (DTA) curves for the heat-treated samples with different indigo/clay mineral ratios. The major difference between the samples with different indigo/clay mineral ratios is the mass-loss step initiated at approximately 525 K and 535 K for the palygorskite and sepiolite substrate samples, respectively, in which the mass-loss value and the DTG peak height increase with an increasing amount of indigo. The temperature range of the mass-loss step agrees with that for the sublimation/decomposition of pure indigo crystals (Figure S9). Therefore, it is likely that, in the samples with higher indigo/clay mineral ratios, excess indigo remains unreacted with the clay mineral substrates.

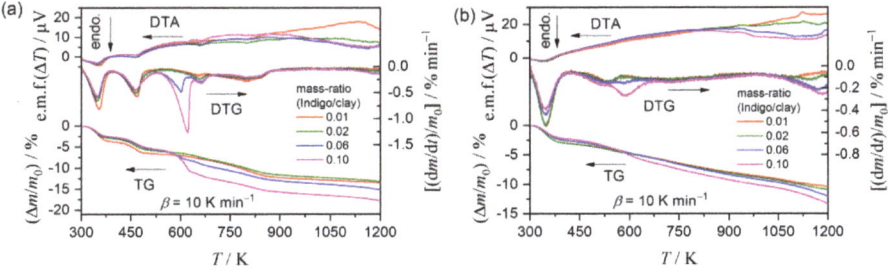

Figure 2. TG–DTG–DTA curves for the heat-treated indigo/clay mineral mixtures synthesized using different initial ratios: (**a**) Indigo/palygorskite and (**b**) indigo/sepiolite.

Figure 3 compares the TG–DTG–DTA curves for the heat-treated indigo/palygorskite sample and those further treated with $Na_2S_2O_4$(aq) for removing excess indigo. The TG–DTG–DTA curves were recorded under an atmosphere of flowing N_2 (Figure 3a) and clearly indicate that the mass-loss step initiated at approximately 525 K disappears after samples are treated with $Na_2S_2O_4$(aq). For the sample treated with $Na_2S_2O_4$(aq), several mass-loss steps that occur at the higher temperatures also disappeared in the TG–DTG–DTA curves recorded in flowing air (Figure 3b). The disappeared mass-loss steps are expected to be attributed to either oxidation or combustion of the thermal decomposition product of indigo. The comparable results of $Na_2S_2O_4$(aq) treatment were also observed for the heat-treated indigo/sepiolite samples, as shown in Figure S10. These results indicate that the removal of excess indigo from the MB samples was successful. Figure S11 compares the sample coloration before and after the $Na_2S_2O_4$(aq) treatment. The faded color that results after the treatment arises from the removal of excess indigo. The color fading is characterized by a decrease in the ultraviolet-visible (UV-Vis) absorption in the wavelength range of 425–600 nm and the appearance of a maximum absorption at approximately 650 nm, as illustrated in Figure S12. Figure S13 presents the SEM images of the samples treated with $Na_2S_2O_4$(aq). The appearance of the synthesized MB samples was not significantly different from those of the palygorskite and sepiolite samples (Figure 1).

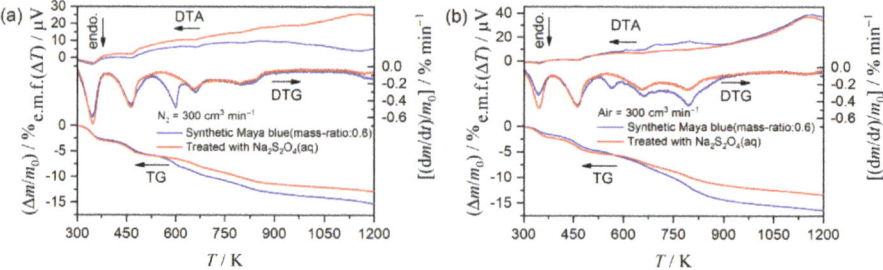

Figure 3. Comparison of the TG–DTG–DTA curves for the heat-treated indigo/palygorskite samples before and after treatment with $Na_2S_2O_4$(aq) recorded in (**a**) flowing N_2 and (**b**) flowing air.

2.2. Kinetic Analysis of the Thermal Decomposition of Clay Minerals

Figure 4 presents TG–DTG curves for the purified clay minerals, recorded at different heating rates (β) in the flow of N_2 gas. The thermal decomposition processes of both clay minerals are multistep processes comprised of three and four distinguishable DTG peaks for palygorskite (Figure 4a) and sepiolite (Figure 4b), respectively. The systematic shift of all the distinguishable DTG peaks to higher temperatures with increasing β was observed for both samples. This is a normal feature for the kinetic process. The average value for the total mass loss during heating the samples to 1223 K was 13.6 ± 0.3% and 9.9 ± 0.1% for palygorskite and sepiolite, respectively. The overall thermal decomposition behaviors of palygorskite and sepiolite approximately agree with those previously reported [34,46,47]. The first to third DTG peaks in both the samples correspond to the thermal dehydration of zeolite water, the first coordinated water, and the second coordinated water. The fourth DTG peak in the thermal decomposition of sepiolite is attributed to the thermal dehydration of structural water.

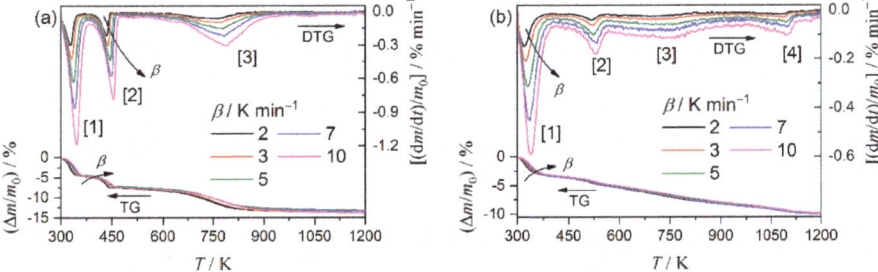

Figure 4. TG–DTG curves for the thermal decomposition of the purified clay minerals recorded at different β in a flow of N_2 gas: (**a**) Palygorskite and (**b**) sepiolite.

As part of the preliminary kinetic approach to the multistep thermal decomposition process, the isoconversional kinetic analysis was examined for the overall thermal decomposition. For the ideal single-step reaction, Equation (2) can be used as the fundamental kinetic equation [48].

$$\frac{d\alpha}{dt} = A \exp\left(-\frac{E_a}{RT}\right) f(\alpha) \qquad (2)$$

Taking the natural logarithm of both sides of Equation (2), one can obtain the following equation:

$$\ln\left(\frac{d\alpha}{dt}\right) = \ln[Af(\alpha)] - \frac{E_a}{RT} \qquad (3)$$

At the selected α, the plot of the left-hand side of Equation (3) versus the reciprocal temperature should exhibit a linear correlation when the value of $\ln[Af(\alpha)]$ is constant. The apparent E_a values at different α can be calculated from the slope of the plot, known as a Friedman plot [49]. The application of the Friedman plot to the overall kinetic data of the multistep thermal decomposition process recorded at different β is not supported by theory, because more than one reaction step overlaps at each α. Even so, some possibility of finding α region characterized by a relevant E_a values or a specific trend of the E_a variation is still anticipated. Figure 5 illustrates the E_a values at different α for the overall thermal decomposition. For the thermal decomposition of palygorskite, four distinguishable reaction steps are expected from the constant E_a regions and the region that exhibits specific E_a variation trends (Figure 5a). The regions assigned as (1), (3), and (4) correspond to the major reaction steps observed as distinguishable DTG peaks. Five distinguishable α regions were found for the thermal decomposition of sepiolite (Figure 5b), in which the regions assigned as (1), (3), (4), and (5) correspond to the major DTG peaks.

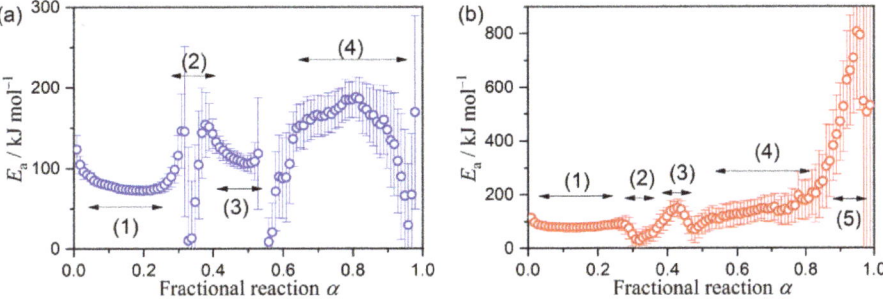

Figure 5. The apparent E_a values at different α with reference to the total mass-loss during the overall thermal decomposition of the purified clay minerals, determined by the Friedman plot: (**a**) Palygorskite and (**b**) sepiolite.

Based on the results of the empirical application of the isoconversional method to the multistep thermal decomposition process, the component reaction steps were deconvolved based on the DTG curves by mathematical deconvolution analysis (MDA); that is, the statistical shape analysis assuming overlapping independent peaks are present [18,50,51].

$$\frac{dm}{dt} = \sum_{i=1}^{N} F_i(t) \qquad (4)$$

In Equation (4), N is the number of component peaks. The value $F_i(t)$ is the statistical function used to satisfactorily fit the component peak i. As the component DTG peaks typically have an asymmetric shape, one of the statistical functions that are applicable to symmetric peaks, such as the Weibull and Frazer–Suzuki functions, is favorable for MDA [18,50,51]. According to the number of distinguishable regions of α observed in the results of the isoconversional kinetic analysis (Figure 5), MDA for the thermal decomposition of palygorskite and sepiolite was carried out by setting $N = 4$ and $N = 5$, respectively. The Weibull function (Equation (S1)) was applied to fit all the component peaks.

Figure 6 illustrates a typical result of the MDA. The second DTG peaks in both samples are described by the partial overlapping of two peaks in the MDA results. The primary outcome from the MDA is the rough estimation of the contribution c_i of each reaction step i with reference to the overall reaction. Table S1 lists the contribution of each component step i. In both the samples, the first and fourth deconvolved steps, which correspond to the dehydration of zeolite water and the second coordinated water, have been indicated as significant contributions to the overall thermal decomposition.

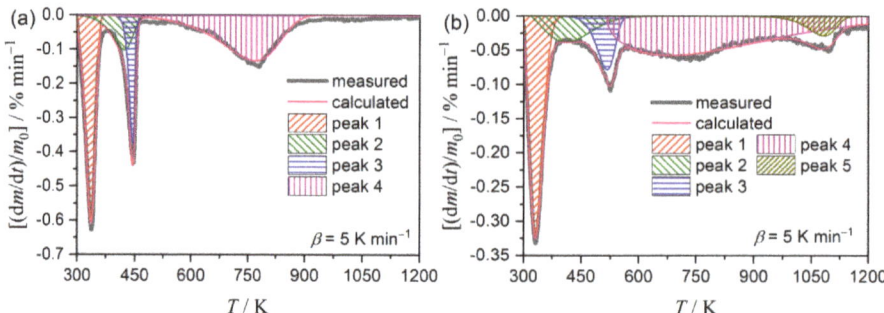

Figure 6. Typical results of MDA applied to the multistep thermal decomposition of the clay minerals: (a) Palygorskite and (b) sepiolite.

The other outcome from MDA is the separated kinetic curves for each reaction step. The features of the separated kinetic curves and the formal kinetic analyses for these kinetic curves are described in Section S2 in the Supplementary Materials.

Based on the results obtained by the preliminary kinetic approaches using the conventional isoconversional method and MDA, the overall kinetic curves were analyzed by assuming that the overlapping multistep process was comprised of independent reaction steps. In this case, the cumulative kinetic equation in Equation (1) is applicable [17–19]. For the kinetic model function $f_i(\alpha_i)$ for each reaction step i, an empirical kinetic model that accommodates different types of the physico-geometrical reaction mechanisms and those that deviate are needed to obtain the sophisticated fit for the calculated kinetic curve to the experimental kinetic curve. The Šesták–Berggren (SB) model with three kinetic exponents [52–54], SB(m, n, p), is one such empirical kinetic model with the high flexibility needed for the fitting, as follows:

$$f(\alpha) = \alpha^m (1-\alpha)^n [-\ln(1-\alpha)]^p \qquad (5)$$

The nonlinear least squares analysis for fitting the calculated kinetic curve to the calculated kinetic curve while simultaneously optimizing all the kinetic parameters in Equations (1) and (5) is a typical procedure of KDA [17–19]. For reliable KDA, appropriate initial values of all the kinetic parameters that will be optimized through KDA are necessary. For the thermal decomposition of palygorskite and sepiolite, the initial c_i and $E_{a,i}$ values were adapted from the results of the MDA (Table S1). The initial kinetic exponents in the SB model were set to SB(0, 1, 0), which is the first-order kinetic model. Then, the order of A_i values was determined graphically by monitoring the fit of the calculated kinetic curve to the experimental kinetic curve. After inputting all the initial values, KDA was run to optimize the values through nonlinear least squares analysis to minimize the sum of squares of the differences between the experimental and calculated kinetic curves, as follows:

$$F = \sum_{j=1}^{M}\left[\left(\frac{d\alpha}{dt}\right)_{\exp,j} - \left(\frac{d\alpha}{dt}\right)_{\text{cal},j}\right]^2 \quad (6)$$

where M is the total number of data points in the experimental kinetic curve at a β value.

Figure 7 illustrates typical results of the KDA for the thermal decomposition of the purified palygorskite (Figure 7a) and sepiolite (Figure 7b). Regardless of the kinetic curve recorded at different β values, the calculated kinetic curve was fit to the experimental kinetic curve with a determination coefficient for the nonlinear least squares analysis (R^2) better than 0.99. The optimized kinetic parameters for each reaction step for the thermal decompositions of the purified palygorskite and sepiolite are summarized in Table 1. The contributions for the first reaction step, attributed to the thermal dehydration of zeolite water, were comparable between the two samples. This was also true for the thermal dehydration of the second coordinated water, which appeared as the fourth reaction step for the thermal decompositions of palygorskite and sepiolite. The optimized E_a values for each reaction step did not significantly change from the initial values in both samples. The rate behavior of each reaction step was simulated from the SB(m_i, n_i, p_i) model with the optimized kinetic exponents, as illustrated in Figure 8. The first reaction step, i.e., the thermal dehydration of zeolite water, exhibits nearly linear deceleration in both samples. The linear deceleration behavior was also seen for the third reaction step of the thermal decomposition of sepiolite. The corresponding reaction step for the thermal dehydration of the first coordinated water in the thermal decomposition of palygorskite exhibited zero-order-like behavior in a wide α_3 range. For the other reaction steps in both samples, deceleration behavior characterized by concaved shapes was observed, possibly indicating that the process was controlled by diffusional removal of evolved water vapor.

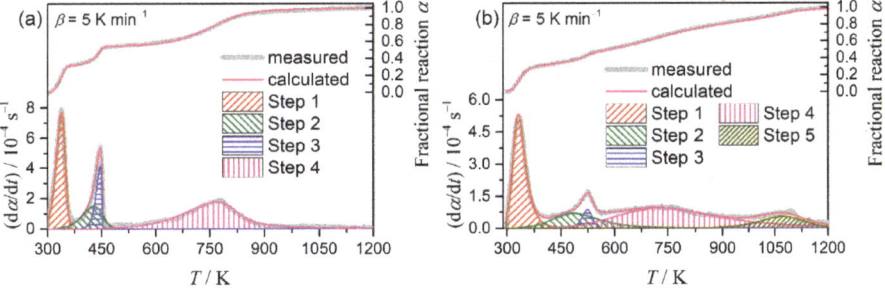

Figure 7. Typical results of KDA for the thermal decomposition of the purified clay minerals: (**a**) Palygorskite and (**b**) sepiolite.

Table 1. The average kinetic parameters optimized by KDA for each reaction step of the thermal decomposition of the purified clay minerals.

Sample	i	c_i	$E_{a,i}/\text{kJ mol}^{-1}$	A_i/s^{-1}	$f_i(\alpha_i) = \alpha_i^m (1-\alpha_i)^n [-\ln(1-\alpha_i)]^p$			R^2
					m	n	p	
Palygorskite	1	0.32 ± 0.01	63.7 ± 0.2	(4.35 ± 0.07) × 10^7	−0.63 ± 0.04	1.26 ± 0.03	0.73 ± 0.03	0.99 ± 0.01
	2	0.12 ± 0.02	90.5 ± 1.7	(2.94 ± 0.01) × 10^8	−0.33 ± 0.02	1.13 ± 0.09	−0.35 ± 0.02	
	3	0.11 ± 0.01	114.4 ± 0.2	(2.00 ± 0.02) × 10^{11}	0.03 ± 0.01	0.61 ± 0.07	0.21 ± 0.01	
	4	0.45 ± 0.01	190.0 ± 1.5	(4.79 ± 0.05) × 10^8	−32.8 ± 3.8	13.8 ± 1.7	29.5 ± 3.7	
Sepiolite	1	0.31 ± 0.01	80.4 ± 0.1	(3.25 ± 0.02) × 10^{10}	0.33 ± 0.02	2.34 ± 0.04	−0.27 ± 0.01	0.99 ± 0.01
	2	0.14 ± 0.01	64.6 ± 0.9	(1.28 ± 0.01) × 10^4	−0.01 ± 0.01	2.30 ± 0.05	−0.59 ± 0.01	
	3	0.04 ± 0.01	213.0 ± 1.0	(1.01 ± 0.01) × 10^{19}	−0.02 ± 0.01	1.41 ± 0.02	−0.09 ± 0.01	
	4	0.41 ± 0.01	124.3 ± 1.4	(2.37 ± 0.01) × 10^5	−0.46 ± 0.01	2.93 ± 0.05	−1.43 ± 0.02	
	5	0.10 ± 0.01	738.4 ± 1.6	(9.50 ± 0.01) × 10^{32}	−1.29 ± 0.01	2.30 ± 0.01	−1.78 ± 0.02	

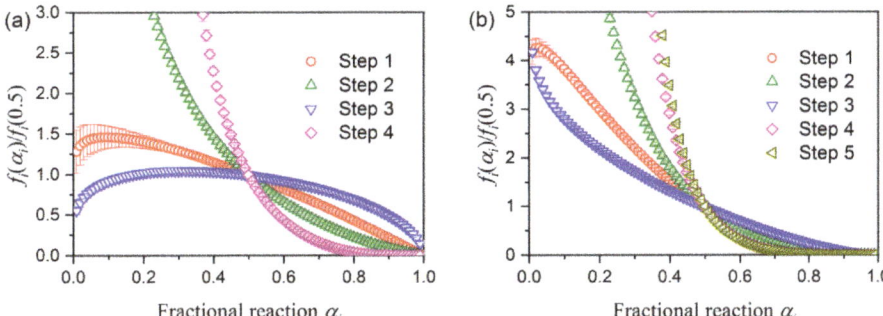

Figure 8. Rate behaviors for each reaction step of the thermal decomposition of (**a**) palygorskite and (**b**) sepiolite, reproduced from SB (m_i, n_i, p_i) optimized by KDA.

2.3. Kinetic Deconvolution Analysis for the Thermal Decomposition of MB

Figure 9 presents the TG–DTG curves recorded at different β values in a flow of air for the thermal decomposition of the synthesized P-MB and S-MB samples. The number of distinguishable DTG peaks was 5 and 6 for the thermal decomposition of P-MB and S-MB samples, respectively. In comparison with the TG–DTG curves for the clay mineral substrates (Figure 4), the third and fifth distinguishable DTG peaks in the thermal decomposition of P-MB appeared in addition to those expected from the thermal decomposition of palygorskite. For S-MB, the third and fourth distinguishable DTG peaks were the additional peaks. These additional peaks can be interpreted as the sublimation/decomposition of indigo incorporated into the clay mineral substrates.

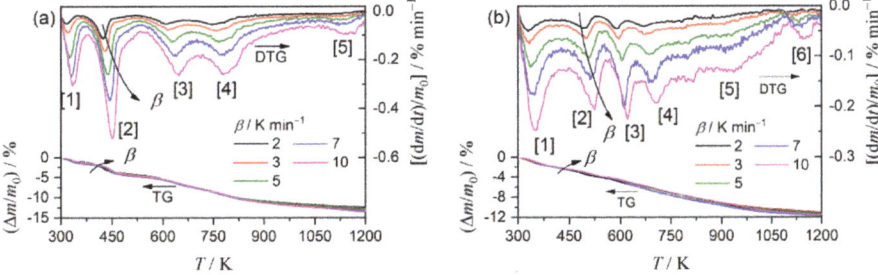

Figure 9. TG–DTG curves for the thermal decomposition of MB recorded at different β in a flow of air: (**a**) P-MB and (**b**) S-MB.

MDA for the thermal decomposition of the MB samples was carried out by adding several minor peaks to the major discernable peaks in the DTG curves (Figure 9). Figure 10 illustrates typical results of MDA carried out by applying a Weibull function to each peak. By comparing the results of MDA for the thermal decomposition of clay mineral substrates, four additional peaks were revealed in both the P-MB and S-MB samples. These additional peaks (that is, the fourth, fifth, seventh, and eighth peaks for P-MB and the fourth, sixth, seventh, and eighth peaks for S-MB) are attributed to the thermal decomposition of the indigo molecules incorporated into the clay mineral matrix. The details of the analysis of the mathematically separated peaks are provided in Section S3 in the Supplementary Materials.

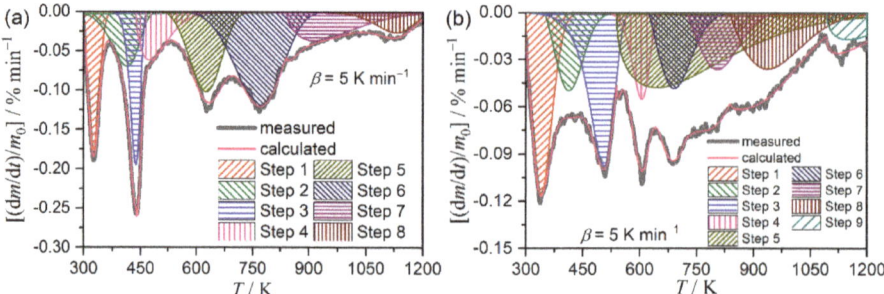

Figure 10. Typical results of MDA for the multistep thermal decomposition of MB samples: (**a**) P-MB and (**b**) S-MB.

Further, the overall thermal decomposition of the MB samples was analyzed by KDA. The initial values of c_i were substituted in for the values determined by MDA (Table S2). For the reaction steps attributed to the thermal decomposition of the clay mineral substrates, the initial values of the kinetic parameters were substituted with values from the results of the KDA for thermal decomposition of the substrates (Table 1). For the reaction steps attributed to the thermal decomposition of indigo, the apparent $E_{a,i}$ values determined for the corresponding steps using MDA (Table S2) were used as the initial values. The $f_i(\alpha_i)$ for the reaction steps of the thermal decomposition of indigo were set to be SB(0, 1, 0) as the initial setting. Then, the order of apparent A_i values was determined by graphically comparing the fit of the calculated curves with the experimental kinetic curve.

Figure 11 illustrates typical results of KDA for the thermal decomposition of the MB samples. The nearly perfect fit using Equation (1) with SB(m, n, p) in Equation (5) as the kinetic model function was realized by using the optimized kinetic parameters for each reaction step, as listed in Table 2. The optimized kinetic parameters for each reaction step from the overall kinetic curves recorded at different β values were practically invariant, as was determined by the acceptably small standard deviation values of each kinetic parameter. The sums of all the contributions from the reaction steps attributed to the thermal decomposition of indigo, that is, i = 4, 5, 7, and 8 for the P-MB sample and i = 4, 6, 7, and 8 for the S-MB sample, were 0.415 and 0.226, respectively. Compared to the initial sample mass m_0, the fractional mass loss attributed to the thermal decomposition of indigo were calculated to be 5.35% and 2.58% for the P-MB and S-MB samples, respectively. These fractional mass-loss values are smaller than the initial mass ratio of indigo added to the clay mineral, 5.66%, before heat treatment and purification, in both samples. The difference in the mass-loss fractions between both samples indicates that the P-MB sample incorporates twice as much indigo as the S-MB sample.

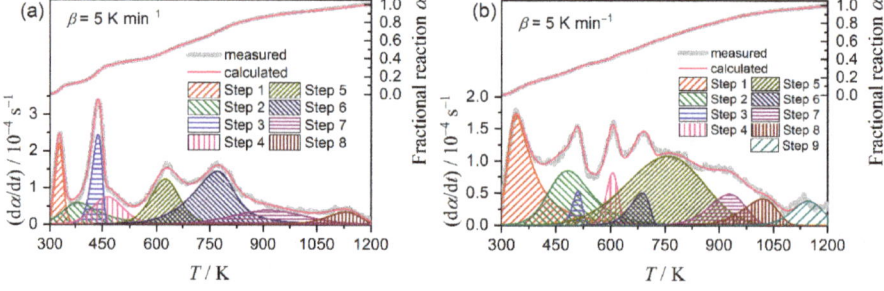

Figure 11. Typical results of KDA for the thermal decomposition of MB: (**a**) P-MB and (**b**) S-MB.

Table 2. Kinetic parameters for each reaction step for the thermal decomposition of MB, as determined by KDA.

Sample	i	c_i	$E_{a,i}$/kJ mol^{-1}	A_i/s^{-1}	$f_i(\alpha_i)=\alpha_i^m(1-\alpha_i)^n[-\ln(1-\alpha_i)]^p$			R^2
					m	n	p	
P-MB	1	0.08 ± 0.01	61.7 ± 0.5	(4.33 ± 0.02) × 10^7	−0.57 ± 0.10	1.69 ± 0.09	0.73 ± 0.05	0.99 ± 0.01
	2	0.10 ± 0.01	46.4 ± 0.3	(1.95 ± 0.01) × 10^3	−0.04 ± 0.01	2.68 ± 0.22	−0.18 ± 0.01	
	3	0.10 ± 0.01	113.0 ± 0.6	(1.70 ± 0.01) × 10^{11}	0.02 ± 0.01	0.85 ± 0.01	−0.07 ± 0.01	
	4	0.09 ± 0.01	43.7 ± 2.0	(1.96 ± 0.02) × 10^2	1.19 ± 0.06	1.05 ± 0.03	−1.10 ± 0.05	
	5	0.15 ± 0.03	143.2 ± 0.3	(1.99 ± 0.02) × 10^9	−0.11 ± 0.01	1.64 ± 0.17	−0.35 ± 0.03	
	6	0.31 ± 0.03	183.2 ± 2.5	(4.99 ± 0.03) × 10^8	−26.61 ± 3.69	11.94 ± 2.27	24.50 ± 4.38	
	7	0.13 ± 0.01	634.3 ± 3.6	(1.70 ± 0.01) × 10^{32}	−2.44 ± 0.19	3.79 ± 0.57	−3.82 ± 0.43	
	8	0.05 ± 0.01	448.3 ± 2.8	(8.84 ± 0.08) × 10^{17}	−0.39 ± 0.03	1.21 ± 0.12	−0.48 ± 0.06	
S-MB	1	0.22 ± 0.01	82.4 ± 0.5	(2.16 ± 0.01) × 10^{10}	0.25 ± 0.01	5.57 ± 0.09	−0.35 ± 0.01	0.98 ± 0.02
	2	0.15 ± 0.01	64.5 ± 0.8	(1.29 ± 0.01) × 10^4	−0.01 ± 0.01	2.45 ± 0.08	−0.40 ± 0.01	
	3	0.02 ± 0.01	206.2 ± 1.9	(9.68 ± 0.01) × 10^{18}	−0.02 ± 0.01	1.26 ± 0.01	−0.11 ± 0.01	
	4	0.05 ± 0.01	160.3 ± 0.4	(4.48 ± 0.02) × 10^{11}	0.06 ± 0.01	1.35 ± 0.08	0.24 ± 0.01	
	5	0.34 ± 0.01	127.9 ± 2.4	(2.33 ± 6.98) × 10^5	−0.46 ± 0.01	1.17 ± 0.02	−1.39 ± 0.05	
	6	0.04 ± 0.01	180.6 ± 0.7	(1.36 ± 0.01) × 10^{11}	−0.20 ± 0.01	0.78 ± 0.01	−0.20 ± 0.01	
	7	0.08 ± 0.01	120.8 ± 3.3	(9.94 ± 0.02) × 10^3	0.06 ± 0.01	1.07 ± 0.01	0.12 ± 0.01	
	8	0.06 ± 0.01	314.0 ± 5.5	(1.51 ± 0.01) × 10^{13}	−0.40 ± 0.01	1.09 ± 0.01	−0.33 ± 0.01	
	9	0.05 ± 0.01	780.9 ± 9.1	(9.65 ± 0.01) × 10^{32}	−0.73 ± 0.01	2.14 ± 0.03	−0.66 ± 0.01	

Figure 12 presents typical optical microscopic views of the P-MB sample as it was heated to different temperatures at $\beta = 10$ K min^{-1} in flowing air. The brilliant blue color of the original P-MB (Figure 12a) was maintained until the sample reached 573 K (Figure 12b), which is a higher temperature than the completion of the fourth reaction step, i.e., the first reaction step of the thermal decomposition of the indigo. This observation supports our previous assumption that the first decomposition step of indigo is the sublimation/decomposition of indigo adsorbed on the surface of the clay mineral substrate, not the indigo incorporated into the clay mineral matrix. Color degradation was observed in the temperature range that corresponds to the fifth reaction step, which is the second decomposition step of incorporated indigo (Figure 12b–d). Thus, the second decomposition step of indigo was interpreted as the decomposition of indigo incorporated into the micropores of the clay mineral substrate. The grayish color of the sample after the fifth reaction step was completed (Figure 12d) was due to the products of indigo decomposition. Upon further heating, the grayish color gradually disappeared after the seventh and eighth reaction steps (Figure 12e,f, respectively), which corresponded to the third and fourth decomposition steps of indigo, respectively. This observation was understood to be the oxidative decomposition of the residues in the flowing air atmosphere.

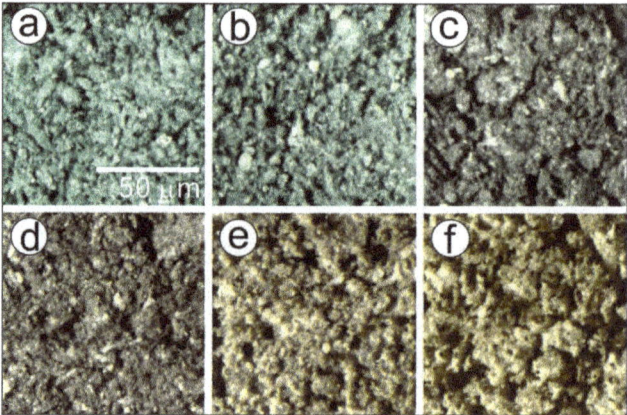

Figure 12. Typical optical microscopic views of the P-MB heated to different temperatures: (**a**) original sample, (**b**) 573 K, (**c**) 713 K, (**d**) 793 K, (**e**) 1003 K, and (**f**) 1223 K.

To compare the thermal decomposition behavior between the indigo incorporated into the MB and pure indigo crystals, thermally induced changes in pure indigo were subjected to a formal kinetic study. Approximately 97.8 ± 0.1% of mass loss was observed during the thermally induced sublimation/decomposition of the pure indigo crystals. The residue (2.2%) was lost by the oxidative decomposition at a higher temperature. The sublimation/decomposition process was characterized by a constant E_a value of 150.7 ± 4.7 kJ mol^{-1} and a phase-boundary controlled model. The details of the kinetic analysis are described in Section S4 in the Supplementary Materials.

Figure 13 compares the extracted kinetic curve ($\beta = 5$ K min^{-1}) and Arrhenius plots for the respective reaction steps, drawn using the optimized Arrhenius parameters listed in Table 2 for the thermal decomposition of indigo incorporated in P-MB ($i = 5$) and S-MB ($i = 6$), with those for the thermally induced sublimation/decomposition of pure indigo crystals. Although the thermal decomposition of indigo in P-MB starts at roughly the same temperature, the reaction proceeds at a slower rate and continues to higher temperatures in comparison to the pure indigo crystals (Figure 13a). Despite the differences in the kinetic data, the Arrhenius plots for these reactions are comparable (Figure 13b), as follows: (E_a/kJ mol^{-1}, A/s^{-1}) values for P-MB ($i = 5$) are (143.2 ± 0.3, (1.99 ± 0.02) × 10^9). Mechanistic differences are a possible reason for the different kinetic behaviors. By contrast, the kinetic data and the Arrhenius plot for the thermal decomposition of the indigo incorporated into S-MB are very different from those of other indigo, having a higher thermal stability and a slower reaction rate. The larger Arrhenius parameters for S-MB ($i = 6$), i.e., $(180.6 \pm 0.7, (1.36 \pm 0.01) \times 10^{11})$, explain the difference between its thermal behavior and that of P-MB.

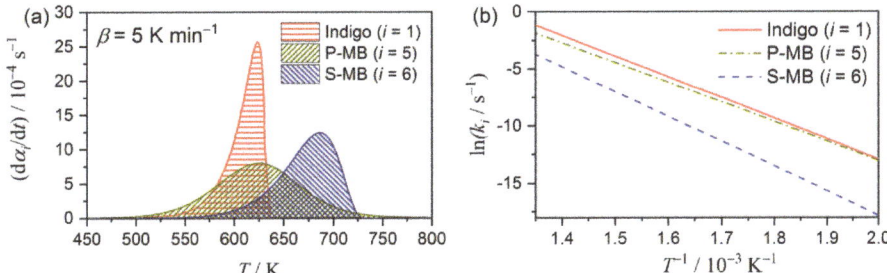

Figure 13. Comparison of the kinetic characteristics for the thermally induced sublimation/decomposition of indigo as crystalline particles, incorporated in P-MB ($i = 5$), and in S-MB ($i = 6$): (**a**) Kinetic curves and (**b**) Arrhenius plots.

Figure 14 is the plot of $f_i(\alpha_i)$ (= SB(m_i, n_i, p_i)) versus α_i for the thermal decomposition step of indigo incorporated into the pores of the substrate minerals, i.e., P-MB ($i = 5$) and S-MB ($i = 6$). In both samples, the curves exhibit a concaved shape, which is characteristic of the deceleration process being controlled by diffusion. The curve was empirically fitted by a model for nucleation and growth controlled by diffusion, i.e., JMA(m) with $m < 1$ [55–58], or for three-dimensional shrinkage of reactant particle controlled by diffusion, i.e., the Jander model, D(3) [59].

$$\text{JMA}(m): f(\alpha) = m(1-\alpha)[-\ln(1-\alpha)]^{1-1/m} \quad (7)$$

$$\text{D}(3): f(\alpha) = \frac{3(1-\alpha)^{2/3}}{2[1-(1-\alpha)^{1/3}]} \quad (8)$$

The removal of gaseous products, which include sublimated indigo, by diffusion from the pores of the substrate mineral is likely the rate-limiting step. A reaction mechanism that is controlled by diffusion is very different from that of the sublimation/decomposition of pure indigo crystals (Figure S22c), which is also one reason for the thermal stability of the indigo incorporated into MB samples.

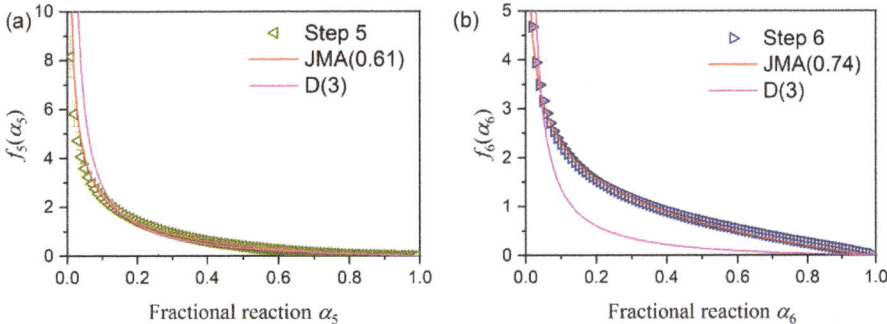

Figure 14. The plot of $f_i(\alpha_i)$ (= SB(m_i, n_i, p_i)) versus α_i for the thermal decomposition step of indigo incorporated into the pores of the MB samples: (**a**) P-MB ($i = 5$) and (**b**) S-MB ($i = 6$).

3. Materials and Methods

3.1. Sample Preparation

The fibrous clay minerals, i.e., palygorskite and sepiolite, were purchased from SEPIO Japan. Each clay mineral was ground using an agate mortar and a pestle. The ground sample was sieved to different particle sizes using a series of sieves with different mesh sizes and an electric shaking apparatus (MVS-1, AS ONE). The sample particles sieved to 170–200 μm in diameter were used to synthesize MB. The clay mineral samples were immersed in 1 M-HCl(aq) and the slurry was stirred for approximately 10 h to remove the $CaCO_3$ impurity included in the clay mineral [60]. The precipitate was filtered and washed repeatedly until the chloride ions were not detected in the filtrates against $AgNO_3$(aq). The separated clay minerals were dried in an electric oven (DK240S, YAMATO) at 343 K for 24 h.

MB was prepared following previously reported procedures [31,33,35,40,41,43,61]. The clay minerals treated with HCl(aq) were used as the substrates for synthesizing MB. Indigo (≥95.0%, NACALAI Tesque) was used as the pigment. Indigo and the clay mineral substrate were mixed in the following mass ratios: 0.01, 0.02, 0.06, or 0.10. Approximately 2.5 g of the mixed sample was placed into a ceramic crucible and covered with a ceramic lid. Samples were heated in an electric furnace (KDF P-70, DENKEN) at 368 K for 24 h and subsequently at 413 K for 24 h.

Excess indigo that was not reacted with the clay mineral substrate was removed as follows [62]. Approximately 0.5 g of the heat-treated sample was dispersed into 20 mL of a 0.25 M-NaOH(aq) solution, to which 0.5 g of sodium dithionate ($Na_2S_2O_4$) had been dissolved. The samples were kept at 348 K and stirred using a magnetic stirrer for 5 min. After filtration, the separated solid was washed repeatedly with water and dried in air.

3.2. Sample Characterization

The clay mineral samples were identified using powder X-ray diffractometry (XRD) and FT-IR. Samples were press-fitted to sample holders to carry out XRD measurement on a diffractometer (RINT-2200V, Rigaku, Tokyo, Japan) with a radiation source (Cu-Kα, 40 kV, 20 mA) in the 2θ range of 5–60° at a scan speed of 4° min^{-1}. FT-IR spectra were measured in a wavenumber range of 400–4600 cm^{-1} using the diffuse reflectance method in a spectrophotometer (FT-IR8400S, Shimadzu, Kyoto, Japan) after diluting the sample with KBr. The clay mineral substrates and MB samples were subjected to simultaneous TG–DTA, UV-Vis spectroscopy, and morphological observation using a SEM. Approximately 10 mg of each sample was weighed into a platinum pan (5 mm in diameter and 2.5 mm in height) and heated at a β of 10 K min^{-1} from room temperature to 1223 K in flowing N_2 (flowrate: 300 cm^3 min^{-1}) for recording TG–DTA curves using an instrument (STA7300, Hitachi High-Tech. Sci., Tokyo, Japan). The UV-Vis spectra of the samples press-fitted to a glass slide were recorded in a wavelength range of 400–700 nm using a spectrophotometer (V-560, JASCO, Tokyo, Japan) equipped with an integrating sphere. For SEM observations, the sample was coated with a thin Pt layer by sputtering (30 mA, 40 s, JFC-1600, JEOL, Tokyo, Japan) and observed using an instrument (JSM-6510, JEOL).

3.3. Tracking of the Thermal Decomposition Process

Thermal behaviors of the clay minerals received and those that were treated with HCl(aq) were investigated by TG/DTA–mass spectrometry (MS). Approximately 10 mg of each sample was weighed into a platinum pan (5 mm in diameter and 2.5 mm in height) and TG–DTA measurements were carried out using an instrument (Thermoplus TG-8120, Rigaku). The sample was heated from room temperature to 1223 K in flowing He (200 cm^3 min^{-1}). During the TG–DTA measurement, the outlet gas from the instrument was transferred into a MS instrument (M-200QA, Anelva, Kanagawa, Japan) through a capillary tube (0.007 mm in inner diameter and 0.8 m in length) that was heated at 500 K.

MS measurements (EMSN: 1.0 A; SEM: 1.0 kV) for the outlet gas were continuously repeated in a m/z range of 10–50.

Using the TG–DTA instrument (STA7300), approximately 10 mg of MB samples were heated to different temperatures at $\beta = 10$ K min^{-1} in flowing air (300 cm^3 min^{-1}). After the sample was cooled to room temperature, the partially decomposed samples were observed using an optical microscope (SZX7, Olympus, Tokyo, Japan) for recording the color of the sample.

3.4. Measurement of the Kinetic Data for the Thermal Decomposition

To record the kinetic data for the thermal decomposition of the clay mineral substrates treated with HCl(aq) and the MB samples, TG–DTA measurements were carried out using the STA7300 instrument. Approximately 5 mg of each sample was weighed into a platinum pan (5 mm in diameter and 2.5 mm in height) and heated from room temperature to 1223 K at various β values between 2 and 10 K min^{-1} in flowing N$_2$ or air (flowrate: 300 cm^3 min^{-1}). The TG–DTA instrument was previously calibrated in view of mass change values and temperature using standard procedures.

4. Conclusions

Thermal decompositions of palygorskite and sepiolite were characterized by three and four distinguished DTG peaks, respectively, which were attributed to the dehydration of zeolite, coordinating, and structural water. Kinetically, the second mass-loss process in both palygorskite and sepiolite was further separated into two reaction steps. Consequently, the thermal decompositions of palygorskite and sepiolite were kinetically separated into four and five reaction steps, respectively. The processes occurring in MB samples, i.e., P-MB and S-MB, were separated by KDA into eight and nine reaction steps, respectively. The additional four reaction steps that were observed for MB samples are attributed to the thermal decomposition of the indigo incorporated into the clay mineral substrates. Discoloration of MB samples occurred during the second reaction step of the thermal decomposition of the indigo incorporated into the clay mineral substrates. The second reaction step is expected to be directly correlated to the thermal stability of the MB samples as a pigment. The second step of the thermal decomposition of indigo in P-MB started at approximately the same temperature as the thermally induced sublimation/decomposition of pure indigo crystals. In addition, the apparent Arrhenius parameters evaluated for these reaction processes were also comparable. However, the second step in the thermal decomposition of indigo in P-MB occurs at a slower rate and continues to higher temperatures compared to pure indigo crystals. This is explained by different physico-geometrical reaction mechanisms; the thermally induced sublimation/decomposition of pure indigo crystals is a phase boundary-controlled reaction and the second reaction step of indigo decomposition in P-MB is a diffusion-controlled reaction. Even though the second reaction step of indigo decomposition in S-MB started at a higher temperature in comparison to that in P-MB, both reactions ended at around the same temperature. The difference in starting temperatures can be explained by the larger Arrhenius parameters for S-MB. Although the stability of MB is commonly discussed in connection with the strength of the chemical bonds between indigo and the clay mineral substrate, the present study indicates that the physico-geometrical kinetic behavior of the thermal decomposition of indigo incorporated into the clay mineral substrate is another important factor in discussing the thermal stability of MB.

Supplementary Materials: The following are available online, S1: Sample preparation and characterization (Figures S1–S13), S2: Kinetic analysis for the thermal decomposition of clay mineral substrate (Table S1, Figures S14–S16), S3: Kinetic analysis for the thermal decomposition of Maya blue (Table S2, Figures S17–S20), S4: Kinetic analysis for the thermally induced sublimation/decomposition of indigo (Figures S21 and S22).

Author Contributions: Conceptualization, N.K.; Methodology, N.K.; Software, N.K.; Formal Analysis, Y.Y.; Investigation, Y.Y.; Data Curation, Y.Y.; Writing-Original Draft Preparation, Y.Y.; Writing-Review & Editing, N.K.; Visualization, Y.Y.; Supervision, N.K.; Project Administration, N.K.; Funding Acquisition, N.K.

Funding: This research was funded by JSPS KAKENHI grant number 17H00820.

Conflicts of Interest: The authors declare no conflict of interest.

References

1. Galwey, A.K.; Brown, M.E. *Thermal Decomposition of Ionic Solids*; Elsevier: Amsterdam, The Netherlands, 1999; ISBN 9780444824370.
2. Galwey, A.K. Structure and order in thermal dehydrations of crystalline solids. *Thermochim. Acta* **2000**, *355*, 181–238. [CrossRef]
3. Koga, N.; Tanaka, H. A physico-geometric approach to the kinetics of solid-state reactions as exemplified by the thermal dehydration and decomposition of inorganic solids. *Thermochim. Acta* **2002**, *388*, 41–61. [CrossRef]
4. Kitabayashi, S.; Koga, N. Thermal decomposition of tin(II) oxyhydroxide and subsequent oxidation in air: Kinetic deconvolution of overlapping heterogeneous processes. *J. Phys. Chem. C* **2015**, *119*, 16188–16199. [CrossRef]
5. Nakano, M.; Wada, T.; Koga, N. Exothermic behavior of thermal decomposition of sodium percarbonate: Kinetic deconvolution of successive endothermic and exothermic processes. *J. Phys. Chem. A* **2015**, *119*, 9761–9769. [CrossRef]
6. Muravyev, N.V.; Koga, N.; Meerov, D.B.; Pivkina, A.N. Kinetic analysis of overlapping multistep thermal decomposition comprising exothermic and endothermic processes: Thermolysis of ammonium dinitramide. *Phys. Chem. Chem. Phys.* **2017**, *19*, 3254–3264. [CrossRef] [PubMed]
7. Koga, N.; Kameno, N.; Tsuboi, Y.; Fujiwara, T.; Nakano, M.; Nishikawa, K.; Iwasaki-Murata, A. Multistep thermal decomposition of granular sodium perborate tetrahydrate: A kinetic approach to complex reactions in solid-gas systems. *Phys. Chem. Chem. Phys.* **2018**, *20*, 12557–12573. [CrossRef] [PubMed]
8. Koga, N.; Kodani, S. Thermally induced carbonation of $Ca(OH)_2$ in a CO_2 atmosphere: Kinetic simulation of overlapping mass-loss and mass-gain processes in a solid-gas system. *Phys. Chem. Chem. Phys.* **2018**, *20*, 26173–26189. [CrossRef]
9. Noda, Y.; Koga, N. Phenomenological kinetics of the carbonation reaction of lithium hydroxide monohydrate: Role of surface product layer and possible existence of a liquid phase. *J. Phys. Chem. C* **2014**, *118*, 5424–5436. [CrossRef]
10. Koga, N.; Suzuki, Y.; Tatsuoka, T. Thermal dehydration of magnesium acetate tetrahydrate: Formation and in situ crystallization of anhydrous glass. *J. Phys. Chem. B* **2012**, *116*, 14477–14486. [CrossRef]
11. Koga, N.; Yamada, S.; Kimura, T. Thermal decomposition of silver carbonate: Phenomenology and physicogeometrical kinetics. *J. Phys. Chem. C* **2013**, *117*, 326–336. [CrossRef]
12. Wada, T.; Koga, N. Kinetics and mechanism of the thermal decomposition of sodium percarbonate: Role of the surface product layer. *J. Phys. Chem. A* **2013**, *117*, 1880–1889. [CrossRef] [PubMed]
13. Yoshikawa, M.; Yamada, S.; Koga, N. Phenomenological interpretation of the multistep thermal decomposition of silver carbonate to form silver metal. *J. Phys. Chem. C* **2014**, *118*, 8059–8070. [CrossRef]
14. Wada, T.; Nakano, M.; Koga, N. Multistep kinetic behavior of the thermal decomposition of granular sodium percarbonate: Hindrance effect of the outer surface layer. *J. Phys. Chem. A* **2015**, *119*, 9749–9760. [CrossRef] [PubMed]
15. Tsuboi, Y.; Koga, N. Thermal decomposition of biomineralized calcium carbonate: Correlation between the thermal behavior and structural characteristics of avian eggshell. *ACS Sustainable Chem. Eng.* **2018**, *6*, 5283–5295. [CrossRef]
16. Kameno, N.; Koga, N. Heterogeneous kinetic features of the overlapping thermal dehydration and melting of thermal energy storage material: Sodium thiosulfate pentahydrate. *J. Phys. Chem. C* **2018**, *122*, 8480–8490. [CrossRef]
17. Koga, N. Physico-Geometric Approach to the Kinetics of Overlapping Solid-State Reactions. In *Handbook of Thermal Analysis and Calorimetry*, 2nd ed.; Vyazovkin, S., Koga, N., Schick, C., Eds.; Elsevier: Amsterdam, The Netherlands, 2018; Volume 6, pp. 213–251. [CrossRef]
18. Koga, N.; Goshi, Y.; Yamada, S.; Pérez-Maqueda, L.A. Kinetic approach to partially overlapped thermal decomposition processes. *J. Therm. Anal. Calorim.* **2013**, *111*, 1463–1474. [CrossRef]
19. Sánchez-Jiménez, P.E.; Perejón, A.; Criado, J.M.; Diánez, M.J.; Pérez-Maqueda, L.A. Kinetic model for thermal dehydrochlorination of poly(vinyl chloride). *Polymer* **2010**, *51*, 3998–4007. [CrossRef]

20. Kitabayashi, S.; Koga, N. Physico-geometrical mechanism and overall kinetics of thermally induced oxidative decomposition of tin(II) oxalate in air: Formation process of microstructural tin(IV) oxide. *J. Phys. Chem. C* **2014**, *118*, 17847–17861. [CrossRef]
21. Ogasawara, H.; Koga, N. Kinetic modeling for thermal dehydration of ferrous oxalate dihydrate polymorphs: A combined model for induction period-surface reaction-phase boundary reaction. *J. Phys. Chem. A* **2014**, *118*, 2401–2412. [CrossRef]
22. Fukuda, M.; Koga, N. Kinetics and mechanisms of the thermal decomposition of copper(II) hydroxide: A consecutive process comprising induction period, surface reaction, and phase boundary-controlled reaction. *J Phys Chem C* **2018**, *122*, 12869–12879. [CrossRef]
23. Nishikawa, K.; Ueta, Y.; Hara, D.; Yamada, S.; Koga, N. Kinetic characterization of multistep thermal oxidation of carbon/carbon composite in flowing air. *J. Therm. Anal. Calorim.* **2017**, *128*, 891–906. [CrossRef]
24. Hara, D.; Nishikawa, K.; Koga, N. Characterization of carbon/carbon composites by kinetic deconvolution analysis for a thermal oxidation process: An examination using a series of mechanical pencil leads. *Ind. Eng. Chem. Res.* **2018**, *57*, 14460–14469. [CrossRef]
25. Kikuchi, S.; Koga, N.; Yamazaki, A. Comparative study on the thermal behavior of structural concretes of sodium-cooled fast reactor. *J. Therm. Anal. Calorim.* **2019**. [CrossRef]
26. Koga, N.; Kikuchi, S. Thermal behavior of perlite concrete used in a sodium-cooled fast reactor: Multistep reaction kinetics and melting for safety assessment. *J. Therm. Anal. Calorim.* **2019**. [CrossRef]
27. Nakano, M.; Fujiwara, T.; Koga, N. Thermal decomposition of silver acetate: Physico-geometrical kinetic features and formation of silver nanoparticles. *J. Phys. Chem. C* **2016**, *120*, 8841–8854. [CrossRef]
28. Fujiwara, T.; Yoshikawa, M.; Koga, N. Kinetic approach to multistep thermal behavior of Ag_2CO_3–graphite mixtures: Possible formation of intermediate solids with Ag_2O–Ag and Ag_2CO_3–Ag core–shell structures. *Thermochim. Acta* **2016**, *644*, 50–60. [CrossRef]
29. Koga, N.; Kasahara, D.; Kimura, T. Aragonite crystal growth and solid-state aragonite–calcite transformation: A physico–geometrical relationship via thermal dehydration of included water. *Cryst. Growth Des.* **2013**, *13*, 2238–2246. [CrossRef]
30. Koga, N.; Nishikawa, K. Mutual Relationship between solid-state aragonite–calcite transformation and thermal dehydration of included water in coral aragonite. *Cryst. Growth Des.* **2014**, *14*, 879–887. [CrossRef]
31. Polette-Niewold, L.A.; Manciu, F.S.; Torres, B.; Alvarado, M., Jr.; Chianelli, R.R. Organic/inorganic complex pigments: Ancient colors Maya blue. *J. Inorg. Biochem.* **2007**, *101*, 1958–1973. [CrossRef]
32. Giustetto, R.; Seenivasan, K.; Bordiga, S. Spectroscopic characterization of a sepiolite-based Maya blue pigment. *Period. Mineral.* **2010**, *79*, 21–37. [CrossRef]
33. Chiari, G.; Giustetto, R.; Druzik, J.; Doehne, E.; Ricchiardi, G. Pre-columbian nanotechnology: Reconciling the mysteries of the maya blue pigment. *Appl. Phys. A* **2007**, *90*, 3–7. [CrossRef]
34. Ovarlez, S.; Giulieri, F.; Delamare, F.; Sbirrazzuoli, N.; Chaze, A.-M. Indigo–sepiolite nanohybrids: Temperature-dependent synthesis of two complexes and comparison with indigo–palygorskite systems. *Microporous Mesoporous Mater.* **2011**, *142*, 371–380. [CrossRef]
35. Giustetto, R.; Seenivasan, K.; Bonino, F.; Ricchiardi, G.; Bordiga, S.; Chierotti, M.R.; Gobetto, R. Host/guest interactions in a sepiolite-based Maya blue pigment: A Spectroscopic Study. *J. Phys. Chem. C* **2011**, *115*, 16764–16776. [CrossRef]
36. Hubbard, B.; Kuang, W.; Moser, A.; Facey, G.A.; Detellier, C. Structural study of Maya blue: Textural, thermal and solidstate multinuclear magnetic resonance characterization of the palygorskite-indigo and sepiolite-indigo adducts. *Clays Clay Miner.* **2003**, *51*, 318–326. [CrossRef]
37. Sánchez-Ochoa, F.; Cocoletzi, G.H.; Canto, G. Trapping and diffusion of organic dyes inside of palygorskite clay: The ancient Maya blue pigment. *Microporous Mesoporous Mater.* **2017**, *249*, 111–117. [CrossRef]
38. Giustetto, R.; Wahyudi, O. Sorption of red dyes on palygorskite: Synthesis and stability of red/purple Mayan nanocomposites. *Microporous Mesoporous Mater.* **2011**, *142*, 221–235. [CrossRef]
39. Tartaglione, G.; Tabuani, D.; Camino, G. Thermal and morphological characterisation of organically modified sepiolite. *Microporous Mesoporous Mater.* **2008**, *107*, 161–168. [CrossRef]
40. Zhou, W.; Liu, H.; Xu, T.; Jin, Y.; Ding, S.; Chen, J. Insertion of isatin molecules into the nanostructure of palygorskite. *RSC Adv.* **2014**, *4*, 51978–51983. [CrossRef]
41. Giustetto, R.; Llabres, I.X.F.X.; Ricchiardi, G.; Bordiga, S.; Damin, A.; Gobetto, R.; Chierotti, M.R. Maya blue: A computational and spectroscopic study. *J. Phys. Chem. B* **2005**, *109*, 19360–19368. [CrossRef]

42. Ovarlez, S.; Giulieri, F.; Chaze, A.M.; Delamare, F.; Raya, J.; Hirschinger, J. The incorporation of indigo molecules in sepiolite tunnels. *Chem. Eur. J.* **2009**, *15*, 11326–11332. [CrossRef]
43. Manciu, F.S.; Reza, L.; Polette, L.A.; Torres, B.; Chianelli, R.R. Raman and infrared studies of synthetic Maya pigments as a function of heating time and dye concentration. *J. Raman Spectrosc.* **2007**, *38*, 1193–1198. [CrossRef]
44. Ovarlez, S.; Chaze, A.-M.; Giulieri, F.; Delamare, F. Indigo chemisorption in sepiolite. Application to Maya blue formation. *C. R. Chim.* **2006**, *9*, 1243–1248. [CrossRef]
45. Tilocca, A.; Fois, E. The color and stability of Maya blue: TDDFT Calculations. *J. Phys. Chem. C* **2009**, *113*, 8683–8687. [CrossRef]
46. Post, J.E.; Heaney, P.J. Synchrotron powder X-ray diffraction study of the structure and dehydration behavior of palygorskite. *Am. Mineral.* **2008**, *93*, 667–675. [CrossRef]
47. Nagata, H. On dehydration of bound water of sepiolite. *Clays Clay Miner.* **1974**, *22*, 285–293. [CrossRef]
48. Koga, N.; Šesták, J.; Simon, P. Some Fundamental and Historical Aspects of Phenomenological Kinetics in the Solid State Studied by Thermal Analysis. In *Thermal Analysis of Micro, Nano- and Non-Crystalline Materials*; Šesták, J., Simon, P., Eds.; Springer: Dordrecht, The Netherlands, 2013; pp. 1–28.
49. Friedman, H.L. Kinetics of thermal degradation of cha-forming plastics from thermogravimetry, application to a phenolic plastic. *J. Polym. Sci., Part C* **1964**, *6*, 183–195. [CrossRef]
50. Perejón, A.; Sánchez-Jiménez, P.E.; Criado, J.M.; Pérez-Maqueda, L.A. Kinetic analysis of complex solid-state reactions. A new deconvolution procedure. *J. Phys. Chem. B* **2011**, *115*, 1780–1791. [CrossRef] [PubMed]
51. Svoboda, R.; Málek, J. Applicability of Fraser–Suzuki function in kinetic analysis of complex crystallization processes. *J. Therm. Anal. Calorim.* **2013**, *111*, 1045–1056. [CrossRef]
52. Šesták, J.; Berggren, G. Study of the kinetics of the mechanism of solid-state reactions at increasing temperatures. *Thermochim. Acta* **1971**, *3*, 1–12. [CrossRef]
53. Šesták, J. Diagnostic limits of phenomenological kinetic models introducing the accommodation function. *J. Therm. Anal.* **1990**, *36*, 1997–2007. [CrossRef]
54. Šesták, J. Rationale and fallacy of thermoanalytical kinetic patterns. *J. Therm. Anal. Calorim.* **2011**, *110*, 5–16. [CrossRef]
55. Avrami, M. Kinetics of phase change. I. General theory. *J. Chem. Phys.* **1939**, *7*, 1103–1112. [CrossRef]
56. Avrami, M. Kinetics of phase change. II. Transformation-time relations for random distribution of nuclei. *J. Chem. Phys.* **1940**, *8*, 212–223. [CrossRef]
57. Avrami, M. Kinetics of phase change. III. Granulation, phase change, and microstructure. *J. Chem. Phys.* **1941**, *9*, 177–184. [CrossRef]
58. Johnson, W.A.; Mehl, K.F. Reaction kinetics in processes of nucleation and growth. *Trans. Am. Inst. Mining. Metall. Eng.* **1939**, *135*, 416–458.
59. Jander, W. Reaktionen im festen zustande bei höheren temperaturen. Reaktionsgeschwindigkeiten endotherm verlaufender umsetzungen. *Z. Anorg. Allg. Chem.* **1927**, *163*, 1–30. [CrossRef]
60. Cui, W.; Zhang, H.; Xia, Y.; Zou, Y.; Xiang, C.; Chu, H.; Qiu, S.; Xu, F.; Sun, L. Preparation and thermophysical properties of a novel form-stable $CaCl_2 \cdot 6H_2O$/sepiolite composite phase change material for latent heat storage. *J. Therm. Anal. Calorim.* **2017**, *131*, 57–63. [CrossRef]
61. Leitão, I.M.V.; Seixas de Melo, J.S. Maya blue, an ancient guest–host pigment: Synthesis and models. *J. Chem. Educ.* **2013**, *90*, 1493–1497. [CrossRef]
62. Boykin, D.W. A convenient apparatus for small-scale dyeing with indigo. *J. Chem. Educ.* **1998**, *75*, 769. [CrossRef]

Sample Availability: Samples of the compounds are available from the authors.

© 2019 by the authors. Licensee MDPI, Basel, Switzerland. This article is an open access article distributed under the terms and conditions of the Creative Commons Attribution (CC BY) license (http://creativecommons.org/licenses/by/4.0/).

Article

Kinetic Processes in Amorphous Materials Revealed by Thermal Analysis: Application to Glassy Selenium

Jiří Málek * and Roman Svoboda

Department of Physical Chemistry, Faculty of Chemical Technology, University of Pardubice, Studentská 95, 53210 Pardubice, Czech Republic
* Correspondence: jiri.malek@upce.cz; Tel.: +420-466-036-553

Academic Editor: Sergey Vyazovkin
Received: 1 July 2019; Accepted: 22 July 2019; Published: 26 July 2019

Abstract: It is expected that viscous flow is affecting the kinetic processes in a supercooled liquid, such as the structural relaxation and the crystallization kinetics. These processes significantly influence the behavior of glass being prepared by quenching. In this paper, the activation energy of viscous flow is discussed with respect to the activation energy of crystal growth and the structural relaxation of glassy selenium. Differential scanning calorimetry (DSC), thermomechanical analysis (TMA) and hot-stage infrared microscopy were used. It is shown that the activation energy of structural relaxation corresponds to that of the viscous flow at the lowest value of the glass transition temperature obtained within the commonly achievable time scale. The temperature-dependent activation energy of crystal growth, data obtained by isothermal and non-isothermal DSC and TMA experiments, as well as direct microscopic measurements, follows nearly the same dependence as the activation energy of viscous flow, taking into account viscosity and crystal growth rate decoupling due to the departure from Stokes–Einstein behavior.

Keywords: glass; structural relaxation; crystallization; viscosity; thermal analysis

---

1. Introduction

Glasses are amorphous materials lacking the periodic atomic arrangement typical for crystalline substances. Structurally, they resemble supercooled liquids but behave mechanically like solids. Figure 1 shows the specific volume or enthalpy as a function of temperature for a typical glass-forming liquid. Upon slow cooling from high temperatures, a liquid may crystallize at T_m forming a stable crystalline material. However, if the cooling through this temperature range is fast enough to avoid nucleation and subsequent crystal growth, a metastable supercooled liquid state is attained.

When a supercooled liquid is cooled by a cooling rate of q^+_1 to lower temperatures, the internal molecular motion slows down and its viscosity significantly increases. At the glass transition temperature (T_g), the time needed for molecular rearrangement becomes comparable to the experimental time scale. At a lower cooling rate ($q^+_2 < q^+_1$) the supercooled liquid stays in metastable equilibrium until lower temperatures. Therefore, the glass transition is not a true thermodynamic phase transition depending on procedural variables such as cooling rate. As a consequence, there is not a single glassy state, and the properties of the glass depend upon how it was obtained [1].

Clear evidence of the non-existence of a single glassy state is a slow, gradual approach of volume or enthalpy towards the extrapolated supercooled liquid equilibrium line that has been called "structural relaxation". This process, associated with a slow molecular rearrangement, is experimentally observable in the glass transition range (see Figure 1). It seems that structural relaxation is strongly affected by the supercooled liquid dynamics. An understanding of clues between long time-scale structural relaxation and short time-scale molecular dynamics of corresponding supercooled liquids is of fundamental importance for glass science.

On reheating of a glass, the structural relaxation peak is usually observed just above T_g. At higher temperatures the crystallization process takes place [2]. These processes can be followed by thermal analysis methods such as differential thermal analysis (DTA) or differential scanning calorimetry (DSC). Both these methods are quite frequently used to study the structural relaxation and crystallization behavior in glasses. Usually, the kinetic parameters of such activation energies are extracted from experiments taken at different heating rates. In this paper, the physical meaning of these parameters determined for structural relaxation and crystallization of glassy selenium is analyzed and discussed with respect to the viscosity behavior of supercooled selenium.

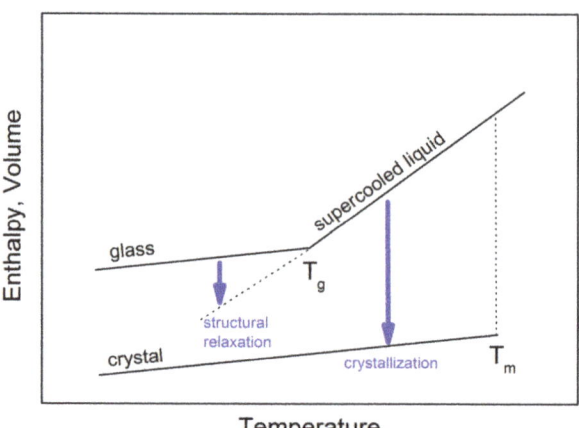

Figure 1. Schematic diagram showing the change in enthalpy and volume during glass formation, structural relaxation and crystallization.

2. Results

This section provides a concise description of the structural relaxation experiments as well as the crystallization of glassy selenium obtained by differential scanning calorimetry (DSC) and dilatometry.

2.1. Volume and Enthalpy Relaxation

The isothermal volume relaxation is typically studied by a temperature down-jump experiment. In this experimental set-up the classical mercury-filled dilatometer containing a selenium glass is first equilibrated in a thermal bath at temperature T_0. Then it is quickly transferred to another temperature bath at temperature $T < T_0$ and the time-dependent volume contraction is recorded. The temperature up-jump experiment starts by equilibration of a dilatometer at a temperature T_0. Then it is quickly transferred to another thermal bath at temperature $T > T_0$ and the time dependent volume expansion is recorded. Isothermal volume relaxation can be expressed as a relative departure of an actual volume V from the equilibrium volume V_∞:

$$\delta = \frac{V - V_\infty}{V_\infty}. \tag{1}$$

The relaxation response immediately after the temperature jump from temperature T_0 to T is defined by the following equation:

$$\delta_0 = \Delta\alpha(T_0 - T), \tag{2}$$

where $\Delta\alpha$ corresponds to the difference between the thermal expansion coefficient in a selenium supercooled liquid and selenium glass [3]. Figure 2 shows the isothermal volume relaxation response of selenium glass subjected to the temperature down-jump (T_0 = 37 °C, T = 32 °C) and up-jump (T_0 = 27 °C, T_B = 32 °C). Open circles represent experimental data. The logarithmic time axis is normalized to the initial time that corresponds to the thermal equilibration of dilatometer after the temperature jump (t_i = 70 s) [3].

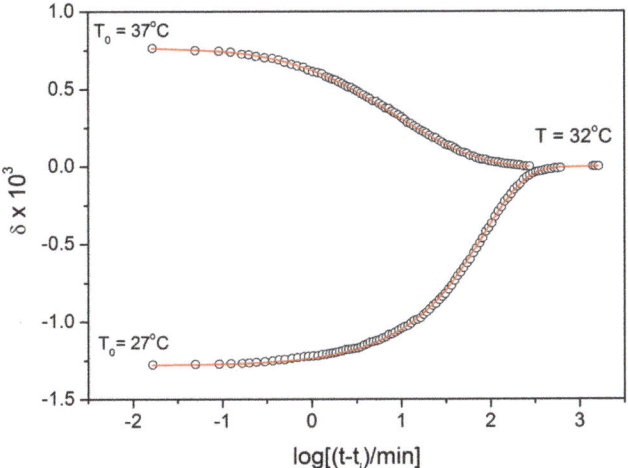

Figure 2. The volume change of Se glass subjected to a temperature down-jump and up-jump ±5 °C. Points correspond to experimental data obtained by dilatometry. Full lines were calculated for the Tool–Narayanaswamy–Moynihan (TNM) model (parameters in text).

It is clearly seen that both relaxation responses after temperature down-jump and up-jump are non-exponential and non-linear. Such behavior can be described by a non-exponential decay function including the reduced time integral [4]:

$$\delta(t) = \delta_0 \exp\left[-\left[\int_0^t \frac{dt}{\tau(T, T_f)}\right]^\beta\right], \qquad (3)$$

where β is the non-exponentiality parameter ($0 < \beta \leq 1$), inversely proportional to the width of the spectrum of relaxation times. It is assumed that τ depends on temperature T, as well as on the instantaneous structure of amorphous material characterized by the fictive temperature T_f [5]. The relaxation time can be expressed by the following equation [6]:

$$\tau(T, T_f) = A_{rel} \cdot \exp\left[x \frac{E_{rel}}{RT} + (1-x) \frac{E_{rel}}{RT_f}\right], \qquad (4)$$

where A_{rel} is the pre-exponential constant, x is the non-linearity parameter ($0 < x \leq 1$) and E_{rel} is the effective activation energy of the relaxation process. The time-dependent fictive temperature for this Tool–Narayanaswamy–Moynihan (TNM) model can be expressed as

$$T_f = \frac{\delta(t)}{\delta_0}[T_0 - T] + T. \qquad (5)$$

Full lines in Figure 2 were calculated by using Equations (3)–(5) for the following set of parameters: β = 0.58 ± 0.05, x = 0.42 ± 0.05, ln (A_{rel}/s) = −133.0 ± 0.5 and E_{rel} = 355 ± 2 kJ·mol^{-1} [3].

Enthalpy relaxation cannot be measured directly in a similar way. However, the DSC heating scans exhibit a typical relaxation overshot just above the glass-transition temperature. It can be shown that normalized heat capacity (C_p^N) measured by DSC can be expressed [6] as the first derivative of the fictive temperature dT_f/dT:

$$C_p^N = \frac{C_p - C_{pg}}{C_{pl} - C_{pg}} = \frac{dT_f}{dT},\qquad(6)$$

where C_p is measured heat capacity, C_{pg} is heat capacity of a glass and C_{pl} is heat capacity of a supercooled liquid.

The temperature dependent plots of C_p^N are shown in Figure 3. Open circles represent experimental data for heating scans at $q^+ = 10$ K·min^{-1} taken immediately after the cooling scans performed at the indicated rates. Full lines in Figure 3 were calculated by using Equations (3)–(6) for the following set of parameters: $\beta = 0.65 \pm 0.05$, $x = 0.52 \pm 0.05$, $\ln(A_{rel}/s) = -133.0 \pm 0.5$ and $E_{rel} = 355 \pm 2$ kJ·mol^{-1} [3]. The parameters $\ln(A_{rel}/s)$ and E_{rel} are identical as for the volume relaxation. However, the values of non-exponentiality and non-linearity parameters are higher than those obtained for volume relaxation.

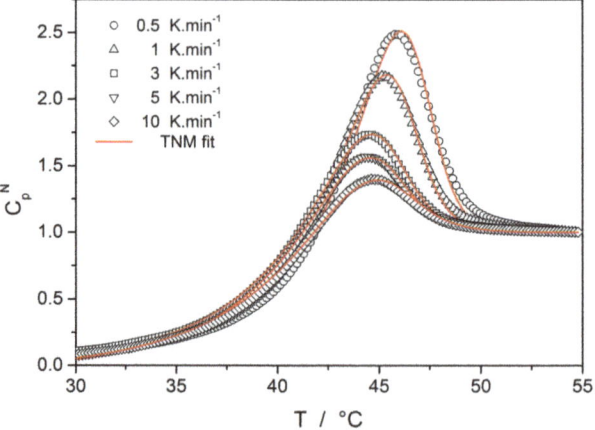

Figure 3. The normalized heat capacity of Se glass reflecting the structural relaxation in the glass transition range. Points correspond to selected experimental data-curves obtained by differential scanning calorimetry (DSC). Full lines were calculated for the TNM model (parameters in text).

Very similar results are also obtained for the thermal expansion coefficient measured by thermomechanical analysis for As$_2$Se$_3$ glass [7,8]. The agreement between the experiment and the TNM model is very good below T_g. Nevertheless, some deviations are observed in the supercooled liquid above T_g due to viscous flow deformation of the sample. Such effect is not relevant in classical mercury dilatometry or DSC experiments described above.

2.2. Nucleation and Crystal Growth

On further reheating of the supercooled liquid above T_g, the nucleation process takes place being followed by crystal growth. In a selenium supercooled liquid, crystals grow from a relatively low-density nuclei population. Well defined and compact spherulitic structures grow from these centers. It seems that the nucleation has negligible effects during the crystal growth. The crystal growth is usually visible on a microscopic level and therefore it can directly be observed by microscopy methods using hot stage.

Temperature-dependent data for crystal growth velocity u [9,10] and viscosity η [10–12] in a selenium supercooled liquid are shown in Figure 4. These two kinetic processes are closely bound together as will be discussed later.

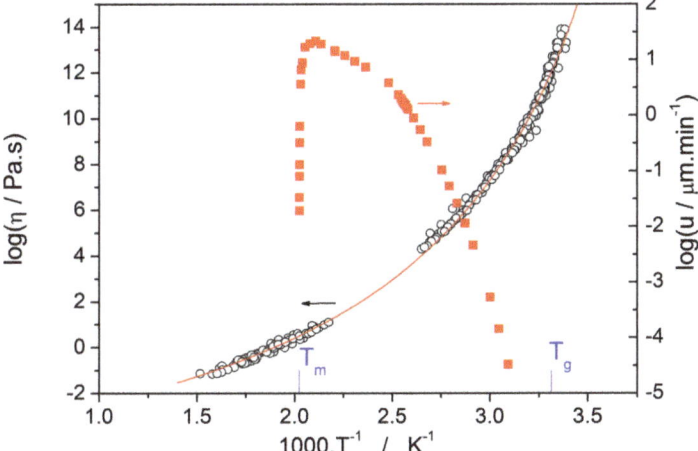

Figure 4. The crystal growth rate of isothermally grown spherulitic crystals and viscosity of a selenium supercooled liquid as a function of reciprocal temperature.

We can assume that in a narrow temperature range the crystal growth velocity can be described by a simple exponential dependence on reciprocal temperature u ~ exp(−E_G/RT), that should be linear on a logarithmic scale. The activation energy of crystal growth E_G can then be obtained from the slope of such linear dependence,

$$\frac{d \log(u)}{d(1/T)} = -\frac{E_G}{2.303 \cdot R}, \qquad (7)$$

where the coefficient 2.303 in Equation (7) comes from the conversion from the natural to the decimal logarithm.

From the crystal growth rate data shown in Figure 4 it is clearly seen that the activation energy is gradually decreasing from a relatively high value just above T_g (≅250 kJ·mol^{-1}) to a considerably lower value (≅40 kJ·mol^{-1}) just below the maximum growth rate.

The heat evolved during the crystal growth can easily be recorded by DSC. Figure 5 shows such heat flow at different scanning rates ranging from 1 to 30 K·min^{-1}.

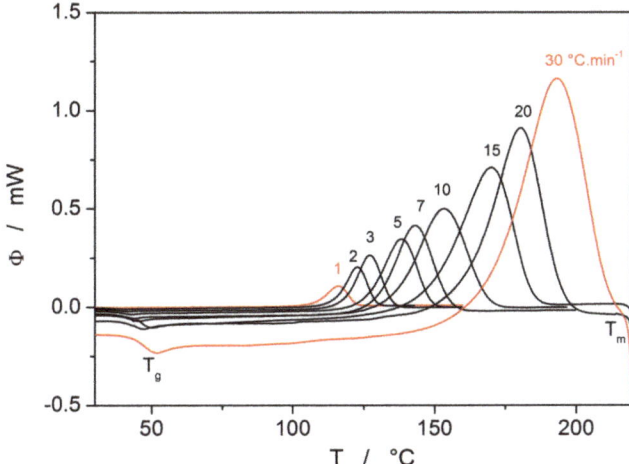

Figure 5. The DSC measurement of bulk selenium glass at different heating rates.

The crystallization peaks shown in Figure 5 involve the whole crystallization process including nucleation and crystal growth. More detailed calorimetric studies [13–16] indicate a complex behavior involving both bulk and surface crystal growth with nucleation possibly affected by internal stresses. Nevertheless, nucleation and surface crystal growth are negligible for the bulk selenium sample.

The DSC curves can easily be converted to kinetic data. It is assumed that the fraction crystallized, α, can be obtained by partial integration of non-isothermal heat flow, ϕ, after baseline subtraction:

$$\alpha = \frac{1}{\Delta H q^+} \int_{T_{on}}^{T} \phi \cdot dT, \qquad (8)$$

where ΔH_c correspond to enthalpic change of crystallization, q^+ is the heating rate and T_{on} is the starting point of baseline approximation. The heat flow due to the crystallization process can then be written as

$$\phi = \Delta H_c A \cdot \exp(-E_c/RT) \cdot f(\alpha), \qquad (9)$$

where A is the preexponential factor and E_c is the activation energy of the crystallization process. The $f(\alpha)$ function corresponds to the Johnson–Mehl–Avrami–Kolmogorov (commonly abbreviated as JMA) model of nucleation-growth process:

$$f(\alpha) = m(1-\alpha)[-\ln[1-\alpha]]^{1-1/m}. \qquad (10)$$

The development of this equation is described in [8] and references quoted in.

Analysis of DSC curves shown in Figure 5 is complicated due to the strong temperature dependence of heat flow as follows from Equation (9). Nevertheless, it has been shown [17,18] that if the measured heat flow is multiplied by T^2 and plotted as a function of α, all data taken at different heating rates collapse to one master curve defined as

$$z(\alpha) = f(\alpha) \int_0^\alpha \frac{d\alpha}{f(\alpha)} \cong \phi \cdot T^2. \qquad (11)$$

This function can be expressed for the JMA model as follows:

$$z(\alpha) = m(1-\alpha)[-\ln[1-\alpha]]. \qquad (12)$$

Figure 6 shows the $z(\alpha)$ function obtained by transformation of all crystallization peaks from Figure 5 by Equations (8) and (11). These plots are scaled within the $0 < z(\alpha) \leq 1$ range for easier comparison of different data sets (points). The $z(\alpha)$ function (full line) calculated by Equation (12) fits the experimental data quite well. This confirms the applicability of the JMA model for the description of non-isothermal crystallization kinetics in selenium glass. Another method to test the validity of this model is the shape analysis of DSC curve [19]. Slight data scatter might indicate variation in thermal contacts between the sample and DSC sensor.

The activation energy of the non-isothermal crystallization process E_c can be determined by the Kissinger method [20] from the shift of the maximum of the DSC peak T_p with heating rate q^+:

$$\frac{d\ln(q^+/T_p^2)}{d(1/T_p)} = -\frac{E_c}{R}. \qquad (13)$$

In case of the isothermal data, the Friedman method [21] is usually used, where $(d\alpha/dt)_\alpha$ and T_α are the conversion rate and temperature corresponding to arbitrarily chosen values of conversion α:

$$\ln([d\alpha/dt]_\alpha) = -\frac{E}{RT_\alpha} + const.. \quad (14)$$

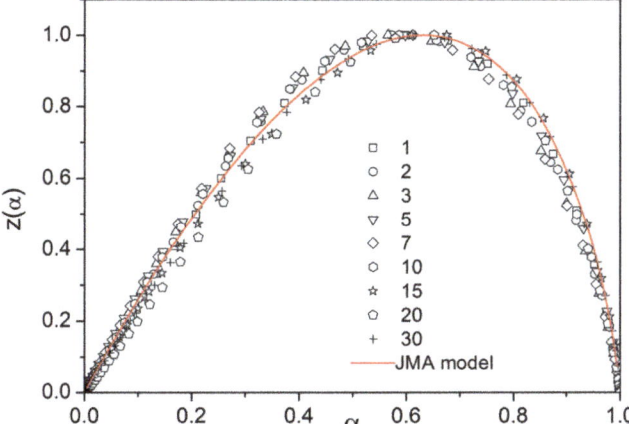

Figure 6. The $z(\alpha)$ function obtained by transformation of DSC data shown in Figure 5. Points were calculated by using Equation (11) (numbers indicate heating rate). The full line was calculated by Equation (12).

The Kissinger and Friedman (for $\alpha = 0.50$) plots are shown in Figure 7 for all the analyzed non-isothermal and isothermal DSC data, respectively. Similarly, as shown above for crystal growth experiments, also here it is seen that the activation energy significantly changes from relatively low values at a higher temperature ($\cong 40$ kJ·mol^{-1}) to higher values at a lower temperature ($\cong 117$ kJ·mol^{-1}).

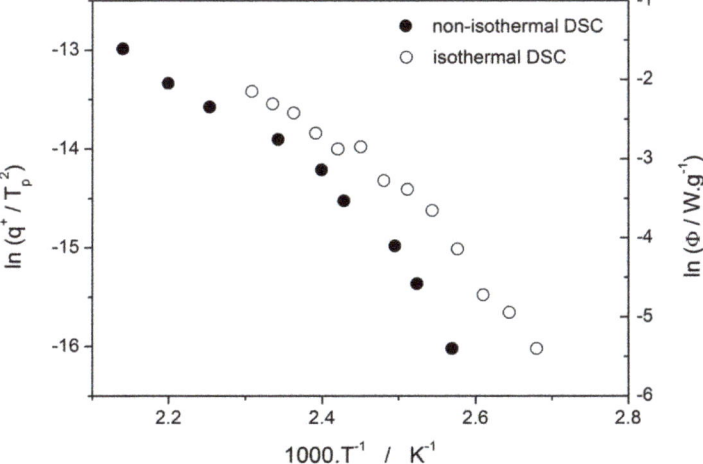

Figure 7. Determination of the crystallization activation energy for selenium glass by the Kissinger method (non-isothermal data) and the Friedman method (isothermal data).

Such important variation of the activation energy makes difficult further analysis of experimental data. The next step of kinetic analysis should be an estimation of the kinetic exponent m in Equation (10). Equation (9) can be rewritten in a somewhat different form,

$$y(\alpha) = \phi \cdot \exp(E_c/RT) = \Delta H_c A \cdot f(\alpha). \tag{15}$$

Figure 8 shows the $y(\alpha)$ function obtained by transformation of all crystallization peaks from Figure 5 by Equation (15) for $E_c = 105$ kJ·mol^{-1}. These plots are scaled within the $0 < y(\alpha) \leq 1$ range for easier comparison of different data sets (points). Full lines shown in Figure 8 were calculated by Equation (10) for two different values of kinetic exponent ($m = 1.5$ and 2.5). It is clearly visible that a higher value of the kinetic exponent better fits data taken at lower heating rates, and lower values of m does the same for data taken at higher heating rates. However, this effect is artificial being just a consequence of important changes in activation energy discussed above.

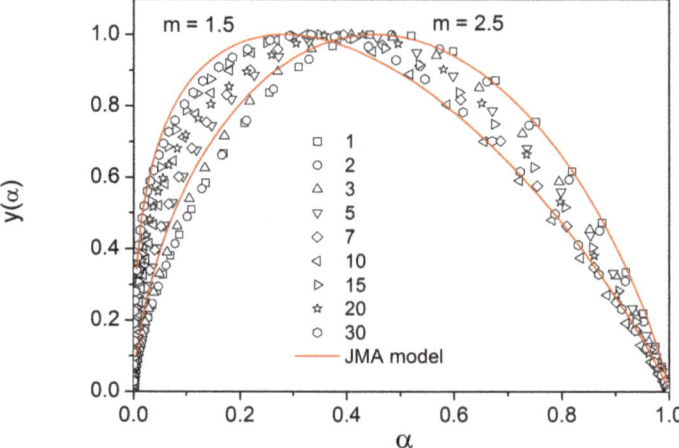

Figure 8. The $y(\alpha)$ function obtained by transformation of DSC data shown in Figure 5. Points were calculated by using Equation (15) (numbers indicate heating rate). Full lines were calculated by Equation (10).

Crystallization in glassy selenium was also measured by means of thermomechanical analysis (TMA). The example curve obtained at 0.2 °C·min^{-1} is shown in Figure 9 as the temperature dependence of sample height decreased. As was shown in [22], one of most reliable and reproducible characteristic temperatures associated with the TMA crystallization measurements is the extrapolated endset temperature T_e—its evaluation is suggested in the figure. The minimum achieved sample height is the point at which the sample deformation caused by viscous flow is ceased by the rigid crystalline structure formed within the sample. The occasional increase of sample height occurring during further heating is the consequence of the ongoing outwards surface crystal growth building up on the stiffened sample profile. The extrapolated endset temperatures can be utilized [23] in a similar way as the characteristic temperatures obtained via calorimetry, i.e., e.g., using the Kissinger equation (Equation (13)). The resulting dependence is for the glassy selenium depicted by the red data and axes in Figure 9. Note the curvature of the Kissinger plot data, similar to the one observed for the DSC crystallization data in Figure 7.

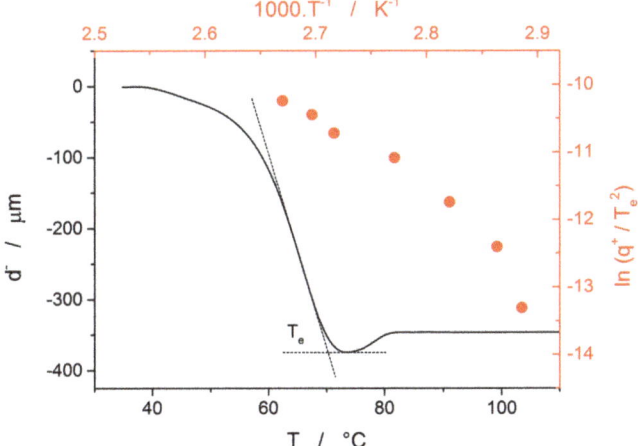

Figure 9. Black-based data and axes show an example thermomechanical analysis (TMA) crystallization curve (obtained at 0.2 °C·min^{-1}); evaluation of the extrapolated endset is illustrated. Red-based data and axes show the Kissinger plot for the TMA crystallization data.

3. Discussion

Thermo-kinetic data described in the previous section imply that the activation energy of most kinetic processes observable in glassy materials by thermal analysis varies with temperature. This gives an interesting opportunity to compare the apparent activation energies among the particular processes for the wide range of experimental conditions. In such a situation, the equilibrium viscosity can be seen as the overarching quantity offering the most sensible comparison for all other data. The activation energy of viscous flow E_η can be obtained from the tangent slope for the viscosity data depicted in Figure 4:

$$\frac{d \log(\eta)}{d(1/T)} = \frac{E_\eta}{2.303 \cdot R}. \qquad (16)$$

The temperature dependence of E_η is shown in Figure 10 (solid line).

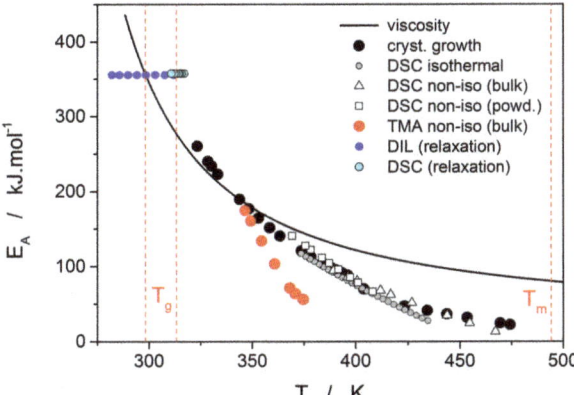

Figure 10. Activation energies of viscous flow (solid line), microscopic crystal growth, macroscopic crystallization observed by DSC (two data series, for bulk and powdered samples) and TMA, as well as structural relaxation observed by DSC and dilatometry.

Starting with the structural relaxation process, it is somewhat surprising that the activation energy of this process remains constant throughout the whole measured temperature range (9–45 °C). Note that structural relaxation is generally considered to be very closely interlinked with viscous flow. Nonetheless, the constant value of E_{rel} in the above-given temperature range was unambiguously confirmed from the dilatometric data [3] (see, e.g., Figure 2), where E_{rel} was evaluated by non-linear optimization as well as by the linearization method [23], and also from the calorimetric data [3,24] (see, e.g., Figure 3), where the non-linear optimization was complemented by the newly developed methodology [25] based on the shift of the relaxation peak. Values of E_{rel} are shown in Figure 10 for dilatometry (dark blue) and calorimetry (light blue). It is well apparent that the relaxation data exhibit a constant value of E_{rel}, not only in the glass transition temperature range commonly revealed via the non-isothermal measurement techniques (indicated by the red dashed lines), but also well below these temperatures. In the temperature window where the relaxation times considerably exceed the time-scale characteristic for the non-isothermal measurements, long-term isothermal annealing experiments need to be performed. Interestingly, the intersection of the E_{rel} and E_η well corresponds to the lowest value of T_g obtained during cooling within the commonly achievable time scale. The constancy of E_{rel} is a priori given by the definition of the TNM model, where the potential temperature-dependent component of E_{rel} is replaced by the T_f-based term on the right-hand side of Equation (4). Direct incorporation of the $E_{rel}(T)$ dependence into Equation (4) might, however, be the way towards solving the occasionally raised questions [26] associated with the universality of TNM formalism—as such it is certainly worth of further exploration.

The second process occurring during further heating of the glassy materials is crystal growth, observable either microscopically (see the crystal growth data in Figure 4) or macroscopically, usually via calorimetric methods (see, e.g., Figure 5). Although the two approaches and the temperature ranges of their applicability differ to a great extent, the observed process is essentially the same and the corresponding activation energies should exhibit unified temperature dependence. In order to verify this hypothesis, E_G was determined from the crystal growth data depicted in Figure 4 by using Equation (7) (via the direct tangential approach), and E_c was determined from both non-isothermal and isothermal DSC measurements by using the Kissinger (Equation (13)) and Friedman (Equation (14)) methods, respectively (see Figure 7 for the two overall dependences). Again, the temperature-resolved tangential approach to the determination of E_c was adopted. The values of E_G and E_c are then compared in Figure 10, showing a very good agreement and confirming the universal nature of the activation energy for the crystal growth process. The non-isothermal DSC crystallization measurements were performed (in addition to the bulk samples) also for a finely powdered glassy selenium. Despite the different crystallization mechanism occurring in case of the fine Se powders [13,14], the corresponding $E_c(T)$ dependence also falls on the crystal growth master-curve depicted in Figure 10, further confirming the universality of this aspect of the crystallization process kinetics.

With regard to the relation between the crystal growth rate and viscosity, it results from the Turnbull–Cohen formula where u is inversely proportional to η. However, it has been shown [27] that for a number of materials the so-called decoupling of these two quantities occurs, breaking the Stokes–Einstein formalism [28]. Formally, the decoupling is described by the apparent decoupling parameter ξ_a:

$$\xi_a = \frac{d \log(u)}{d \log(\eta)} \cong \frac{E_G}{E_\eta}. \qquad (17)$$

This rather empirical expression can be further corrected by accounting only for the rate at which the structural entities (atoms, molecules) present in the liquid phase are attached to the growth liquid-crystal interface. This correction is based on the elimination of the term f_p from the expression

References

1. Ediger, M.D.; Angell, C.A.; Nagel, S.R. Supercooled liquids and glasses. *J. Phys. Chem.* **1996**, *100*, 13200–13212. [CrossRef]
2. Debenedetti, P.G. *Metastable Liquids. Concepts and Principles*; Princeton University Press: Princeton, NJ, USA, 1996; ISBN 9780691085951.
3. Málek, J.; Svoboda, R.; Pustková, P.; Čičmanec, P. Volume and enthalpy relaxation of a-Se in the glass transition region. *J. Non Cryst. Solids* **2009**, *355*, 264–272. [CrossRef]
4. Naraynaswamy, O.S. A Model of Structural Relaxation in Glass. *J. Am. Ceram. Soc.* **1971**, *54*, 491–498. [CrossRef]
5. Tool, A.Q. Relation between inelastic deformability and thermal expansion of glass in its annealing range. *J. Am. Ceram. Soc.* **1946**, *29*, 240–253. [CrossRef]
6. DeBolt, M.A.; Easteal, A.J.; Macedo, P.B.; Moynihan, C.T. Analysis of Structural Relaxation in Glass Using Rate Heating Data. *J. Am. Ceram. Soc.* **1976**, *59*, 16–21. [CrossRef]
7. Málek, J.; Svoboda, R. Structural Relaxation and Viscosity Behavior in Supercooled Liquids at the Glass Transition. In *Thermal Analysis of Micro, Nano- and Non-Crystalline Materials*; Šesták, J., Šimon, P., Eds.; Hot Topics in Thermal Analysis and Calorimetry; Springer Science+Business Media: Dordrecht, The Netherlands, 2013; Volume 9, pp. 147–173. ISBN 978-90-481-3149-5.
8. Málek, J.; Shánělová, J. Crystallization Kinetics in Amorphous and Glassy Materials. In *Thermal Analysis of Micro, Nano- and Non-Crystalline Materials*; Šesták, J., Šimon, P., Eds.; Hot Topics in Thermal Analysis and Calorimetry; Springer Science+Business Media: Dordrecht, The Netherlands, 2013; Volume 9, pp. 291–324. ISBN 978-90-481-3149-5.
9. Ryschenkow, G.; Faivre, G. Bulk crystallization of liquid selenium—Primary nucleation, growth kinetics and modes of crystallization. *J. Cryst. Growth* **1988**, *87*, 221–235. [CrossRef]
10. Málek, J.; Barták, J.; Shánělová, J. Spherulitic Crystal Growth Velocity in Selenium Supercooled Liquid. *Cryst. Growth Des.* **2016**, *16*, 5811–5821. [CrossRef]
11. Bernatz, K.; Echeveria, I.; Simon, S.; Plazek, D. Characterization of the molecular structure of amorphous selenium using recoverable creep compliance measurements. *J. Non Cryst. Solids* **2002**, *307*, 790–801. [CrossRef]
12. Košťál, P.; Málek, J. Viscosity of selenium melt. *J. Non Cryst. Solids* **2010**, *356*, 2803–2806. [CrossRef]
13. Svoboda, R.; Málek, J. Crystallization kinetics of a-Se, part 1: Interpretation of kinetic functions. *J. Therm. Anal. Calorim.* **2013**, *114*, 473–482. [CrossRef]
14. Svoboda, R.; Málek, J. Crystallization kinetics of a-Se, part 2: Deconvolution of a complex process—The final answer. *J. Therm. Anal. Calorim.* **2014**, *115*, 81–91. [CrossRef]
15. Svoboda, R.; Málek, J. Crystallization kinetics of a-Se, part 3: Isothermal data. *J. Therm. Anal. Calorim.* **2015**, *119*, 1363–1372. [CrossRef]
16. Svoboda, R.; Gutwirth, J.; Málek, J. Crystallization kinetics of a-Se, part 4: Thin films. *Philos. Mag.* **2014**, *94*, 3036–3051. [CrossRef]
17. Málek, J. The Kinetic Analysis of Non-Isothermal Data. *Thermochim. Acta* **1992**, *200*, 257–269. [CrossRef]
18. Málek, J. The applicability of Johnson-Mehl-Avrami model in the thermal analysis of the crystallization kinetics of glasses. *Thermochim. Acta* **1995**, *267*, 61–73. [CrossRef]
19. Málek, J.; Criado, J.M. The Shape of a Thermoanalytical Curve and Its Kinetic Information Content. *Thermochim. Acta* **1990**, *164*, 199–209. [CrossRef]
20. Kissinger, H.E. Reaction kinetics in differential thermal analysis. *Anal. Chem.* **1957**, *29*, 1702–1706. [CrossRef]
21. Friedman, H.L. Kinetics of thermal degradation of char-forming plastics from thermogravimetry. Application to a phenolic plastic. *J. Polym. Sci. Part C* **1964**, *6*, 183–195. [CrossRef]
22. Zmrhalová, Z.; Pilný, P.; Svoboda, R.; Shánělová, J.; Málek, J. Thermal properties and viscous flow behavior of As_2Se_3 glass. *J. Alloys Compd.* **2016**, *655*, 220–228. [CrossRef]
23. Málek, J. Rate-determining factors for structural relaxation in non-crystalline materials II. Normalized volume and enthalpy relaxation rate. *Thermochim. Acta* **1998**, *313*, 181–190. [CrossRef]
24. Svoboda, R.; Pustková, P.; Málek, J. Relaxation behavior of glassy selenium. *J. Phys. Chem. Sol.* **2007**, *68*, 850–854. [CrossRef]

As can be seen, most datapoints can be reasonably described by using the JMA kinetic exponent $m = 3$, which may correspond to the assumption that for bulk selenium samples the crystal growth starts dominantly via formation of three-dimensional volume-located crystallites.

4. Materials and Methods

Glassy selenium was prepared from pure elements (5N, Sigma-Aldrich, Prague, Czech Republic) by melt-quenching. Elementary Se was melted in an evacuated fused silica ampoule, which was then let to cool in air. The glassy material was crushed in agate mortar and sieved through defined mesh so that the various particle size fractions were obtained. Powder DSC crystallization data reported in this paper were obtained for the 20–50 µm fraction [13,14]. Pieces of glass with a diameter larger than 1 mm were used for the DSC relaxation and bulk crystallization measurements [3,13]. Melt-quench in thin ampoules was utilized to prepare cylindrical Se samples. The ampoules were quenched vertically to obtain a glassy ingot, which was then sawed into samples with the following diameters (d_m) and heights (h_m): For viscosity measurements $d_m = 6$ mm and $h_m = 2.5$ mm [12], for microscopic crystal growth measurements $d_m = 4$ mm and $h_m = 2$ mm [10] and for TMA crystallization measurements $d_m = 4$ mm and $h_m = 1$ mm.

Experimental setups and details of most measurements were already published in the respective papers: Structural relaxation by DSC and dilatometry in [3,24], viscosity in [12], microscopic crystal growth in [10] and non-isothermal powder and bulk crystallization by DSC in [13,14]. The new, previously unpublished data are those for isothermal bulk crystallization measured by DSC and non-isothermal crystallization measured by TMA.

The isothermal DSC data were obtained using a Q2000 DSC (TA Instruments, Prague, Czech Republic) equipped with a cooling accessory, autolid, autosampler and T-zero Technology. Dry nitrogen was used as the purge gas at a rate of 50 cm^3·min^{-1}. The calorimeter was calibrated using In, Zn and H$_2$O. The stability of the DSC signal was checked daily. Open T-zero low-mass pans were used. Regarding the applied temperature program, the sample (8–10 mg) was first subjected to a 5 min isotherm at 45 °C and then heated at 100 °C·min^{-1} to a selected temperature T_i, where the sample was allowed to isothermally crystallize until the crystallization process was complete. The isothermal crystallization temperatures utilized in the case of each particle size fraction were 100, 105, 110, 115, 120, 125, 130, 135, 140, 145, 150, 155 and 160 °C. In order to obtain a baseline for the isothermal measurement, each DSC pan with a crystalline sample of glass was kept in the DSC cell and the above-described temperature procedure was repeated (in this way the data subtracted from the isothermal crystallization signal truly surrogated the presence of an inert material with similar heat capacity, mass, grain size and positioning in the DSC pan/cell). Perfect flatness of the baseline and reproducibility of the crystallization measurements were confirmed. For the extensive testing of the suitability and repeatability of the initial 100 °C·min^{-1} heating ramp, see [15].

The thermomechanical measurements were realized by using a TMA Q400EM (TA Instruments), where the cylindrical samples were compressed in-between two alumina plates, and the force applied to the sample was 30 mN. A linear heating rate was applied to study the effect of crystal growth suppressing the decrease of sample height caused by viscous flow. The following heating rates were applied between 35 and 170 °C: 0.2, 0.5, 1, 2, 3, 4 and 5 °C·min^{-1}.

Author Contributions: Conceptualization, J.M.; methodology, J.M.; validation, R.S. and J.M.; formal analysis, J.M. and R.S.; investigation, R.S.; resources, J.M. and R.S.; data curation, R.S.; writing—original draft preparation, J.M. and R.S.; supervision, J.M.; project administration, R.S.; and funding acquisition, R.S.

Funding: This research was funded by Czech Science Foundation grant number 17-11753S.

Conflicts of Interest: The authors declare no conflict of interest.

The last data set depicted in Figure 10 is the one corresponding to the TMA crystallization measurements. Evaluation of activation energy from the TMA data was done similarly as in the case of non-isothermal DSC, i.e., by the tangential approach of the Kissinger dependence (Equation (13)), where the extrapolated endset T_e (see Figure 9) was used as the characteristic temperature. It was shown for several other chalcogenide glassy systems [31,32] that the activation energy evaluated in this way from the TMA measurements is in a good agreement with E_c from DSC, but the data are (due to the choice of the extrapolated characteristic temperature) shifted to lower temperatures; that is, the crystal growth process is seemingly observed "in advance" and the E_A–T dependence gets shifted to lower temperatures. From this point of view, it is the high-temperature points shown in Figure 10 that in this dependence represent the commonly observed behavior. Contrary to what was observed for most other studied chalcogenides, E_A rapidly increases at lower temperatures. The exact position of T_e depends on many factors (including applied force, sample geometry, nucleation density, location and morphology of forming crystallites, etc.) but essentially can be understood as the competition between the viscous flow and crystal growth rate (which also depends on viscosity). In case of amorphous selenium, the position of T_e appears to be driven more by viscous flow (in combination with surface tension) at low temperatures. Thus, the activation energy determined from the TMA measurements at low q^+ gets closer to E_η.

There are of course many consequences associated with the temperature variation of crystallization activation energies. In the last part of the Discussion section we will focus on the model-based master-plot evaluation method utilizing the characteristic kinetic functions $z(\alpha)$ and $y(\alpha)$. The former can be expressed as the product $f(\alpha) \cdot g(\alpha)$, see Equation (11), and as such is invariable with E_A. On the other hand, function $y(\alpha)$ is proportional to $f(\alpha)$, which utilizes E_A during the transformation of experimental data, see Equation (15). It naturally suggests itself to use the full $E_c(T)$ dependence (see Figure 10) in Equation (15). However, the $y(\alpha)$ function is too sensitive to the value of activation energy, and the large variation of E_c throughout each non-isothermal DSC measurement effectively results in $y(\alpha)$ distortions reminiscent of output obtained for the JMA formalism with the sub-unity kinetic exponent m. It is therefore reasonable to replace the full $E_c(T)$ dependence by constant values of E_c, selected for each measurement individually based on the arbitrarily determined characteristic temperature point. If the value of E_c for the temperature corresponding to the maximum transformation rate (maximum of the DSC peak) is used, the resulting values of JMA kinetic exponents are too large to be physically meaningful. Nevertheless, if we consider that the dimensionality of the formed crystallites is being already set at the start of the crystallization process, and, correspondingly, we utilize E_c values corresponding to $\alpha = 0.10$, the data depicted in Figure 12 are obtained.

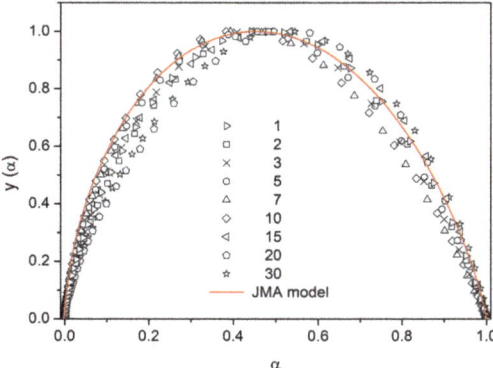

Figure 12. The $y(\alpha)$ functions obtained by transformation of DSC data shown in Figure 5. Points were calculated by using Equation (15) (numbers indicate heating rate); E_c utilized in the calculations were determined from the corresponding dependence shown in Figure 10 for the temperatures matching $\alpha = 0.1$ of each respective DSC data curve. The solid red line was calculated by Equation (10) for m = 3.

for the crystal growth rate u, where f_p is the probability of the structural entity, newly attached to the crystal growth interface, remaining within the crystalline phase:

$$\xi = \frac{d[\log[u] - f_p]}{d\log(\eta)} = \frac{d[\log[u] - \log[1 - \exp[-\Delta G_{lc}/RT]]]}{d\log(\eta)} = \frac{d\log(u_{kin})}{d\log(\eta)} \cong \frac{E_{G,kin}}{E_\eta}. \quad (18)$$

ΔG_{lc} is the difference between the Gibbs energies of supercooled liquid and crystalline phases. Both forms of the decoupling parameter were essentially calculated from the ratio of the respective activation energies; ΔG_{lc} was calculated based on the standard thermodynamic expression [29] from the selenium enthalpy and entropy of fusion and the heat capacity data published in [30].

Temperature dependences of both forms of the decoupling parameter are shown in Figure 11. Values of ξ_a and ξ indicate that at low temperatures/growth rates almost no decoupling between u and η can be found (in fact, slight negative decoupling occurs below 340 K). However, as the temperature increases the decoupling becomes more prominent and very well recognizable at temperatures above 370 K. Increasing temperature also results in a rising difference between the two forms of decoupling parameters ξ and ξ_a. This difference is negligible up to ~365 K but then rapidly increases with the exponential increase of the correction term f_p. Note that the melting entropy of crystalline selenium is $\Delta S_m \approx 1.5R$ [10]. Such a relatively small value in combination with low supercooling causes relatively large difference between ξ and ξ_a at higher temperatures. A similar effect may also bring N-type heat capacity of supercooled selenium [30], which effectively lowers the f_p contribution. Interestingly, both decoupling dependences exhibit a sudden step-like decrease of the decoupling parameter at 400 K. This is the consequence of the change in preferential morphology of the growing crystallites (transition from spherulitic form B to spherulitic form A, as described in [9]). The step-like change very well agrees with the break on the temperature dependences of the corresponding integral data utilized for calculation of E_G and E_c, as depicted in Figure 11, where the dashed vertical line indicates the break point. It is noteworthy that the corrected rate of growth u_{kin} for the high-temperature spherulitic form A does not show further increase in decoupling (which would be represented by the decrease of ξ) and is close to $\xi = 0.62$.

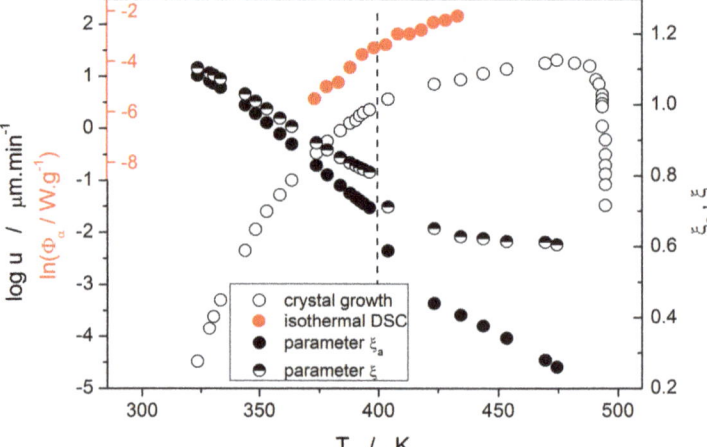

Figure 11. Temperature dependences of decoupling parameters ξ_a and ξ (right-hand axis), crystal growth rate u (black outer left-hand axis) and logarithm of heat flow corresponding to $\alpha = 0.50$ obtained during isothermal DSC experiments (red inner left-hand axis). The vertical dashed line indicates the transition between the spherulitic B and A crystallite forms.

25. Svoboda, R. Novel equation to determine activation energy of enthalpy relaxation. *J. Therm. Anal. Calorim.* **2015**, *121*, 895–899. [CrossRef]
26. Kovacs, A.J. Transition vitreuse dans les polymères amorphes. Etude phénoménologique. *Fortschr. Hochpolym. Forsch.* **1963**, *3*, 394–507. [CrossRef]
27. Ediger, M.D.; Harrowell, P.; Yu, L. Crystal growth kinetics, exhibit a fragility-dependent decoupling from viscosity. *J. Chem. Phys.* **2008**, *128*, 034709. [CrossRef] [PubMed]
28. Gutzow, I.S.; Schmelzer, J.W.P. *The Vitreous State: Thermodynamics, Structure, Rheology, and Crystallization*; Springer: Berlin/Heidelberg, Germany, 2013; ISBN 978-3-642-34632-3.
29. Busch, R.; Kim, Y.J.; Johnson, W.L. Thermodynamics and kinetics of the undercooled liquid and the glass transition of the $Zr_{41.5}Ti_{13.8}Cu_{12.5}Ni_{10.0}Be_{22.5}$ alloy. *J. Appl. Phys.* **1995**, *77*, 4039–4043. [CrossRef]
30. Svoboda, R.; Málek, J. Thermal behavior in Se-Te chalcogenide system: Interplay of thermodynamics and kinetics. *J. Chem. Phys.* **2014**, *141*, 224507. [CrossRef] [PubMed]
31. Svoboda, R.; Brandová, D.; Chromčíková, M.; Setnička, M.; Chovanec, J.; Černá, A.; Liška, M.; Málek, J. Se-doped $GeTe_4$ glasses for far-infrared optical fibers. *J. Alloys Compd.* **2017**, *695*, 2434–2443. [CrossRef]
32. Svoboda, R.; Brandová, D.; Chromčíková, M.; Liška, M. Thermokinetic behavior of Ga-doped $GeTe_4$ glasses. *J. Non Cryst. Solids* **2019**, *512*, 7–14. [CrossRef]

Sample Availability: Samples of the amorphous selenium are not available from the authors.

© 2019 by the authors. Licensee MDPI, Basel, Switzerland. This article is an open access article distributed under the terms and conditions of the Creative Commons Attribution (CC BY) license (http://creativecommons.org/licenses/by/4.0/).

Article

Critical Appraisal of Kinetic Calculation Methods Applied to Overlapping Multistep Reactions

Nikita V. Muravyev [1,*], Alla N. Pivkina [1] and Nobuyoshi Koga [2]

1. Energetic Materials Laboratory, Semenov Institute of Chemical Physics, Russian Academy of Sciences, 119991 Moscow, Russia; alla_pivkina@mail.ru
2. Chemistry Laboratory, Department of Science Education, Graduate School of Education, Hiroshima University, 1-1-1 Kagamiyama, Higashi-Hiroshima 739-8524, Japan; nkoga@hiroshima-u.ac.jp
* Correspondence: n.v.muravyev@ya.ru

Academic Editor: Sergey Vyazovkin
Received: 23 May 2019; Accepted: 19 June 2019; Published: 21 June 2019

Abstract: Thermal decomposition of solids often includes simultaneous occurrence of the overlapping processes with unequal contributions in a specific physical quantity variation during the overall reaction (e.g., the opposite effects of decomposition and evaporation on the caloric signal). Kinetic analysis for such reactions is not a straightforward, while the applicability of common kinetic calculation methods to the particular complex processes has to be justified. This study focused on the critical analysis of the available kinetic calculation methods applied to the mathematically simulated thermogravimetry (TG) and differential scanning calorimetry (DSC) data. Comparing the calculated kinetic parameters with true kinetic parameters (used to simulate the thermoanalytical curves), some caveats in the application of the Kissinger, isoconversional Friedman, Vyazovkin and Flynn–Wall–Ozawa methods, mathematical and kinetic deconvolution approaches and formal kinetic description were highlighted. The model-fitting approach using simultaneously TG and DSC data was found to be the most useful for the complex processes assumed in the study.

Keywords: kinetic analysis; overlapping reactions; TG; DSC; kinetic deconvolution; isoconversional analysis; formal kinetic analysis

1. Introduction

Thermally-induced transformations in heterogeneous system usually do not obey the idealized single-step kinetic pattern but are comprised of consecutive or concurrent reaction steps. The reaction behavior is further complicated by the contribution of some physical phenomena (melting, diffusion, etc.) [1–3]. Thermal decomposition of solids represents a typical example of such complex process [4–8]. Determination of the kinetic parameters for the thermal decomposition is not a straightforward, but large experimental efforts should be paid for finding a possible way to attain the rigorous solution [9–11]. Alongside, examination of the applicability of different kinetic calculation methods through analyzing the mathematically simulated kinetic curves assuming the specific complex process is the possible method to obtain a guideline for the successful kinetic calculations [12–14]. For example, Svoboda et al. [15] analyzed the simulated process with independent reactions and revealed that the commonly used Kissinger method [16,17] provides the good estimate for the activation energy of the dominant reaction. Vyazovkin et al. applied isoconversional methods to the simulated parallel independent [18] and consecutive [19] reactions, and summarized the typical shapes of the activation energy dependency on the conversion degree [20]. Some examples of the complex reactions where the isoconversional hypothesis is not fulfilled were also reported [21–23]. Burnham [24] proposed the formal kinetic methods to describe the complex processes, but usually some preliminary insights are necessary to give the close initial guesses for many fitted parameters. The attempt to consider simultaneously various kinetic calculation methods to underline their advantages and shortcomings was performed by the Kinetic committee of ICTAC (International Confederation for Thermal Analysis and Calorimetry) [25]. Since then several thermokinetic techniques have been proposed [26,27], however, the critical assessment of the applicability of the modern thermokinetic methods for analyzing the complex processes is still missing.

If a partially overlapping multistep reaction is considered, the contributions of each component reaction step to the overall process determined by the different thermoanalytical (TA) techniques are not equal in general, because of the different physical quantities subjected to individual TA measurements. In some cases, the respective reaction steps can bring the opposite contributions to the overall process as measured using a TA technique. Experimental studies concerning the overall reaction composed of more than one reaction step with oppositely signed TA signals have been reported, i.e., partially overlapping mass loss and mass gain [3,28,29], or exothermic and endothermic events [30–32]. In practice, such complex processes are observed during thermal decomposition of energetic materials [5] and high-temperature operations with ionic liquids [33]. This particular situation, e.g., two overlapping mass-loss processes corresponding to endothermic and exothermic effects, respectively, educe that the apparent kinetic curves recorded using differential scanning calorimetry (DSC) can be different from that obtained using thermogravimetry (TG). The phenomenon is caused by different nature of these TA signals and was observed experimentally for isopropylammonium nitrate decomposition [30] and kerogen pyrolysis [34]. A sophisticated kinetic deconvolution technique have been applied recently to this type of solid-state reactions, where the kinetic parameters obtained for the constituted single processes have been linked with the physicochemical and physicogeometrical features of the transformations [31,32].

The aim of the present study is to evaluate the applicability of the available kinetic calculation methods to the simulated complex process characterized as the partially overlapping multistep reaction and to provide a guideline of successful kinetic analysis to the processes of this type. The kinetic curves were simulated mathematically (thus, correct kinetic parameters are available for verification), and processed using different kinetic calculation methods including Kissinger method [16], isoconversional methods [12,35–37], mathematical [27] and kinetic [31,38] deconvolution methods, and formal kinetic analysis [39,40]. Through the kinetic calculation using different methods, the resulting apparent kinetic parameters determined for simulated TG and DSC curves are mutually compared for finding the possible way to obtain relevant kinetic information of the real-life complex reaction process, for which one has limited information a priori.

2. Theoretical

2.1. Simulation of Successive Reactions

The thermoanalytical data were simulated for the following theoretical process comprising two successive reaction steps:

$$A(s) \xrightarrow{k_1} B(s) + C(g), \ B(s) \xrightarrow{k_2} D(s) + E(g) \tag{1}$$

To model the existence ratio of solid components – A(s), B(s), and D(s) at a time, a set of differential kinetic equations has to be solved. When TA signals under linear nonisothermal conditions at a heating rate β are simulated, a set of kinetic equations is expressed using conversion degrees, α_1 and α_2, for the respective component reaction steps [41]:

$$\frac{d\alpha_1}{dT} = \frac{A_1}{\beta} e^{-E_{a1}/(RT)} f_1(\alpha_1), \ \frac{d\alpha_2}{dT} = \frac{A_2}{\beta} e^{-E_{a2}/(RT)} f_2(\alpha_1, \alpha_2), \tag{2}$$

where A and E_a are the apparent Arrhenius parameters and $f(\alpha)$ is the kinetic model function. The subscripts 1 or 2 indicate the first and second component reaction steps. Considering the derivation of the kinetic rate data from derivative TG (DTG), the normalized overall reaction rate, $(d\alpha/dt)_{DTG}$, is expressed:

$$\left(\frac{d\alpha}{dt}\right)_{DTG} = \eta \frac{d\alpha_1}{dt} + (1-\eta)\frac{d\alpha_2}{dt}, \qquad (3)$$

where α is the conversion degree for the overall reaction and η is the ratio of the total mass-loss value during the first reaction process, Δm_1, to the total mass-loss for the overall reaction, Δm_Σ: $\eta = \Delta m_1/\Delta m_\Sigma$. On the other hand, the normalized overall reaction rate derived from DSC, $(d\alpha/dt)_{DSC}$, is represented as follows [19]:

$$\left(\frac{d\alpha}{dt}\right)_{DSC} = \gamma \frac{d\alpha_1}{dt} + (1-\gamma)\frac{d\alpha_2}{dt}, \qquad (4)$$

where γ is the ratio of the total thermal effect for the first reaction, Q_1, to the total thermal effect for the overall reaction, Q_Σ: $\gamma = Q_1/Q_\Sigma$. It must be noted from Equations (3) and (4) that DTG and DSC curves for the overall process are generally different because of different definitions of η and γ.

For simplicity, a series of kinetic curves were simulated mathematically by assuming a successive two first-order reactions; therefore, $f_1(\alpha_1) = 1 - \alpha_1$ and $f_2(\alpha_1, \alpha_2) = \alpha_1 - \alpha_2$. In addition, equivalent total mass loss for the first and second reaction processes was assumed, thus $\eta = 0.5$. The thermal effects of the component reaction processes were assumed to have opposite signs, and the absolute value of the total thermal effect of the first reaction process (exothermic) was set to be twice as large as that for the second reaction process (endothermic), i.e., $\gamma = 2$. Based on the aforementioned assumptions, four different cases of the temperature dependences of the rate constants of each component reaction step were simulated:

- Case 1: the rate constant for the first reaction, k_1, is larger than that of the second reaction, k_2, and this difference decreases with increasing temperature, T;
- Case 2: the value of k_1 is larger than k_2, and its ratio, k_1/k_2, is constant independent of T;
- Case 3: the value of k_1 is smaller than k_2, and this difference increases with T;
- Case 4: the value of k_1 is smaller than k_2, and k_1/k_2 is constant independent of T.

Variation of k_1/k_2 with temperature is plotted for cases 1–4 in Figure S1, and the apparent Arrhenius parameters assumed for simulating the kinetic curves are listed in Table 1. The simulation of the kinetic curves at different β values were performed using Mathcad (ver. 15.0), in which the fourth-order Senum and Yang approximation [42] was employed for the approximation of exponential temperature integral.

2.2. Kinetic Calculation Methods

Kinetic calculations for the mathematically simulated kinetic curves were performed using self-developed thermokinetic code THINKS [43], which comprises various calculation methods including Kissinger (ASTM E698 [17]), Friedman [35], Flynn–Wall–Ozawa (FWO) [36], Starink [44], and Vyazovkin [45] methods, combined kinetic analysis (CKA) [26], and model-fitting approaches, kinetic deconvolution [31,38] and formal kinetic analysis [39,40]. For analyzing the partially overlapping thermal events, the statistical peak separation technique, i.e., mathematical deconvolution [27], was applied. The fundamentals of these kinetic calculation methods can be found in original papers, the basic ideas relevant to the present study are implemented in the text.

3. Results and Discussion

3.1. Apparent Features of the Simulated Kinetic Datas

The simulated kinetic curves at $\beta = 2$ K min^{-1} for the Cases 1–4 are shown in Figure 1. All the simulated kinetic curves at different β values for each case are presented in Figures S2–S5. Although various patterns of overlapping two mass-loss processes with exothermic and endothermic thermal effects are expected, the kinetic curves simulated by assuming four different cases in this study are believed to cover typical apparent features of the kinetic curves possibly obtained by TG and DSC measurements.

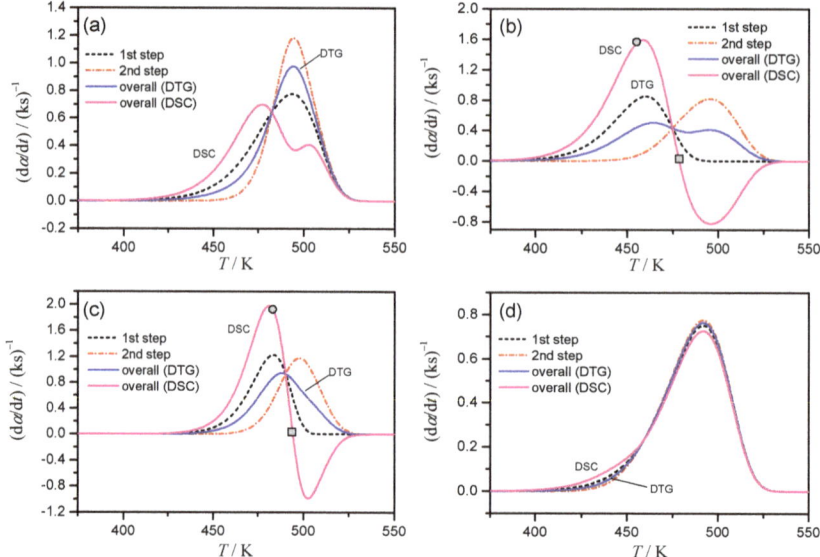

Figure 1. Comparison of the simulated kinetic curves at $\beta = 2$ K min^{-1} for each component reaction step calculated according to Equation (2), overall kinetic curve calculated according to Equation (3) (from DTG), and overall kinetic curve calculated according to Equation (4) (from DSC): (**a**) Case 1, (**b**) Case 2, (**c**) Case 3, and (**d**) Case 4.

In Case 1 (Figure 1a), the kinetic curve of DTG indicates a single peak, whereas overlapping two positive peaks with distinguishable peak tops are observed. At first glance, the process is misinterpreted as a single-step reaction from TG/DTG and as overlapping two exothermic reactions from DSC. However, the kinetic features assumed to simulate the kinetic rate can be deduced by comparing the kinetic curves derived from DTG and DSC: the peak top of DTG and the minimum between the two apparent positive peak tops in DSC appear at the same temperature. The assumed overlapping feature of the complex process is easily deduced in Case 2 (Figure 1b): DTG indicates overlapping of two peaks with well separated peak tops, while DSC indicates one positive and one negative peak. In Case 3 (Figure 1c), the smooth single peak is present on DTG curve. However, because of clearly separated positive and negative peaks on the kinetic curve of DSC, the assumed overlapping feature of the process is easily recognized. Case 4 is the most difficult one for deducing the complex features of the process from DTG and DSC data (Figure 1d). Because the kinetic curves derived from DTG and DSC are practically identical, it is the case usually interpreted from TG and DSC measurements as the single reaction process. In the present case, the situation appears as the result of the practically identical reaction temperature region and the same rate behavior assumed; however, similar situation is also expected when the contribution of a component reaction to the overall mass loss and thermal effect is very limited.

3.2. Kissinger Method

For the kinetic analysis of the single-step reaction, the peak maximum of DTG and DSC curves, i.e., point where $(d^2\alpha/dt^2) = 0$, contain important kinetic information [25] as formalized in the Kissinger method [16]: E_a value is evaluated from the shift of the peak temperature with varying β value. Because of its simplicity and robustness, the Kissinger method is one of the most widely used kinetic calculation methods for the determination of E_a value and also is implemented in ASTM E698-05 [17]. In general, application of the Kissinger method to the complex process as assumed in this study lacks theoretical validity. However, as demonstrated by Svoboda and Málek [15] for the kinetic analysis of the independent overlapping reactions, the results of Kissinger technique often reveal the relevant E_a value for the rate-limiting step.

Table 1 lists the apparent Arrhenius parameters for simulated kinetic data, calculated from the Kissinger plots and assuming the first-order reaction behavior. Irrespective of DTG and DSC data, superficial Arrhenius parameters were obtained for Case 1. Herein, the peak maximums cannot be correlated directly to the constituting single reactions as assumed by Kissinger method, but are determined by the superposition of two processes, which relative contributions are varied with β (Figure 1a, Figure S2). Table 1 also lists the variation range of the conversion degree at peak maximum, $\Delta\alpha_p$, among the curves with different β. This variation should be negligible for rigorous application of Kissinger method [25], so its magnitude can serve as an indicator of the reaction complexity. In Case 2, both DTG and DSC data indicate well separated two peak maxima (Figure 1b) because of limited overlapping of two constituent reactions. The limited influence on each peak maximum from another stage results in nearly constant α_p with β, i.e., negligible $\Delta\alpha_p$ (Table 1). Therefore, the Arrhenius parameters determined by Kissinger method for the first and second peaks of DTG and DSC data fully agree with the correct kinetic parameters. Application of Kissinger method in Case 3 brings the incorrect A, E_a values due to high overlapping of stages, that is recognized by a considerable $\Delta\alpha_p$ values (Table 1). As was seen in Figure 1d, Case 4 is very specific with the same E_a values for stages, but lower rate constant k_1 throughout the process. The $\Delta\alpha_p$ values for DSC and DTG data are negligible supporting the value of Kissinger analysis. As a result, the E_a values obtained from DTG and DSC data closely correspond to that assumed to simulate the kinetic data. The preexponent value calculated from the overall DTG and DSC data correspond to the assumed A value for the first reaction step, because the first reaction step is the rate-limiting step in Case 4.

Summarizing the above findings, the Kissinger method becomes less applicable with increasing the interrelationship between the constituent processes. One should look on the magnitude of peak shift $\Delta\alpha_p$ as a first indicator of the reaction complexity.

Table 1. Kinetic parameters calculated by the Kissinger plot for the simulated successive reactions of Cases 1–4.

Case	Step	Assumed Kinetic Parameters [1]		E_a/kJ mol^{-1} [2]		lg(A/s^{-1}) [2]		$\Delta\alpha_p$ [3]	
		E_a/kJ mol^{-1}	lg(A/s^{-1})	DTG	DSC	DTG	DSC	DTG	DSC
1	1	120	10	138 ± 4	148 ± 1	12.9 ± 0.4	13.7 ± 0.1	0.06	0.19
	2	185	17.5	–	160 ± 10	–	14.0 ± 1.1	–	0.19
2	1	115	10.4	115 ± 1	115 ± 1	10.3 ± 0.1	10.4 ± 0.1	0.002	0.007
	2	115	9.4	115 ± 1	115 ± 1	9.4 ± 0.01	9.4 ± 0.1	0.002	0.003
3	1	185	17.5	203 ± 4	177 ± 2	19.3 ± 0.4	16.7 ± 0.3	0.11	0.11
	2	120	10	–	159 ± 3	–	14.0 ± 0.3	–	0.16
4	1	115	9.5	116 ± 3	116 ± 2	9.6 ± 0.3	9.6 ± 0.2	0.03	0.02
	2	115	11						

[1] Stands for the kinetic parameters used in Equations (2)–(4) to calculate the input data. [2] Correlation coefficients of linear fit in Kissinger plot is higher than 0.999 for all cases. [3] Variation of α_p with β (1–10 K min^{-1}).

3.3. Isoconversional Methods

The kinetic analysis of the TA data using the isoconversional methods is recommended by the kinetic committee of ICTAC [25]. For the ideal single-step reaction recorded under ideal measurement conditions, a constant E_a value should be obtained at different α during the course of reaction, irrespective of the kind of isoconversional methods, i.e., integral method of Flynn-Wall-Ozawa [36], differential method of Friedman [35], and advanced integral method of Vyazovkin [45]. However, in practice, the constant E_a is often not attained even for the relatively simple processes, due to the gradual changes in reaction conditions, reaction geometry, and rate-limiting step as the reaction advances. Therefore, along with the examples of successful applications to complex processes [37], some superficial E_a values and observations of the kinetic parameters variation in an exotic manner during reaction course have been reported [22]. Moreover, when the isoconversional methods are applied to the multistep processes, the physical meaning of conversion degree α should be carefully reconsidered: while, in the fundamental kinetic equation, α is defined as the fractional reaction for the single reaction step [25], for the multistep process the α value is determined experimentally as the fraction of the total changes in the physical quantities during the overall process. Keeping in mind these theoretical limitations, several isoconversional methods were applied to the simulated kinetic data of Cases 1–4.

Application of differential Friedman [35] and integral FWO [36] methods reveals large (up to 35%) discrepancies of calculated E_a values for the same input data (Figure S6). A possible reason of inaccuracy of the Doyle approximation of temperature integral used in FWO method [25] was disregarded after coincidence of the FWO result with the output of Starink isoconversional method [44] that uses more precise representation of the temperature integral (Figures S6 and S7). Thus, the difference is inherent for the integral isoconversional methods and caused by considerable dependency of E_a on α, while in calculation the apparent E_a value is smoothed over 0–α region resulting in systematic error [45]. To prove this, we used the advanced integral method developed by Vyazovkin [45] that is based on the optimization of incremental change in E_a by iterative calculation over small $\alpha - (\alpha + \Delta\alpha)$ and therefore eliminates the "smoothing" effect. Indeed, it reveals the kinetic parameters equal to the output of differential Friedman method (Figure S6).

The results of the isoconversional Friedman analysis applied to the simulated data of Cases 1–4 are shown in Figure 2 along with the assumed values for the constituting steps (dashed lines). For Case 1, the E_a value calculated from DTG indicates the correct value of E_{a1} at the beginning and increases toward E_{a2} in α range characterized by overlapping of two reaction processes but goes back to E_{a1} at the final stage (Figure 2a). As for isoconversional dependency built on DSC data, its variation is superficial with no meaning for the lowering of E_a, although the values at the beginning and end of the reaction correspond to E_{a1} as in the DTG.

For Case 2 that describe two partially overlapping processes with similar E_a values, the E_a values determined from DTG are close to the correct E_a and approximately constant during the course of reaction. Due to distinction between the exothermic and endothermic effects on DSC curve of Case 2, the conversion, derived normally as a partial area under curve, will not only increase but decrease as the reaction advances. This fact results in α value larger than unity or smaller than zero depending on the magnitude relation between the exothermic and endothermic effects (Points designated as circles in Figure 1b,c correspond to the moment when the total conversion, calculated from DSC, is equal to unity). Thus, "normal" isoconversional procedure will treat the truncated to 1 part of DSC data offering the partial and distorted kinetic information. One of possible empirical procedures to deal with the "α > 1" issue is the separate consideration of the exothermic and endothermic parts [46]. The breakpoint between two parts is where DSC signal changes its sign (denoted by square in Figure 1b). To put the isoconversional kinetic parameters on a single plot we adjust the first "exo" part as 0..0.7 (= $Q_{exo}/(Q_{exo} + |Q_{endo}|)$) range of overall conversion, while the second "endo" part – to 0.7..1 region (Figure 2b). The results of the "separation" procedure correspond well to the isoconversional data calculated on DTG and the correct E_a for both stages. Another possible approach to the "α > 1" issue is representing of the overall α as the index of the reaction advancement calculated as the partial area under the absolute magnitude of DSC data. It should be noted that the procedure lacks theoretical validity, because the sum of the absolute values of exothermic and endothermic peak areas possibly changes accompanied by the changes of overlapping degree of the exothermic and endothermic processes with β. Even in such theoretically invalid situation, the apparent E_a values calculated from DSC of Case 2 are close to the correct E_a value (Figure 2b, |DSC| curve).

The results for DTG of Case 3 seem to be more relevant as the barriers for stages are different. Although DTG curve looks like a smooth single peak (Figure 1c), the isoconversional analysis catches the variation of E_a values during the process (Figure 2c). The calculated E_a values at the beginning and the end of the overall reaction indicate the correct E_a values for the first and second mass-loss processes, respectively. For DSC data the "$\alpha > 1$" issue arises due to clear exothermic and endothermic parts of the heat flow. The separate isoconversional analysis of exo- and endothermic parts reveals the activation energy for the first stage about 175 kJ mol^{-1} (exact value is 185 kJ mol^{-1}), for the second endothermic stage E_a value approaches to 120 kJ mol^{-1}, which is the correct value. Treatment of the absolute magnitude of DSC data as in Case 2, seems doubtful due to significant change in the sum of absolute peak areas, i.e., at $\beta = 10$ K min^{-1} it is larger in 1.23 times than at $\beta = 1$ K min^{-1}. However, the E_a values calculated from |DSC| data indicate the similar variation as the reaction advances with that from DTG (Figure 2c).

Case 4 represents two stages that are overlapping in a smooth single peak either in DSC or DTG (Figure 1d). Because the E_a values assumed for the first and second reaction steps are identical, the results of Friedman method represent the close correspondence to the correct E_a value during the course of reaction both for DTG and DSC data (Figure 2d). The multistep reaction behavior cannot be deduced from the TA curves and the results of isoconversional analysis in this instance.

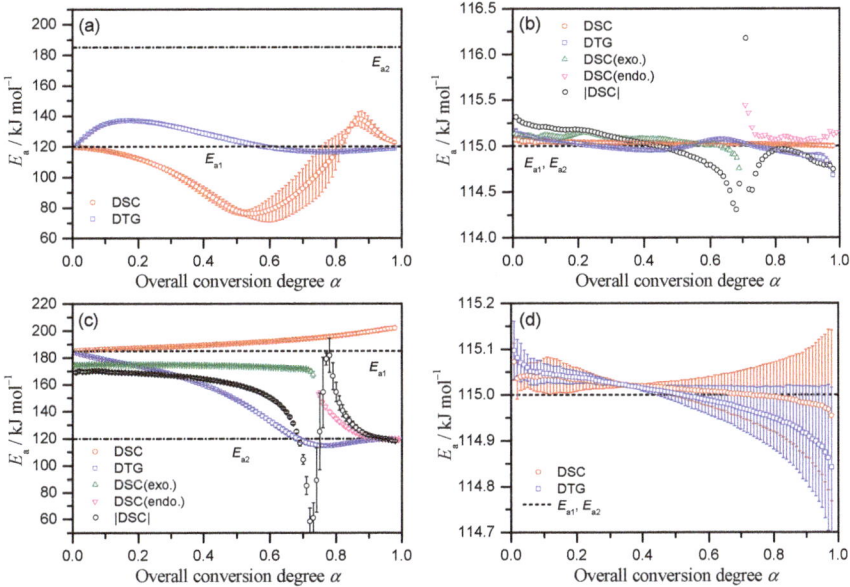

Figure 2. Isoconversional Friedman analysis of the simulated DTG and DSC data at β of 1–10 K min^{-1}: (**a**) Case 1, (**b**) Case 2 (for clarity the uncertainties are not shown), (**c**) Case 3, and (**d**) Case 4.

These results show the strength of isoconversional technique in catching the change of E_a in course of reaction and its weakness in case of significant overlapping. To evaluate the isoconversional results further, we compare the prediction of thermal behavior based on isoconversional kinetic parameters with the output of the exact model. Indeed, the results of the isoconversional analysis appear as the mathematical representation of the process. They can be easily extended outside of the temperature region where the kinetic parameters have been determined, i.e., for prediction purposes (e.g., [47]). What are the limits of this procedure, and does the isoconversional parameters derived for considered complex process allow successful prediction? Answering these questions, we compare the model output (exact solution) for heating rates from 10^{-9} to 10^9 K min^{-1} with the prediction based on isoconversional results (obtained with data at 1–10 K min^{-1} rates). In Table S1 the results of the isoconversional-based simulation and exact solution are compared in terms of the peak temperatures of the conversion rate, Figure S8 depicts the reaction profiles. For Cases 2 and 4, i.e., where the E_a values for both stages are the same, the isoconversional prediction based on TG data is fully consistent with the true outputs from the assumed model. Isoconversional results for Case 2, where the overall α calculated from DSC goes above unity, reveal that for 0–1 region the prediction is correct (e.g., Figure S8c). To account the rest of the process, we apply two modified approaches, i.e., analysis of separate exo- and endothermic parts, and absolute magnitude of DSC, both results in inaccurate prediction. For Cases 1 and 3 with more complex kinetic pattern the isoconversional technique gives the results differing from the exact output of the assumed model in terms of the peak temperatures up to 260 K using DSC data and 75 K using TG data. Therefore, in case of high reaction complexity, the predictions based on the isoconversional kinetic parameters have to be performed with great caution.

3.4. Mathematical Deconvolution and Subsequent Combined Kinetic Analysis

An alternative approach is to extract (deconvolute) the constituent peaks from the complex profile for their subsequent kinetic analysis. This approach is based on the mathematical deconvolution—the empirical peak separation by summing up mathematically fitted component peaks using a fitting function, $F(t)$ [48]:

$$\frac{d\alpha}{dt} = \sum_{i=1}^{N} F_i(t), \quad (5)$$

where N is the total number of component reaction steps. The fitting function with the asymmetric peak shape is generally recommended, e.g., Fraser–Suzuki function [27]:

$$F(t) = a_0 \cdot \exp\left(-\ln 2 \left[\frac{\ln\left(1 + 2a_3 \frac{t-a_1}{a_2}\right)}{a_3}\right]^2\right), \quad (6)$$

where a_0, a_1, a_2, a_3 are the amplitude, position, half-width and asymmetry of the curve, respectively. This approach has been successfully applied for real-life complex processes [49,50].

The deconvolution procedure for DSC data of Case 1, where the endothermic event is not apparent, yields the superficial results as shown in Figure 3a. The kinetic data derived from DSC (thick line) is equally well, with the correlation coefficient higher than 0.997, described by the superposition of two exothermic processes (dotted lines) or superposition of exothermic and endothermic events (dashed lines). Moreover, for the latter (correct choice) the deconvoluted exothermic and endothermic peaks are largely different from the assumed contribution of stages, i.e., γ ($d\alpha_1/dt$) and $(1-\gamma)$ ($d\alpha_2/dt$) (solid lines, Figure 3a). The problem of selecting the peak combination is easily eliminated for Cases 2 and 3, where the overall process clearly represents the combination of the exothermic and endothermic steps. Truly, in Case 2, where both stages have the same E_a value, but different A values, the deconvoluted peaks closely resemble the true reactions conversions (Figure S9a). However, in more realistic Case 3, when the E_a values for component reactions differ, the deconvoluted peaks although being exothermic and endothermic, not follow true reaction rates (Figure S9b). For DTG data, one is caught between deconvolution with two peaks (dashed lines, Figure 3b) or single peak (dotted line). However, the deconvoluted curves again are different from the true ones, i.e., $\eta(d\alpha_1/dt)$ and $(1-\eta)(d\alpha_2/dt)$ (solid lines, Figure 3b).

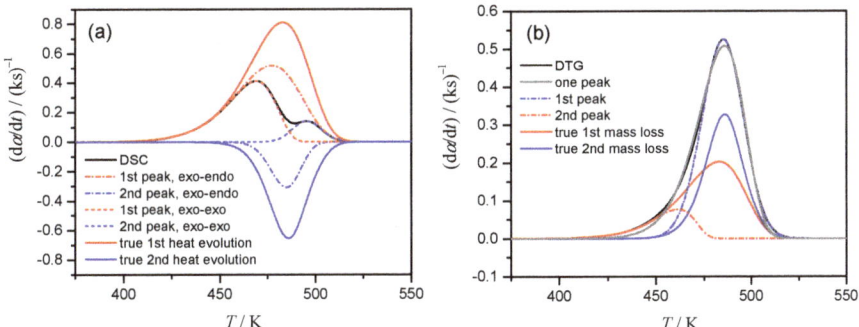

Figure 3. Selection of the peak combinations for Case 1, $\beta = 1$ K min^{-1}: (**a**) superposition of two exothermic processes (dotted lines) or exothermic and endothermic ones (dashed lines) results in perfect correlation with DSC signal (black solid) but completely different from true heat evolution data, (**b**) DTG peaks are described well both by two peaks (dashed lines) and by single peak (dotted line), but both are different from true signals.

In the practical kinetic analysis, usually it is not possible to recognize the superficial mathematical deconvolution as was seen in Figure 3. For critical evaluation of the results we investigate how these superficial deconvoluted reactions for Cases 1–4 (Figure 3, Figure S9) behave being further subjected to the kinetic analysis. The kinetic analysis assumed a single-step reaction, so called combined kinetic analysis (CKA) [26], is based on the optimization of the linearized equation:

$$\ln\frac{d\alpha}{dt} - \ln[\alpha^m(1-\alpha)^n] = \ln(cA) - \frac{E_a}{RT}, \qquad (7)$$

During the optimization we use for m and n the ranges of $-1 \le m \le 2$ and $0 \le n \le 2$ as the reasonable limits [10]. The calculated linear dependency of Equation (7) is shown in Figure 4a for the first reaction in Case 1, deconvoluted as the combination of two exothermic processes on DSC (incorrect scheme!). Note, that the nice linearity not allows one to suspect the superficial nature of analyzed data. After the determination of the optimum m and n values, the reaction type have been discriminated by comparing the reduced $f(\alpha)/f(0.5)$ function with that for ideal reaction types. The analyzed reaction follows three-dimensional diffusion of Ginstling–Brounshtein model (designated as D4 in Figure 4b)!

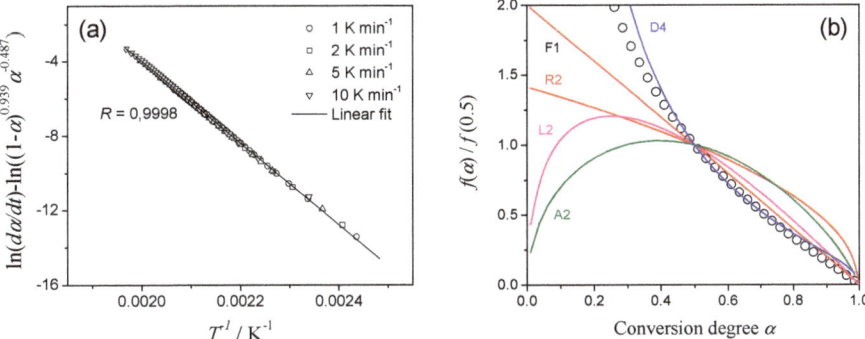

Figure 4. CKA analysis of the first reaction step of Case 1 deconvoluted from DSC curve by assuming two successive exothermic processes. (**a**) CKA plot, (**b**) Comparison of theoretical master plots of different kinetic models with calculated data (circles).

Table S3 summarizes the apparent kinetic results obtained via the mathematical peak deconvolution and subsequent CKA treatment. Although the kinetic calculation using CKA method is superficially successful as was seen in Figure 4, the correct kinetic parameters were obtained only for Cases 2 and 4 with the same E_a for both component reactions. Friedman isoconversional analysis of the deconvoluted peaks (Figure S10) supports the conclusion: improper deconvolution disturbs the results, although some kinetic features of the initial kinetic model can still be noticed. Overall, the advantage of the mathematical deconvolution—high flexibility and excellent fit quality—appears as its drawback when the superficial peaks are reconstructed from the reaction profiles of the complex process.

3.5. Kinetic Deconvolution Analysis

Another deconvolution approach – kinetic deconvolution analysis (KDA) [10,48] –represents the input data as a superposition of the independent stages, i, with contributions c_i:

$$\frac{d\alpha}{dt} = \sum_i^n c_i A_i \exp\left(-\frac{E_{a,i}}{RT}\right) f_i(\alpha_i). \tag{8}$$

An empirical reaction model $f(\alpha)$ is usually taken in a flexible Šesták–Berggren form [51]:

$$f(\alpha) = \alpha^m \cdot (1-\alpha)^n \cdot [-\ln(1-\alpha)]^p. \tag{9}$$

Nonlinear regression is carried out for fitting the calculated data based on Equation (8) to the experimental data giving the optimized kinetic parameters c_i, E_a, A, m, n, p for each reaction step.

Whereas the mathematical deconvolution-based analysis and the KDA are based on the similar idea of superposition with several peaks, these two approaches are largely different in the light of methodological procedures. KDA accepts the "kinetic" shape of peaks according to Equations (8), (9), and this restriction results in more sound kinetic results (vide infra). However, the considerable number of optimized parameters imposes heavy demands on the reliability of the initial values for kinetic parameters. Therefore, the detailed kinetic information collected using a range of physicochemical and microscopic techniques is necessary to the successful KDA, in addition to preliminary kinetic calculations for kinetic curves recorded systematically [3,7,32].

To assess the value of the KDA for analyzed data and allow the nonlinear regression to converge, we take the Šesták-Berggren model in truncated form, i.e., with $p = 0$, and use the recently proposed approach for scheme of two partially overlapping reactions [31]. Therein, combining Equations (3) and (4), one obtains:

$$P(T) = \frac{\left(\frac{d\alpha}{dt}\right)_{DSC}}{\left(\frac{d\alpha}{dt}\right)_{DTG}} = \frac{\gamma \frac{d\alpha_1}{dt} + (1-\gamma)\frac{d\alpha_2}{dt}}{\eta \frac{d\alpha_1}{dt} + (1-\eta)\frac{d\alpha_2}{dt}}. \tag{10}$$

Analyzing the $P(T)$ dependency, we look at the beginning of the process, when $d\alpha_2/dt \approx 0$ and $P \approx \gamma/\eta$. In fact, in all Cases 1–4, the $P(T)$ curve starts from value of 4 (Figure 5a). In Cases 2 and 3, where at the end of the process k_1 exceeds k_2, we obtain $P \approx (1-\gamma)/(1-\eta) = -2$ since $d\alpha_1/dt \approx 0$. Therefore, in specific instances when the contribution of either reaction step is negligible at the beginning and the end of the process, values of γ and η can be calculated from starting and final P values. These γ and η values were used as initial guesses in regression according to Equation (8).

Results of KDA show its advantage and robustness in revealing the correct reaction model, i.e., first-order reaction, using either DSC or DTG data (Table S3). As for kinetic pairs (E_a, lg$A$) the correct results were obtained for simple Cases 2 and 4 where the ratio between the rate constants are independent on temperature. Considering the Cases 1 and 3, the statistical and visual (Figure 5b) quality of the fit is still high, but the E_a values differ from that in exact model. However, the A values compensate this difference in accordance with the kinetic compensation effect (KCE). It seems that this KCE is caused by the conceptual feature of the above deconvolution techniques: the conversion rate in form of Equation (8) in general is not equal to that parameter for the successive reactions due to unaccounted dependency of the component kinetic processes as considered in Equations (2) and (3) by the dependence of $d\alpha_2/dt$ on α_1 or in case of parallel processes because of the same interdependence of the conversion degrees for the components.

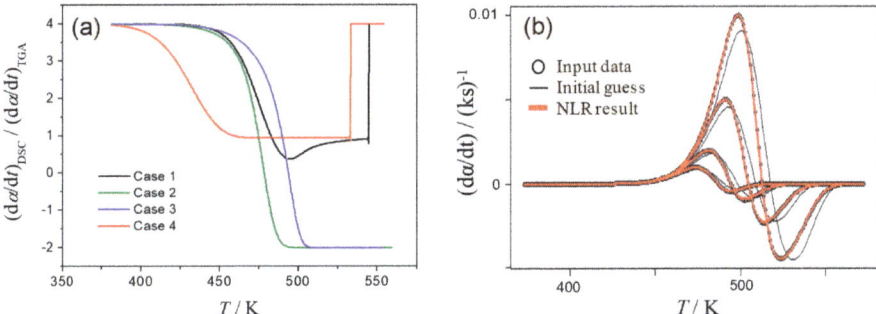

Figure 5. Kinetic deconvolution analysis: (**a**) Ratio of the conversion rates from DSC to DTG dependences on temperature for cases 1–4 at $\beta = 2$ K min^{-1}, (**b**) Fit of DSC data for Case 3 under various heating rates (points) with Equation (8) with optimized parameters (thick red line).

3.6. Formal Kinetic Analysis

The last considered method is formal kinetic analysis. It combines the strength of the model-fitting techniques with allowing for various connections between the process stages (e.g., successive, parallel and independent steps). This approach seems to be the most advantageous for the complex multistep process especially with exothermic and endothermic reactions in line with observations by Burnham [40]. The kinetic calculations have been performed with THINKS software by assuming single and various combinations of two-step reactions.

While performing the formal kinetic analysis, one faces with even higher (than for KDA) demands of the appropriate initial guesses and a need of the selection of the reaction scheme prior to calculation. Till now the selection of the reaction pattern depends on the experience of the researcher and the supplementary physicochemical insights to the process that are available. Recently, Muravyev and Pivkina propose to use the artificial neural networks to make a "recognition" of the kinetic scheme, however, the results only for the single-step reactions have been obtained [52]. Herein, we first tried the correct scheme with two subsequent first-order reactions to obtain the kinetic parameters and compare them with exact values. Overall, the kinetic parameters that were optimized throughout nonlinear regression based on DSC or DTG are close to that in assumed model (Table S4). But can we show that this kinetic scheme is indeed better than the above KDA model with two independent reactions, or model with the single-step reaction (e.g., for single DTG peak, Case 1)? From our calculations, all these schemes offer equally high correlation coefficients in certain cases and to select statistically best model we use the recently proposed advanced tool, the Bayes information criteria (BIC) [53,54]. Table 2 and Table S5 list the BIC values with the minimal one corresponding to the statistically best model. Evidently, true kinetic scheme can be identified by this kind of analysis. Also, we perform the calculations on joint DSC and DTG input data as was proposed recently by authors [31], i.e., the same reaction scheme but various contributions of stages for gravimetric and caloric data. Results show that using both datasets simultaneously increases the accuracy in kinetic parameters determination and allows easier distinguishing of the appropriate kinetic scheme (Figures S11 and S12).

Table 2. Bayes information criteria for the formal kinetic analysis of Case 1 data with several reaction schemes.

Kinetic Scheme	DSC Data	DTG Data	DSC+DTG Data
Single-step reaction	−4671	−6339	−23010
Two parallel reactions	−4888	−5355	−18186
Two consecutive reactions	−9873	−10733	−32343
Two independent reactions (KDA)	−6791	−7611	−24676

High performance of the model-fitting analysis is evidenced by these results; however, the analyzed kinetic curves were drawn by assuming overlapping two-step reactions that are characterized by homogeneous-like first-order reactions with a simple interaction between the reaction steps as expressed by $f_1(\alpha_1) = 1 - \alpha_1$ and $f_2(\alpha_1, \alpha_2) = \alpha_1 - \alpha_2$. Further careful examinations with more complex input data are required to further assess the strength of the formal kinetic approach and its applicability to the real heterogeneous reactions.

4. Summary and Outlook

The kinetic parameters calculated by above methods for both reaction steps of Case 1 are brought together in Figure 6 as lg$A$ dependency on E_a. The exact values (used to simulate the data) are marked as red circles on the plot. Results of Kissinger method, kinetic parameters for deconvoluted peaks and the output of the formal kinetic analysis fall onto the straight line drawn through the exact values of (lg$A$, E_a). Same trend is observed for Case 3 (Figure S13). This phenomenon is known as kinetic compensation effect and it has been extensively discussed in literature [55–57]. Application of the incorrect reaction model $f(\alpha)$ or inappropriate computational technique to fit the data will manifest in results as KCE [57–61]. Apparently, our results shown in Figure 6 and Figure S13 reveal this, mathematical, type of KCE.

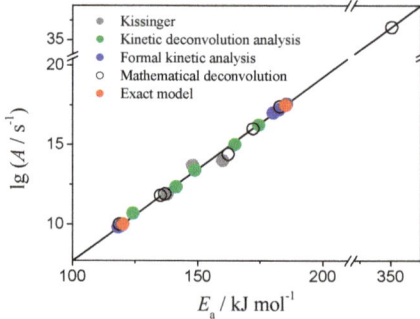

Figure 6. Kinetic compensation effect on the kinetic parameters calculated for Case 1 data.

Summarizing the findings of present study, we can highlight the guidelines for designing the kinetic approach to the considered type of complex processes. First of all, it should be noted that the kinetic data resolved using TG and DSC are generally different for the multistep reactions and the observation of both signals is advantageous. Application of the isoconversional method can be useful for preliminary kinetic assessment as it often reflects some kinetic features of the process, e.g., the starting/final E_a values. However, the resulting changes of E_a during reaction course can easily become superficial when significant overlapping of stages takes place. The situation will propagate in isoconversional prediction; thus, it has to be performed with great precautions. The mathematical deconvolution technique shows high flexibility in catching of the reaction profile, while the flip side is extremely unstable kinetic results. Kinetic deconvolution analysis uses the same idea of superposition, but its "kinetic shape" of peaks increases robustness of the approach drastically. KDA method is recommended for the preliminary evaluation of the kinetic parameters, number of stages, and the reaction types. The formal kinetic analysis is shown to be the most useful in present evaluation since it provides the correct parameters for each reaction step in considered process. Its main problem is selection of an appropriate reaction scheme, must be solved in each particular case with the help of the preliminary kinetic analysis and additional physicochemical insights to the process.

There appears to be no single approach leading to the correct kinetic triplet in all situations but having the clear idea about the considered process and having analyzed the whole available experimental data (DSC and TG, structure changes, evolved gases etc.) will help one to select an appropriate way to perform the kinetic analysis and obtain correct results. As a general framework that eases the selection of particular kinetic calculation method, we use the following logical approach (Figure 7). The synthetic kinetic data that reveal the specific features of the real data was first generated. Then possible kinetic calculation methods are applied and justified by comparison of their output with exact kinetic parameters used in model. Last step is performing the kinetic analysis using selected kinetic calculation method over the real experimental data (i.e., with unknown answer). Present study and previous experimental work [31] taken altogether show the value of recommended approach.

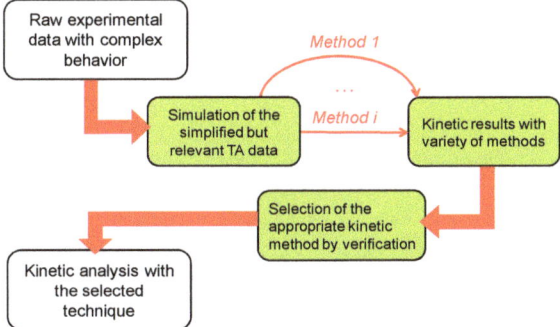

Figure 7. Proposed framework for the kinetic analysis of complex multistep processes.

Supplementary Materials: The following are available online, Figures S1–S13. Tables S1–S5.

Author Contributions: Conceptualization, N.V.M. and N.K.; methodology, N.V.M. and N.K.; software, N.V.M.; validation, N.V.M., A.N.P., and N.K.; formal analysis, N.V.M. and A.N.P.; investigation, N.V.M., resources, A.N.P.; data curation, N.V.M.; writing—original draft, N.V.M.; writing—review & editing, N.V.M. and N.K., visualization, N.V.M. and N.K., project administration, N.V.M. and A.N.P.; supervision, A.N.P.

Funding: This research was funded by the State research project 0082-2018-0002, AAAA-A18-118031490034-6 to Semenov Institute of Chemical Physics.

Conflicts of Interest: The authors declare no conflicts of interest.

References

1. Kaisersberger, E.; Opfermann, J. Kinetic evaluation of exothermal reactions measured by DSC. *Thermochim. Acta* **1991**, *187*, 151–158. [CrossRef]
2. Burnham, A.K.; Weese, R.K.; Weeks, B.L. A Distributed Activation Energy Model of Thermodynamically Inhibited Nucleation and Growth Reactions and Its Application to the β–δ Phase Transition of HMX. *J. Phys. Chem. B* **2004**, *108*, 19432–19441. [CrossRef]
3. Kitabayashi, S.; Koga, N. Thermal Decomposition of Tin(II) Oxyhydroxide and Subsequent Oxidation in Air: Kinetic Deconvolution of Overlapping Heterogeneous Processes. *J. Phys. Chem. C* **2015**, *119*, 16188–16199. [CrossRef]
4. Galwey, A.K.; Brown, M.E. *Thermal Decomposition of Ionic Solids: Chemical Properties and Reactivities of Ionic Crystalline Phases*; Elsevier: Burlington, NJ, USA, 1999; ISBN 978-0-08-054279-9.

5. Long, G.T.; Vyazovkin, S.; Brems, B.A.; Wight, C.A. Competitive Vaporization and Decomposition of Liquid RDX. *J. Phys. Chem. B* **2000**, *104*, 2570–2574. [CrossRef]
6. Muravyev, N.V.; Monogarov, K.A.; Asachenko, A.F.; Nechaev, M.S.; Ananyev, I.V.; Fomenkov, I.V.; Kiselev, V.G.; Pivkina, A.N. Pursuing reliable thermal analysis techniques for energetic materials: Decomposition kinetics and thermal stability of dihydroxylammonium 5,5'-bistetrazole-1,1'-diolate (TKX-50). *Phys. Chem. Chem. Phys.* **2017**, *19*, 436–449. [CrossRef] [PubMed]
7. Koga, N.; Kameno, N.; Tsuboi, Y.; Fujiwara, T.; Nakano, M.; Nishikawa, K.; Iwasaki Murata, A. Multistep thermal decomposition of granular sodium perborate tetrahydrate: A kinetic approach to complex reactions in solid–gas systems. *Phys. Chem. Chem. Phys.* **2018**, *20*, 12557–12573. [CrossRef] [PubMed]
8. Kameno, N.; Koga, N. Heterogeneous Kinetic Features of the Overlapping Thermal Dehydration and Melting of Thermal Energy Storage Material: Sodium Thiosulfate Pentahydrate. *J. Phys. Chem. C* **2018**, *122*, 8480–8490. [CrossRef]
9. Opfermann, J.; Hädrich, W. Prediction of the thermal response of hazardous materials during storage using an improved technique. *Thermochim. Acta* **1995**, *263*, 29–50. [CrossRef]
10. Sánchez-Jiménez, P.E.; Perejón, A.; Criado, J.M.; Diánez, M.J.; Pérez-Maqueda, L.A. Kinetic model for thermal dehydrochlorination of poly(vinyl chloride). *Polymer* **2010**, *51*, 3998–4007. [CrossRef]
11. Koga, N.; Yamada, S.; Kimura, T. Thermal Decomposition of Silver Carbonate: Phenomenology and Physicogeometrical Kinetics. *J. Phys. Chem. C* **2013**, *117*, 326–336. [CrossRef]
12. Flynn, J.H.; Wall, L.A. General treatment of the thermogravimetry of polymers. *J. Res. Natl. Bur. Stand. Phys. Chem.* **1966**, *70A*, 487. [CrossRef]
13. Vyazovkin, S.; Linert, W. The Application of Isoconversional Methods for Analyzing Isokinetic Relationships Occurring at Thermal Decomposition of Solids. *J. Solid State Chem.* **1995**, *114*, 392–398. [CrossRef]
14. Agrawal, R.K. Analysis of non-isothermal reaction kinetics: Part 2. Complex reactions. *Thermochim. Acta* **1992**, *203*, 111–125. [CrossRef]
15. Svoboda, R.; Málek, J. Is the original Kissinger equation obsolete today? *J. Therm. Anal. Calorim.* **2014**, *115*, 1961–1967. [CrossRef]
16. Kissinger, H.E. Reaction Kinetics in Differential Thermal Analysis. *Anal. Chem.* **1957**, *29*, 1702–1706. [CrossRef]
17. ASTM. *ASTM E698-05 Standard Test Method for Arrhenius Kinetic Constants for Thermally Unstable Materials*; ASTM International: West Conshohocken, PA, USA, 2005.
18. Vyazovkin, S.V.; Goryachko, V.I.; Lesnikovich, A.I. An approach to the solution of the inverse kinetic problem in the case of complex processes. Part III. Parallel independent reactions. *Thermochim. Acta* **1992**, *197*, 41–51. [CrossRef]
19. Vyazovkin, S. Conversion dependence of activation energy for model DSC curves of consecutive reactions. *Thermochim. Acta* **1994**, *236*, 1–13. [CrossRef]
20. Vyazovkin, S.V.; Lesnikovich, A.I. An approach to the solution of the inverse kinetic problem in the case of complex processes. *Thermochim. Acta* **1990**, *165*, 273–280. [CrossRef]
21. Criado, J.M.; Sánchez-Jiménez, P.E.; Pérez-Maqueda, L.A. Critical study of the isoconversional methods of kinetic analysis. *J. Therm. Anal. Calorim.* **2008**, *92*, 199–203. [CrossRef]
22. Cai, J.; Wu, W.; Liu, R. Isoconversional Kinetic Analysis of Complex Solid-State Processes: Parallel and Successive Reactions. *Ind. Eng. Chem. Res.* **2012**, *51*, 16157–16161. [CrossRef]
23. Moukhina, E. Determination of kinetic mechanisms for reactions measured with thermoanalytical instruments. *J. Therm. Anal. Calorim.* **2012**, *109*, 1203–1214. [CrossRef]
24. Burnham, A.K. Application of the Šesták-Berggren Equation to Organic and Inorganic Materials of Practical Interest. *J. Therm. Anal. Calorim.* **2000**, *60*, 895–908. [CrossRef]
25. Vyazovkin, S.; Burnham, A.K.; Criado, J.M.; Pérez-Maqueda, L.A.; Popescu, C.; Sbirrazzuoli, N. ICTAC Kinetics Committee recommendations for performing kinetic computations on thermal analysis data. *Thermochim. Acta* **2011**, *520*, 1–19. [CrossRef]
26. Pérez-Maqueda, L.A.; Criado, J.M.; Sánchez-Jiménez, P.E. Combined Kinetic Analysis of Solid-State Reactions: A Powerful Tool for the Simultaneous Determination of Kinetic Parameters and the Kinetic Model without Previous Assumptions on the Reaction Mechanism. *J. Phys. Chem. A* **2006**, *110*, 12456–12462. [CrossRef]
27. Perejón, A.; Sánchez-Jiménez, P.E.; Criado, J.M.; Pérez-Maqueda, L.A. Kinetic Analysis of Complex Solid-State Reactions. A New Deconvolution Procedure. *J. Phys. Chem. B* **2011**, *115*, 1780–1791. [CrossRef]

28. Noda, Y.; Koga, N. Phenomenological Kinetics of the Carbonation Reaction of Lithium Hydroxide Monohydrate: Role of Surface Product Layer and Possible Existence of a Liquid Phase. *J. Phys. Chem. C* **2014**, *118*, 5424–5436. [CrossRef]
29. Koga, N.; Kodani, S. Thermally induced carbonation of $Ca(OH)_2$ in a CO_2 atmosphere: Kinetic simulation of overlapping mass-loss and mass-gain processes in a solid–gas system. *Phys. Chem. Chem. Phys.* **2018**, *20*, 26173–26189. [CrossRef]
30. Schmid, H.; Eisenreich, N.; Krause, C.; Pfeil, A. Analysis of complex thermoanalytical curves: The thermodynamic and kinetic parameters of isopropylammonium nitrate. *J. Therm. Anal. Calorim.* **1989**, *35*, 569–576. [CrossRef]
31. Muravyev, N.V.; Koga, N.; Meerov, D.B.; Pivkina, A.N. Kinetic analysis of overlapping multistep thermal decomposition comprising exothermic and endothermic processes: Thermolysis of ammonium dinitramide. *Phys. Chem. Chem. Phys.* **2017**, *19*, 3254–3264. [CrossRef]
32. Nakano, M.; Wada, T.; Koga, N. Exothermic Behavior of Thermal Decomposition of Sodium Percarbonate: Kinetic Deconvolution of Successive Endothermic and Exothermic Processes. *J. Phys. Chem. A* **2015**, *119*, 9761–9769. [CrossRef]
33. Heym, F.; Etzold, B.J.M.; Kern, C.; Jess, A. Analysis of evaporation and thermal decomposition of ionic liquids by thermogravimetrical analysis at ambient pressure and high vacuum. *Green Chem.* **2011**, *13*, 1453. [CrossRef]
34. Skala, D.; Sokić, M.; Tomić, J.; Kopsch, H. Kinetic analysis of consecutive reactions using TG and DSC techniques: Theory and application. *J. Therm. Anal. Calorim.* **1989**, *35*, 1441–1458. [CrossRef]
35. Friedman, H.L. Kinetics of thermal degradation of char-forming plastics from thermogravimetry. Application to a phenolic plastic. *J. Polym. Sci. Part C Polym. Symp.* **1964**, *6*, 183–195. [CrossRef]
36. Ozawa, T. A New Method of Analyzing Thermogravimetric Data. *Bull. Chem. Soc. Jpn.* **1965**, *38*, 1881–1886. [CrossRef]
37. Vyazovkin, S. *Isoconversional Kinetics of Thermally Stimulated Processes*; Springer International Publishing: Cham, Switzerland, 2015; ISBN 978-3-319-14174-9.
38. Sánchez-Jiménez, P.E.; Pérez-Maqueda, L.A.; Perejón, A.; Criado, J.M. A new model for the kinetic analysis of thermal degradation of polymers driven by random scission. *Polym. Degrad. Stab.* **2010**, *95*, 733–739. [CrossRef]
39. Opfermann, J. Kinetic Analysis Using Multivariate Non-linear Regression. I. Basic concepts. *J. Therm. Anal. Calorim.* **2000**, *60*, 641–658. [CrossRef]
40. Burnham, A.K.; Dinh, L.N. A comparison of isoconversional and model-fitting approaches to kinetic parameter estimation and application predictions. *J. Therm. Anal. Calorim.* **2007**, *89*, 479–490. [CrossRef]
41. Budrugeac, P.; Homentcovschi, D.; Segal, E. Critical Considerations on the Isoconversional Methods. III. On the evaluation of the activation energy from non-isothermal data. *J. Therm. Anal. Calorim.* **2001**, *66*, 557–565. [CrossRef]
42. Pérez-Maqueda, L.A.; Criado, J.M. The Accuracy of Senum and Yang's Approximations to the Arrhenius Integral. *J. Therm. Anal. Calorim.* **2000**, *60*, 909–915. [CrossRef]
43. Muravyev, N.V. *THINKS—Thermokinetic Software*, Moscow, 2016.
44. Starink, M.J. The determination of activation energy from linear heating rate experiments: A comparison of the accuracy of isoconversion methods. *Thermochim. Acta* **2003**, *404*, 163–176. [CrossRef]
45. Vyazovkin, S. Modification of the integral isoconversional method to account for variation in the activation energy. *J. Comput. Chem.* **2001**, *22*, 178–183. [CrossRef]
46. Muravyev, N.V.; Monogarov, K.A.; Bragin, A.A.; Fomenkov, I.V.; Pivkina, A.N. HP-DSC study of energetic materials. Part I. Overview of pressure influence on thermal behavior. *Thermochim. Acta* **2016**, *631*, 1–7. [CrossRef]
47. Roduit, B.; Borgeat, C.; Berger, B.; Folly, P.; Alonso, B.; Aebischer, J.N. The prediction of thermal stability of self-reactive chemicals: From milligrams to tons. *J. Therm. Anal. Calorim.* **2005**, *80*, 91–102. [CrossRef]
48. Koga, N.; Goshi, Y.; Yamada, S.; Pérez-Maqueda, L.A. Kinetic approach to partially overlapped thermal decomposition processes: Co-precipitated zinc carbonates. *J. Therm. Anal. Calorim.* **2013**, *111*, 1463–1474. [CrossRef]

49. Yan, Q.-L.; Zeman, S.; Zhang, J.-G.; Qi, X.-F.; Li, T.; Musil, T. Multistep Thermolysis Mechanisms of Azido-s-triazine Derivatives and Kinetic Compensation Effects for the Rate-Limiting Processes. *J. Phys. Chem. C* **2015**, *119*, 14861–14872. [CrossRef]
50. Svoboda, R.; Málek, J. Applicability of Fraser–Suzuki function in kinetic analysis of complex crystallization processes. *J. Therm. Anal. Calorim.* **2013**, *111*, 1045–1056. [CrossRef]
51. Šesták, J.; Berggren, G. Study of the kinetics of the mechanism of solid-state reactions at increasing temperatures. *Thermochim. Acta* **1971**, *3*, 1–12. [CrossRef]
52. Muravyev, N.V.; Pivkina, A.N. New concept of thermokinetic analysis with artificial neural networks. *Thermochim. Acta* **2016**, *637*, 69–73. [CrossRef]
53. Bohn, M.A. Assessment of description quality of models by information theoretical criteria based on Akaike and Schwarz-Bayes applied with stability data of energetic materials. In Proceedings of the 46th International Annual Conference of the Fraunhofer ICT, Karlsruhe, Germany, 23–26 June 2015; pp. 6.1–6.23.
54. Roduit, B.; Hartmann, M.; Folly, P.; Sarbach, A.; Baltensperger, R. Prediction of thermal stability of materials by modified kinetic and model selection approaches based on limited amount of experimental points. *Thermochim. Acta* **2014**, *579*, 31–39. [CrossRef]
55. Garn, P.D. An examination of the kinetic compensation effect. *J. Therm. Anal. Calorim.* **1975**, *7*, 475–478. [CrossRef]
56. Agrawal, R.K. On the compensation effect. *J. Therm. Anal. Calorim.* **1986**, *31*, 73–86. [CrossRef]
57. Koga, N. A review of the mutual dependence of Arrhenius parameters evaluated by the thermoanalytical study of solid-state reactions: The kinetic compensation effect. *Thermochim. Acta* **1994**, *244*, 1–20. [CrossRef]
58. Koga, N.; Šesták, J. Kinetic compensation effect as a mathematical consequence of the exponential rate constant. *Thermochim. Acta* **1991**, *182*, 201–208. [CrossRef]
59. Brown, M.E.; Galwey, A.K. The significance of "compensation effects" appearing in data published in "computational aspects of kinetic analysis": ICTAC project, 2000. *Thermochim. Acta* **2002**, *387*, 173–183. [CrossRef]
60. Koga, N.; Šesták, J.; Málek, J. Distortion of the Arrhenius parameters by the inappropriate kinetic model function. *Thermochim. Acta* **1991**, *188*, 333–336. [CrossRef]
61. Koga, N.; Šesták, J. Further aspects of the kinetic compensation effect. *J. Therm. Anal. Calorim.* **1991**, *37*, 1103–1108. [CrossRef]

Sample Availability: Samples of the compounds are available from the authors.

© 2019 by the authors. Licensee MDPI, Basel, Switzerland. This article is an open access article distributed under the terms and conditions of the Creative Commons Attribution (CC BY) license (http://creativecommons.org/licenses/by/4.0/).

Article

Kinetics of Crystallization and Thermal Degradation of an Isotactic Polypropylene Matrix Reinforced with Graphene/Glass-Fiber Filler

Evangelia Tarani [1], George Z. Papageorgiou [2], Dimitrios N. Bikiaris [3] and Konstantinos Chrissafis [1],*

[1] X-ray, Optical Characterization and Thermal Analysis Laboratory, Physics Department, Aristotle University of Thessaloniki, GR541 24 Thessaloniki, Greece; etarani@physics.auth.gr
[2] Chemistry Department, University of Ioannina, P.O. Box 1186, 45110 Ioannina, Greece; gzpap@uoi.gr
[3] Laboratory of Polymer Chemistry and Technology, Department of Chemistry, Aristotle University of Thessaloniki, GR541 24 Thessaloniki, Greece; dbic@chem.auth.gr
* Correspondence: hrisafis@physics.auth.gr; Tel.: +30-2310-99-8188

Academic Editor: Sergey Vyazovkin
Received: 18 April 2019; Accepted: 21 May 2019; Published: 23 May 2019

Abstract: Polypropylene composites reinforced with a filler mixture of graphene nanoplatelet-glass fiber were prepared by melt mixing, while conventional composites containing graphene nanoplatelet and glass fiber were prepared for comparative reasons. An extensive study of thermally stimulated processes such as crystallization, nucleation, and kinetics was carried out using Differential Scanning Calorimetry and Thermogravimetric Analysis. Moreover, effective activation energy and kinetic parameters of the thermal decomposition process were determined by applying Friedman's isoconversional differential method and multivariate non-linear regression method. It was found that the graphene nanoplatelets act positively towards the increase in crystallization rate and nucleation phenomena under isothermal conditions due to their large surface area, inherent nucleation activity, and high filler content. Concerning the thermal degradation kinetics of polypropylene graphene nanoplatelets/glass fibers composites, a change in the decomposition mechanism of the matrix was found due to the presence of graphene nanoplatelets. The effect of graphene nanoplatelets dominates that of the glass fibers, leading to an overall improvement in performance.

Keywords: polypropylene; graphene nanoplatelets; glass fibers; crystallization; kinetics; activation energy

1. Introduction

The continuous demand for high-performing materials in the 20th century has led to the development of polymer composites. Polymer composites usually exhibit advanced mechanical, thermal and electrical performance as a result of the reinforcing nature of the fillers [1]. Polypropylene (PP) is one of the most used semicrystalline polymers. It can be found in various applications, such as heating, piping, sanitary, film and rigid packaging, due to its low cost and good physicochemical properties. Various fillers have already been incorporated in an isotactic PP matrix in an attempt to further improve the properties of PP, and most of those works report a significant enhancement in the final properties of the composite materials [2–5].

Glass fibers (GF) are one of these fillers since they can quite efficiently reinforce most polymeric matrices due to their high stiffness and strength, their outstanding fatigue performance, and low cost [6–8]. For this reason, GF-reinforced composites are already a commodity in the aerospace, automotive, construction, and sporting industries. To achieve the desired properties, a GF filler content of 30–50 wt.% is used. However, the high GF loadings lead to an undesirable increase in density, decreased melt flow and increased brittleness. Carbon-based materials such as carbon fibers (CF), carbon nanotubes (CNT), carbon black (CB), graphene nanoplatelets (GNP), graphene oxide (GO) and graphene have shown excellent potential as reinforcements due to their thermal and electrical conductivity, high mechanical strength and optical properties. For this reason, even a small concentration of the above-mentioned fillers can achieve significant enhancement in the properties of the initial material. GNPs display similar mechanical, electrical, barrier and thermal properties to a few layers of graphene but are significantly cheaper to produce. The GNPs that were used in the current work are produced through an intercalation procedure from graphite flakes, which most often leads to the formation of stacks of a few layers of graphite, and has already been incorporated with success in many polymeric matrices [9–12].

One strategy to further enhance the properties of the PP composites, or counterbalance some of the above-mentioned disadvantages of GF fillers, is the use of a filler mixture where two or more reinforcements are used in combination [13–16]. Pedrazzoli et al. [13] have followed this strategy to produce mechanically robust PP graphene/GF composites. The authors observed an increase in interfacial interactions between the matrix and GF due to the presence of GNP, while increases of approximately 105% for the modulus and 16% for the tensile strength with the hybrid filler were observed too. Papageorgiou et al. [14] combined both GNPs and GF in order to produce a hybrid PP composite. It was found that the modulus of PP–GNP–GF composites was 3-fold that of the matrix, while the thermal conductivity was increased 5-fold. A competitive effect was also found between GF- and GNP-filled PP, since GFs tend to decrease the thermal stability, while GNPs act towards the increase in the decomposition temperature [15].

A further detailed investigation is needed in order to fully evaluate the origins of the interesting phenomenon between GFs and GNPs. On the one hand, an extensive study of thermally stimulated processes such as crystallization and crystallization kinetics is intriguing to control and master the crystallization behavior of PP–GF–GNP composites in order to design materials with desirable properties. On the other hand, the importance of studying the thermal degradation kinetics of PP–GF–GNP composites comes from the need to understand the thermal stability under different conditions because the thermal behavior of plastics can be improved by knowing the parameters of the thermal decomposition process. Many studies on the crystallization kinetics and thermal degradation kinetics of PP have been carried out [17–21]. However, no comprehensive research has been conducted on the isothermal crystallization behavior and thermal degradation kinetics of PP–GF–GNP composites.

The main goal of this study is to investigate the synergistic effects of GFs and GNPs on the thermal properties of PP composites. For this reason, Differential Scanning Calorimetry was employed to investigate the isothermal crystallization behavior of PP–GF–GNP composites. The kinetic constant and half crystallization time were calculated by the Avrami equation, while the surface free energy of folding was calculated by the Lauritzen–Hoffman theory. Moreover, the effect of each filler along with the filler mixture system on kinetic analysis has also been evaluated by Thermogravimetric Analysis. A detailed study has been performed for the calculation of the effective activation energy using Friedman's isoconversional differential method. Finally, the kinetic model and the kinetic parameters of the thermal decomposition process were determined by the multivariate non-linear regression method.

2. Results and Discussion

2.1. Isothermal Crystallization

Understanding the crystallization mechanism is necessary for designing materials with the required properties because the crystallization process influences polymer properties through the crystal structure and morphology established during processing.

Isothermal crystallization experiments for neat PP and the corresponding composites were performed at different temperatures ranging from 125 to 150 °C. The exothermic curves representing the crystallization of the samples were recorded as a function of time (Figure 1a—only the PP–GF10 sample is presented for brevity reasons), while Figure 1b shows a comparative plot of neat PP, PP–GF20, PP–GNP20 and PP–GF–GNP20 composites at 135 °C. The curves shifted to lower values with a decreasing isothermal crystallization temperature, indicating that the crystallization rate was significantly increased.

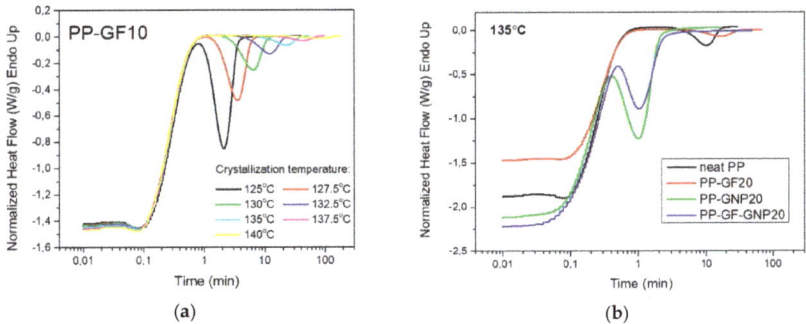

Figure 1. (a) Evolution of the exothermic peaks of PP–GF10 composite during the isothermal crystallization at different temperatures; (b) a comparative plot of neat PP, PP–GF20, PP–GNP20 and PP–GF–GNP20 composites at 135 °C.

The decrease in the time to reach overall crystallization can be used to describe the acceleration of isothermal crystallization. Figure 2 shows the time to the peak for the crystallization exotherm of PP–GF, PP–GNP, and PP–GF–GNP composites. It is obvious that the presence of GNPs accelerated the crystallization (GNPs acted effectively as heterogeneous nucleating agents), while GFs did not enhance the crystallization rates, but on the contrary, retarded the whole crystallization process. Finally, the samples filled with the filler mixture of GF–GNP crystallized at much higher rates due to the presence of GNPs [15]. This anti-nucleation effect caused by GFs is unusual since most inorganic fillers facilitate the heterogeneous nucleation in polymer composites by offering several nucleation sites during the crystallization from the melt. So, a further detailed investigation is needed in order to fully evaluate the origins of the interesting phenomenon between GFs and GNPs.

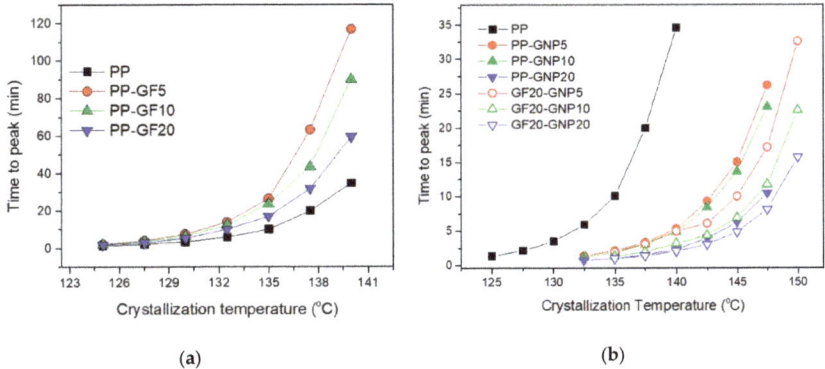

Figure 2. Dependence of the time to peak on the temperature of isothermal crystallization from the melt for (a) PP–GF and (b) PP–GNP and PP–GF–GNP composites.

2.2. Avrami Analysis of Isothermal Crystallization

The relative degree of crystallinity can be obtained if the assumption that the evolution of crystallinity is linearly proportional to the evolution of heat released during the crystallization phenomenon [22]:

$$X(t) = \frac{\int_0^t (dH_c/dt)dt}{\int_0^\infty (dH_c/dt)dt} \quad (1)$$

where dH_c represents the enthalpy of crystallization during an infinitesimal time internal dt, while the limits t and ∞ denote the elapsed time during crystallization and the end of crystallization phenomenon, respectively. Figure 3 shows the relative crystallinity of neat PP and the PP–GF10 composite at different crystallization times in the process of isothermal crystallization. It can be seen that all characteristic sigmoid isotherms shift to the right with increasing isothermal crystallization temperature and the crystallization rate becomes slower.

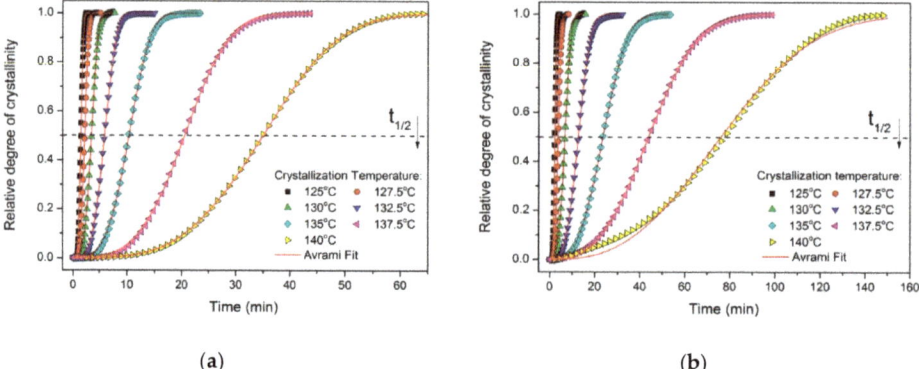

Figure 3. Avrami fit (continuous line) for the data (solid symbols) from the isothermal crystallization of (a) neat PP and (b) PP–GF10 composite.

Then, the isothermal crystallization kinetics of the PP, PP–GF, PP–GNP, and PP–GF-GNP composites were interpreted in terms of the Avrami equation in order to more precisely analyze the origins of the above-mentioned phenomenon, mainly through nucleation and growth process at a fixed crystallization temperature. The Avrami equation is one of the most used macrokinetic theories of the field and, according to this, the relative degree of crystallinity $X(t)$ can be related to the crystallization time according to the expression:

$$X(t) = 1 - exp(kt^n) \text{ or } X(t) = 1 - exp[-(Kt)^n] \quad (2)$$

where n is the Avrami exponent which is related to the nucleation process and k is the growth function which is dependent on the nucleation and crystal growth [23–25]. The composite Avrami form includes K instead of k (where $k = K^n$) [26]. The non-linear curve fitting procedure based on the Marquardt–Levenberg algorithm was employed for the data from isothermal crystallization since it takes into consideration the whole range of crystallization, compared to the linear fitting process [27]. The simulated theoretical lines can be seen in Figure 3 in comparison with the experimental data and it is obvious that the correlation is very high ($R^2 > 0.99$), indicating the efficiency of the Avrami equation for the description of the isothermal crystallization kinetics of the composites. The results from the Avrami analysis are presented in Figure S1 and Table S1. The n values of neat PP were in the vicinity of 2.3–3, similar to most literature reports [28,29], while the Avrami exponent for the GNP-based composites was higher, indicating that there can be an alteration in the growth mode. On the contrary, the GF-based materials presented n values closer to the ones reported in the case of the matrix, but still higher than the matrix. Generally, when n values are close to 2, this is an indication of a two-dimensional growth of the crystals, while when n is close to 3 or higher (as in the case of the composites), this fact is related to heterogeneous nucleation followed by three-dimensional growth. Therefore, all composite materials are exhibiting a 3D growth pattern. Moreover, the increase in the Avrami exponent with increasing T_c is indicative of the sporadic nucleation phenomena which can be observed at higher crystallization temperatures [30]. Regarding the growth rate for the GNP-based composites, the values of K are significantly higher than those of neat PP, while once again the materials with the highest loadings exhibit faster rates because of the vast amount of particles and surface provided for heterogeneous nucleation.

2.3. Lauritzen–Hoffman Analysis

According to the secondary nucleation theory which has been formulated by Hoffman and Lauritzen [31,32], the overall crystallization rate can be controlled by nucleation and transportation of the macromolecules in the melt. The Lauritzen–Hoffman secondary nucleation theory can describe the spherulite growth rate effectively as a function of temperature during isothermal crystallization. Accordingly, G can be expressed by:

$$G = G_0 exp\left[-\frac{U*}{R(T_c - T_\infty)}\right] exp\left[-\frac{K_g}{T_c(\Delta T)f}\right] \quad (3)$$

where G_0 is the pre-exponential factor, $U*$ and T_∞ are the Vogel–Fulcher–Tammann–Hesse (VFTH) parameters describing the transport of the polymer segments across the liquid/crystal interphase, K_g is the nucleation constant and ΔT denotes the undercooling. The first exponential term of the above expression is related to the contribution of the diffusion process to the growth rate, while the second exponential term is the contribution of the nucleation process. The generally accepted VFTH parameters are $U*$ = 1500 cal/mol [31,32] and T_∞ = (T_g-30) K [29,31]. In the present study, the $U*$ parameter and the T_g value of PP was set equal to 1500 cal/mol and 270 K, respectively, [31,33] while the equilibrium melting point was set equal to 485.1 K based on calculations with the non-linear Hoffman–Weeks method [33,34]. Theses substitutions have been widely used in the crystallization study of PP polymer and the corresponding composite systems. It is clear from the above assumptions that the Lauritzen–Hoffman theory provides an approximation and not an absolute value. Thus, the data and calculations presented in this work are mainly for the purpose of qualitative comparison between the PP matrix and its nanocomposites. The calculation of K_g for secondary or heterogeneous nucleation can be obtained by:

$$K_g = \frac{jb_0\sigma\sigma_e T_m^0}{k_B(\Delta h_f)} \quad (4)$$

where j = 4 for regimes I and III and j = 2 for regime II. At a low level of undercooling, crystallization occurs in regime I where the secondary nucleation rate is far less than the surface-spreading rate. At an intermediate level of undercooling, the secondary nucleation rate is comparable to the surface-spreading rate and crystallization takes place in regime II. Regime III emerges at a high level of undercooling where the secondary nucleation rate becomes larger than the surface-spreading rate [31]. The temperature range of crystallization that was selected in this work corresponds to regime III according to Xu et al. [35]. b_0 is the thickness of a single stem on the crystal, σ is the lateral surface free energy, Δh_f is the enthalpy of fusion and k_B is Boltzmann's constant. The approximation that the spherulitic growth rate is proportional to the inverse of the halftime of crystallization ($G = 1/t_{1/2}$) has been commonly used for this type of studies [33]. Therefore, after the logarithmic transformation of Equation (3), the nucleation constant can be calculated from the expression:

$$ln(G) + \frac{U*}{R(T_c - T_\infty)} = ln(G_0) - \frac{K_g}{T_c(\Delta T)f} \quad (5)$$

Plotting the left-hand side of Equation (5) versus $1/T_c(\Delta T)f$ most commonly gives a straight line with a slope and intercept equal to the nucleation constant $-K_g$ and G_0, respectively. The Lauritzen–Hoffman plots for PP and the composites can be seen in Figure 4 and it is obvious that the experimental data are successfully fitted with a linear fitting procedure.

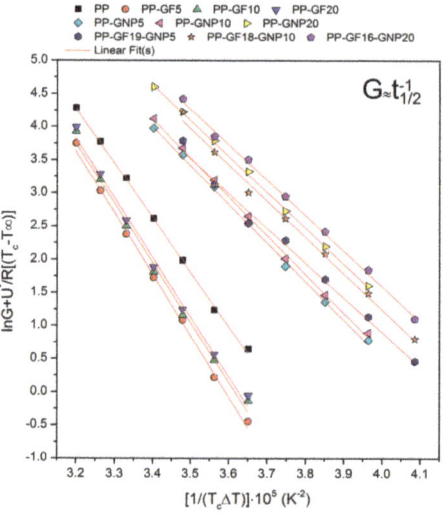

Figure 4. Lauritzen–Hoffman plots of PP and the composites.

The nucleation constant represents the energy that is needed to form a nucleus of a critical size, and it is also related to the lateral and folding surface energy. The results presented in Figure 5 show that the presence of GNPs and their increasing loadings successfully reduces the energy needed for the crystallization of the material. On the contrary, the material filled with GFs demands higher energy values to initiate crystallization and, thus, crystallization is retarded in these samples. The lateral and the fold surface energies are equally important parameters for the crystallization which is governed by secondary nucleation since they are related to both crystal nucleation and growth rates. The lateral surface free energy can be calculated from the empirical equation proposed by Thomas and Staveley [36]:

$$\sigma = \alpha \Delta h_f \sqrt{a_0 b_0} \quad (6)$$

where α is an empirical constant equal to 0.1 and $a_0 b_0$ represents the cross-sectional area of the polymer chains of PP. According to literature, $a_0 = 5.46 \times 10^{-10}$ and $b_0 = 6.26 \times 10^{-10}$ m [37]. Therefore, after obtaining σ, the fold surface energy, σ_e, can be calculated by substituting σ into Equation (4). The results for the PP composites are presented in Figure 5b, and it can be seen that the GNP-filled materials present a lower thermodynamic barrier to chain folding and, as a consequence, to polymer crystallization. Once again, GFs seem to increase the energy needed to create a new surface, along with the critical nucleus size needed for crystal growth and make the crystallization phenomenon development more difficult.

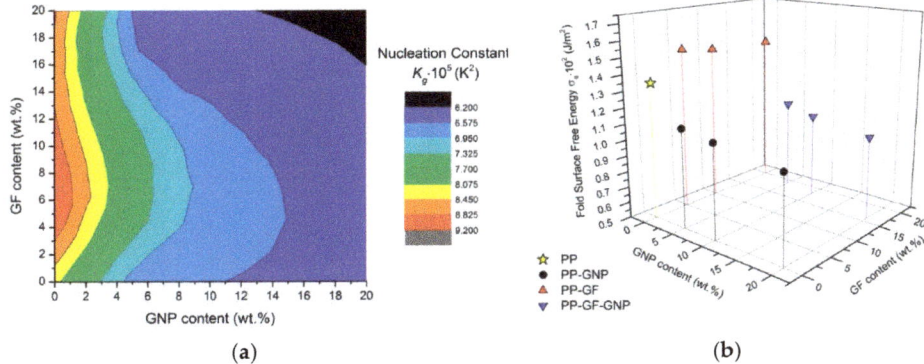

Figure 5. (a) Contour plot of the nucleation constant; (b) 3D graph of the fold surface free energy for PP and the composites.

2.4. Melting Behavior after Isothermal Crystallization

Neat PP, PP–GF10, PP–GNP5, PP–GFNP18–GNP10 composites were subjected to heating after isothermal crystallization. The heating rate applied was 10 °C/min, and the results can be seen in Figure 6a–d. The melting temperatures remain unchanged, but the melting peaks of neat PP and PP–GF10 composite (Figure 6a,b) present an obvious twinning, especially at higher crystallization temperatures. The specific phenomenon has been described in the past by Hikosaka and Seto [38] and other reports [39–41], and it is related to a modification transition mechanism from α_1 to α_2 crystals. The monoclinic crystalline structure of PP (α-phase) is the most stable thermodynamically than the other three crystalline phases (β-, γ- and smectic-phase) and presents two variants: the less stable α_1 phase (phase group C2/c) and the more stable α_2-phase (P2$_1$C) [42]. The generation of α_1 crystals proceeds during fast cooling from the melt, while α_2 crystals are formed during the isothermal procedure at higher temperatures. According to Naiki et al. [43], the origin of the differences between the two variants is the methyl group arrangement, which is in perfect order in the α_2 phase, while it is random in the α_1 phase. Therefore, the duality of the peak represents the transition from situations with a high degree of disorder, to more ordered situations, which are obtained at isothermal processes from lower to higher crystallization temperatures.

On the contrary, no twinning can be seen in the GNP-based samples (Figure 6c,d). This can be attributed to the significantly faster crystallization rates of the material, which do not provide enough time in order to generate the more stable α_2 crystals after the partial melting of α_1 crystals, even though the crystallization temperatures were much higher than the matrix.

2.5. Nucleation Activity

In order to calculate the nucleation activity of the fillers using the method proposed by Dobreva and Gutzow [44], dynamic crystallization experiments were performed for PP and nanocomposite samples at various cooling rates, ranging from 5 to 20 °C/min. The DSC curves were recorded as a function of temperature and they are presented for neat PP and PP–GNP10 composite in Figure 7. It can be seen that the crystallization peak temperature decreased with an increasing cooling rate and that the peaks of the nanocomposite sample can be observed at higher temperatures than neat PP, Figure S2.

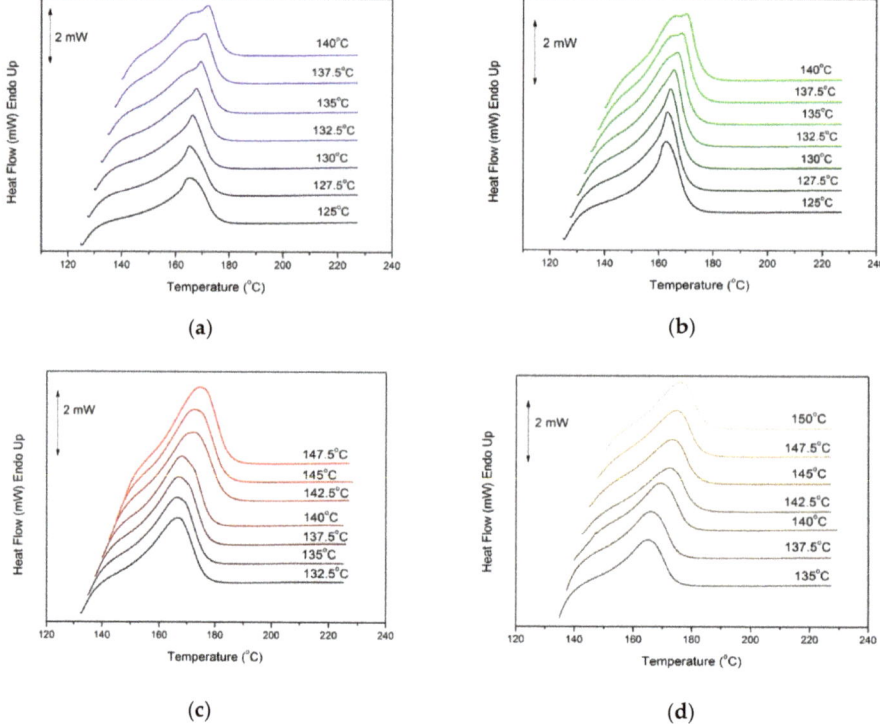

Figure 6. Melting traces of (**a**) PP, (**b**) PP–GF10, (**c**) PP–GNP5, and (**d**) PP–GF18–GNP10 composites after isothermal crystallization at various temperatures.

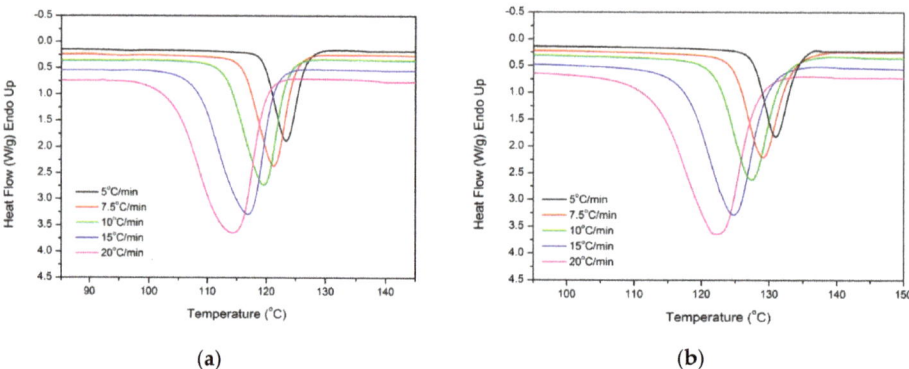

Figure 7. Differential scanning calorimeter (DSC) non-isothermal crystallization exothermic peaks recorded at different cooling rates for (**a**) PP and (**b**) PP–GNP10 composite.

The nucleation activity can be estimated from the ratio $\varphi = B^*/B$, where B^* is a parameter which can be calculated from the following expression:

$$B = \frac{\omega \sigma^3 V_m^2}{3nkT_m^0 \Delta S_m^2} \tag{7}$$

where ω is a geometric factor, σ is specific energy, V_m is the molar volume of the crystallizing substance, n is the Avrami exponent, ΔS_m is the entropy of melting and T_m^0 is the infinite crystal melting temperature. Another way of obtaining B is by plotting $\ln\beta$ versus the inverse squared degree of supercooling $1/(\Delta T_p)^2$ (where $\Delta T_p = T_m - T_p$) and calculating the slope of these plots:

$$\ln\beta = A - \frac{B}{2.303\Delta T_P^2} \quad \text{and} \quad \ln\beta = A - \frac{B^*}{2.303\Delta T_P^2} \tag{8}$$

where A is a constant and B and B^* are the constants related to the homogeneous and heterogeneous nucleation. If the nucleating substance is extremely active, the nucleation activity (B^*/B) will be close to zero, while if it is inert, the nucleation activity will be close to unity. Hence, when the nucleation activity values are higher than 1, this fact indicates an anti-nucleation effect of the filler. Plots of $\ln\beta$ versus $1/(\Delta T_p)^2$ are shown in Figure 8a for the PP–GNP samples, while the results for the nucleation activity are presented in Figure 8b. GNPs in the PP–GNP sample with their large surface area provide increased nucleation activity on the composite samples, which increases with increasing filler content. On the contrary, PP-GFs composites exhibit values of nucleation activity higher than 1, indicating once again their anti-nucleation effect on the PP matrix. Finally, the samples filled with the filler mixture of GF–GNP exhibit a B^*/B ratio very close to their PP–GNP counterparts, signifying that the GF in the specific samples are almost inert during the crystallization from the melt and the GNPs are responsible for creating enough surface to initiate crystallization faster at higher temperatures than the matrix. The high nucleation activity of GNP prevailed despite the high GFs content, and the crystallization rates were significantly enhanced under all conditions compared to the matrix. Interestingly, the crystallization rates in the specific set of samples were slightly higher (1–3 °C) than those observed for PP–GNP samples, even though the GFs should not contribute to the specific phenomenon, from previous observations. Most possibly a de-agglomeration occurred due to the increased shear stress between the various particles (loadings ranging from 24–36 wt.%) and a relatively homogeneous distribution of GNPs enabled higher rates of crystallization. Furthermore, the high interconnectivity of GFs [45] may have enabled the formation of a filler network, which is known to act positively towards the increase in the crystallization rates.

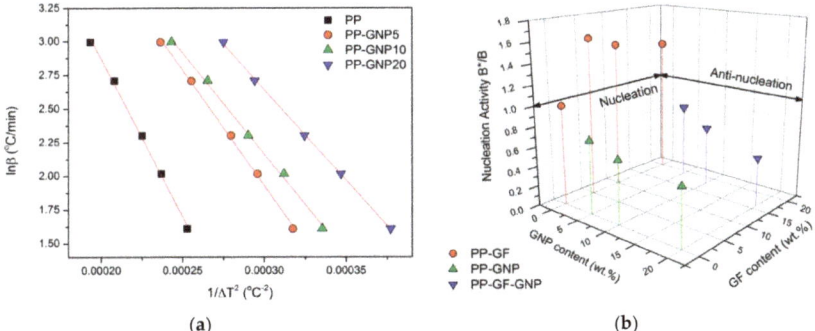

Figure 8. (a) Dobreva plots for the evaluation of the nucleation activity of PP-GNP composites. (b) 3D plot for the calculated B^*/B ratio from Dobreva's method.

3. Kinetic Analysis Based on Thermogravimetric Data

A competitive effect between GFs and GNPs was found for the PP–GF–GNP composite since GFs tend to decrease the thermal stability, while GNPs act towards the increase in the decomposition temperature [15]. A further detailed investigation is needed in order to investigate the reactions which take place during the decomposition process and the mechanisms which describe them. For performing kinetic computations on thermal analysis data, the composites filled with the highest filler content were selected (PP–GF20, PP–GNP20, PP–GF16–GNP20) along with the neat PP. For each sample, four runs were conducted with each one at a different heating rate (5, 10, 15 and 20 °C/min) [46].

3.1. Isoconversional Methods

The kinetic analysis using data from Thermogravimetric Analysis (TGA) is most commonly performed in two steps: isoconversional and model fitting. Isoconversional methods are generally considered accurate for the processing of thermoanalytical data without any assumption on the reaction mechanism and, apart from the E-a dependence, they provide indications on the second step of kinetic analysis, the model-fitting procedure [47,48]. The basis of the isoconversional methods is the assumption that the conversion function $f(\alpha)$ does not change with the variation of the heating rate for the whole range of the degree of conversion a. In the current work, Friedman's differential method was used, which was developed by taking the logarithm of the basic rate equation:

$$\frac{da}{dt} = k(T)f(a) \quad (9)$$

where T is the temperature, t is the time, and $k(T)$ is the rate coefficient which originated from the Arrhenius law $k(T) = Ae^{-E/RT}$. Therefore, Friedman's equation takes the form:

$$ln\frac{d\alpha}{dt} = \left[ln\left(\beta_i \frac{d\alpha}{dT}\right)\right]_{\alpha,i} = ln[A_\alpha f(\alpha)] - \frac{\alpha}{RT_{\alpha,i}} \quad (10)$$

where $\beta = dT/dt$. Subscript i is the ordinal number of an experiment performed at a given heating rate for non-isothermal conditions. This method is rather accurate because it does not include any mathematical approximations. The activation energy values can be obtained by plotting $ln(d\alpha/dT)$ against $1/T$ for a constant α value. The results from the application of Friedman's isoconversional method for the materials filled with the highest filler content, PP–GF20, PP–GNP20, PP–GF16–GNP20 composites can be seen in Figure 9.

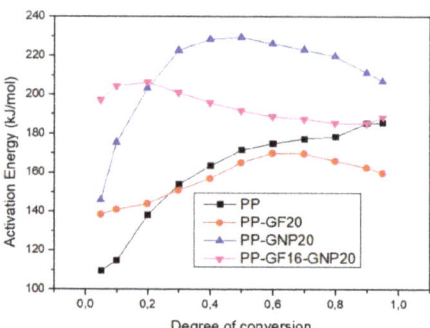

Figure 9. Dependence of the activation energy on the degree of conversion as calculated with Friedman's method for neat PP, and PP–GF20, PP–GNP20, PP–GF16–GNP20 composites.

From Figure 9, it can be seen that both PP–GNP20 and PP–GF16–GNP20 present higher values of activation energy for the whole range of the degree of conversion. On the contrary, the sample filled with GFs (PP–GF20) presents lower values of E compared to neat PP. Moreover, the samples (PP, PP–GF20, PP–GNP20) present a characteristic pattern on the dependence of E on a. The curves can be divided into two regions: the first region is extended up to $a = 0.3$–0.4, where the E increases quite rapidly and corresponds to a small mass loss at the beginning of the decomposition. For a >0.4, the E values remain almost stable for the rest of the reaction and this region corresponds to the main degradation mechanism. The dependence of E on α suggests that the decomposition is a complex reaction and at least two mechanisms should be considered for the description and fitting of these samples. On the contrary, the fact that the E values for the PP–GF16–GNP20 sample remained almost stable throughout the whole range of α is an indication that the decomposition of this sample may be described with a single-step mechanism.

3.2. Model-Fitting Procedure

The second step of a kinetic analysis based on thermogravimetric data involves the use of fitting methods with mathematical models which can describe the decomposition process. During the procedure, the samples were heated under four different rates (5, 10, 15 and 20 °C/min) and the experimental data were fitted with 16 different kinetic models and their combinations (Figure 10). The kinetic triplet has been estimated for each sample on the basis that the quality of fitting was high enough. The first part of the model-fitting procedure involves the attempt of fitting the experimental data with a single-step model, where the mechanism which is described by the model corresponds to the main mass loss. Therefore, the single-step procedure was applied to all samples under study and only the PP–GF16–GNP20 sample was successfully fitted by a single model, which was the n-th order model with autocatalysis $Cn: f(a) = (1-a)^n (1 + K_{cat})X$, where X is the reactants and K_{cat} is a constant. This fact was expected from the isoconversional analysis, since the stable values of the E with increasing α indicated that a simplified procedure could describe the decomposition of the specific sample.

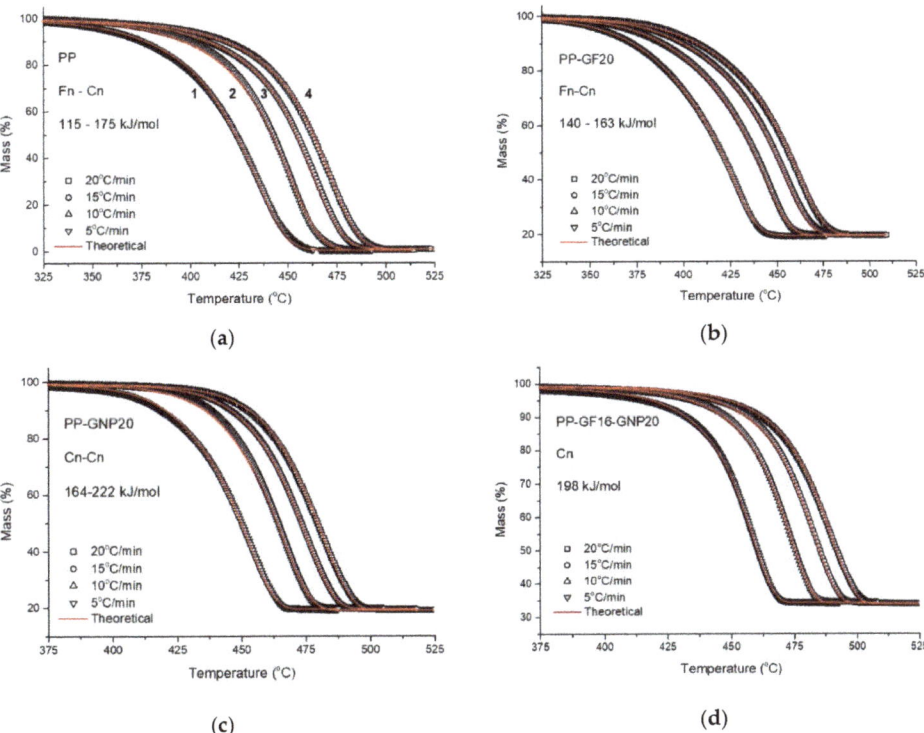

Figure 10. Thermal degradation of (**a**) PP, (**b**) PP–GF20, (**c**) PP–GNP20, and (**d**) PP–GF16–GNP20 composites at different heating rates (**1**) 5 °C/min, (**2**) 10 °C/min, (**3**) 15 °C/min, (**4**) 20 °C/min. The open black symbols represent the experimental data, while the continuous red lines represent the fittings with different models.

The second part of the procedure involves the testing of different combinations of mathematical models (consecutive or parallel) for the simulation of the experimental data that cannot be fitted with a single-step model. The two mechanisms that describe the decomposition were initially considered consecutive and the combinations that gave the most accurate fitting results were Fn-Cn for the neat PP and the PP–GF20 sample, where Fn is an n-th order model: $f(a) = (1-a)^n$, while the PP–GNP20 sample exhibited a different mechanism (Cn-Cn). Once again, the use of the two mechanisms was expected from the E-a dependency that was observed in the isoconversional methods (Figure 9). The results from the model fitting procedure are summarized in Table 1. It is worthwhile noting that the degradation of neat PP was found to take place into two stages in agreement with the literature; the first stage, the initial small mass loss, was simulated with an n-th order model, while the second stage was attributed to the main degradation mechanism and was simulated with an n-th order model with autocatalysis [21].

Table 1. Calculated values of activation energy (E), pre-exponential factor (A), reaction order (n), branching rate constant (K_{cat}) and correlation coefficient (R^2) for PP and the composites.

		1st Step				
Sample	Mechanism	E (kJ/mol)	logA (s^{-1})	N	logK_{cat}	R^2
PP	Fn	115	6.3	0.4	-	0.9998
PP–GF20	Fn	140	8.5	0.6	-	0.9998
PP–GNP20	Cn	163	9.2	0.6	0.6	0.9999
PP–GF16–GNP20	Cn	198	11.4	0.7	0.7	0.9999
		2nd Step				
Sample	Mechanism	E (kJ/mol)	LogA	N	LogK_{cat}	R^2
PP	Cn	175	10.3	1.0	0.3	0.9998
PP–GF20	Cn	163	9.4	0.9	0.6	0.9998
PP–GNP20	Cn	222	13.5	0.8	0.2	0.9999

From the results presented in Table 1, it can be seen that the correlation quality of all fittings (neat PP, PP–GF20, PP–GNP20 and PP–GF16–GNP20 composites) was very high for all combinations. Besides the PP–GF16–GNP20 composite, which exhibited a single value of activation energy and a different decomposition pattern, an increase in the activation energy values was recorded for all samples during the transition from the 1st to the 2nd step. However, the PP–GF20 sample presented lower E than the matrix at the 2nd step and thus affected the thermal stability of the matrix. Moreover, the neat PP and PP–GF20 samples were fitted with the same models (Fn-Cn), which indicates that GFs did not alter the decomposition mechanism. On the contrary, GNPs altered the 1st step of the decomposition for the PP–GNP20 sample (Cn-Fn), while they totally changed the decomposition of PP–GF16–GNP20. The effect of GNP filler is more pronounced, increasing the activation energy of thermal degradation in agreement with the calculated dependence of E on α (Figure 9) and the significant thermal stability enhancement of the PP matrix [15]. This increase in the activation energy of the nanocomposites is associated with the two-dimensional planar structure of GNPs, and it is compatible with the nanoconfinement concept as described by Chen et al., according to which, the presence of GNPs creates areas where the macromolecular chains of the matrix are confined, disturbing their regular coil conformation [15,49–51]. Thus, the mobility of polymer matrix is restricted, and the chemical reactivity of the corresponding chains is lower, increasing the activation energy and eventually further retarding the thermal degradation of the nanocomposites. The nanoconfinement concept can also explain the high activation energy of the PP–GF16–GNP20 composite since the simultaneous presence of GFs and GNPs occupies an extensive area in the volume of the composite, forming a well-distributed filler network and further confining the movement of the PP macromolecules. It should be noted that the density of GNPs is significantly lower than that of GFs and the nanofiller specific surface area is almost two orders of magnitude greater than that of GFs and, thus, GNP surface area dominates.

4. Materials and Methods

4.1. Materials

PP homopolymer was provided by Lyondellbasell with the commercial name Moplen HP501L and exhibited a flow index of 6 g/10min and melt density of 900 kg/m^3. The glass-filled PP was provided by ALBIS under the commercial name Altech PP-H A 2020/159 GF20 CP, which is a PP homopolymer product filled with 20 wt.% GF, and it exhibited a melt density of 1040 kg/m^3. The exfoliated GNPs (xGNP-25) were produced by sulfuric-based intercalated graphite and were obtained from XG Sciences (East Lansing, MI, USA). The nanoplatelets exhibited a mean platelet diameter of 25 μm and an average thickness of 6–8 nm. Their oxygen content was less than 1%, while the residual acid content was less than 0.5 wt.%.

The extrusion process was performed with a twin-screw extruder (Thermo Scientific HAAKE MiniLab micro compounder, Karlsruhe, Germany) at 190 °C and 100 rpm, while the mixing took place for 12 min. Afterwards, the prepared materials were hot pressed in order to prepare films of different thicknesses appropriate for each type of following measurements. The GNP-filled materials were named PP–GNPx throughout the manuscript, where x is the filler content at wt.% (x = 5, 10, 20 wt.%). The same applies to the samples filled with GF (PP–GFx). To produce the PP–GF–GNP composite, the material filled with 20 wt.% GF (PP–GF20) was used as a matrix and the GNPs were added in the melt mixing process. This caused a dilution of the GF content in the final batch and, at increasing GNP content, lowered the amount of GFs. So, three sets of samples were prepared, namely PP–GF19–GNP5, PP–GF18–GNP10, and PP–GF16–GNP20.

4.2. Differential Scanning Calorimetry (DSC)

A TA Q100 differential scanning calorimeter (DSC) from TA Instruments (New Castle, UK), calibrated with indium and zinc standards, was used for the study of crystallization and melting of the composites. Dry nitrogen gas was purged into the DSC cell with a flow rate of 50 mL/min. The weight of the samples was 5 ± 0.2 mg, while they were sealed in aluminium pans and heated at a heating rate 20 °C/min to 240 °C, which is above the equilibrium melting point of PP, for 5 min. The specific procedure was followed in order to erase the thermal history of the samples. For the isothermal crystallization study, the experiments were carried out according to the detailed procedure recommended by Lorenzo et al. [50]. Briefly, the samples were rapidly cooled from 240 °C to the crystallization temperature at a rate of 80 °C/min and then held at the specific temperature for a time period. The crystallization exothermic peak was then recorded. Heating scans were performed at a rate of 10 °C/min. It should be noted that the selected isothermal crystallization temperatures of the GNP-containing samples were not the same for every set of materials because the instrument was not able to equilibrate before reaching the relatively low crystallization temperatures. This happened since the samples crystallized very fast and the apparatus could not settle properly at the specific temperature. For this reason, in order to obtain a well-defined crystallization exotherm and for the avoidance of errors from baseline determination and onset time, higher temperatures were selected to allow for more time for the DSC to stabilize and record the curve properly. The measurement procedure was performed according to the detailed guidelines from Lorenzo et al. [52] for the isothermal crystallization kinetics measurements and the proper use of the Avrami equation to fit the data. For the non-isothermal crystallization, cooling scans were performed at rates of 5, 7.5, 10, 15 and 20 °C/min. A fresh sample was used for each run.

4.3. Thermogravimetric Analysis (TGA)

Thermogravimetric analysis experiments were carried out using a SETARAM SETSYS TG-DTA 16/18 instrument (Caluire, France). Samples of 5 ± 0.5 mg were placed in alumina crucibles, while an empty alumina crucible was used as a reference. According to the recommendations of ICTAC [45] for the kinetic analysis, the matrix and the composite materials filled with the highest filler content

(PP–GF20, PP–GNP20, PP–GF16–GNP20) were heated from room temperature up to 600 °C in a 50 mL/min flow of N_2 at four different rates, namely 5, 10, 15 and 20 °C/min.

5. Conclusions

GNPs and GFs were inserted individually and simultaneously in an isotactic PP matrix; the nucleation, crystallization behavior, and thermal degradation kinetics were studied under a wide variety of conditions. For the samples containing GNPs, the high number of heterogeneous nucleating surfaces along with the geometrical characteristics of the filler enabled the composite samples to crystallize at higher rates in isothermal conditions. On the contrary, the presence of GFs retarded the crystallization procedure. Crystallization at higher temperatures enabled crystalline perfection in the matrix and observation of the melting of both imperfect and perfect crystals of PP and PP–GF samples, while for the PP–GNP and PP–GF–GNP composites, the faster crystallization rates did not allow enough time for the crystals to develop a perfect crystalline structure. The anti-nucleation activity of the GF was also verified from Dobreva's method, and it was attributed to the extensive formation of a rigid amorphous fraction in the material with restricted chain mobility, which did not enhance the crystallization rates. Isoconversional and model fitting methods were successfully employed for the study of the decomposition of PP composites, and the results showed that the presence of GNPs altered the decomposition mechanism of the composites. In detail, the Cn model best described the thermogravimetric data for the PP–GF16–GNP20 composite, while the Fn-Cn models and Cn-Cn models better described the two decomposition stages of the PP–GF20 and PP–GNP20 composites, respectively. The effect of GNPs is more pronounced compared to GFs, increasing the activation energy due to the two-dimensional planar structure of GNPs and the mobility restriction of the segmental movement of the PP matrix. The presence of GNPs creates areas where the macromolecular chains of the matrix are confined, disturbing their regular coil conformation and restricting their movement. In conclusion, the easily produced GNP-based PP composites can be used in a number of advanced applications where good thermal and mechanical performance is needed.

Supplementary Materials: The supplementary materials are available online, Table S1: Avrami parameters of PP and the composite samples at different isothermal crystallization temperatures, Figure S1: (a) Avrami exponent of PP and nanocomposites. The n values of the matrix have been plotted using both symbol and line in order to separate them easier from the values of the composites; (b) growth function K of all materials under study, Figure S2.: Evolution of the crystallization peak temperature as a function of cooling rate for PP and nanocomposites

Author Contributions: Conceptualization, K.C. and D.N.B.; methodology, E.T. and G.Z.P.; investigation, E.T. and G.Z.P.; formal analysis, E.T. and G.Z.P.; writing—original draft preparation, E.T. and G.Z.P.; writing—review and editing, E.T., G.Z.P., D.N.B. and K.C.; supervision, K.C. and D.N.B.

Funding: This research received no external funding.

Conflicts of Interest: The authors declare no conflict of interest.

References

1. Alexandre, M.; Dubois, P. Polymer-layered silicate nanocomposites: Preparation, properties and uses of a new class of materials. *Mat. Sci. Eng. R.* **2000**, *28*, 1–63. [CrossRef]
2. Valles, C.; Abdelkader, A.M.; Young, R.J.; Kinloch, I.A. Few layer graphene-polypropylene nanocomposites: The role of flake diameter. *Faraday Disc.* **2014**, *173*, 379–390. [CrossRef]
3. Kashiwagi, T.; Grulke, E.; Hilding, J.; Groth, K.; Harris, R.; Butler, K.; Shields, J.; Kharchenko, S.; Douglas, J. Thermal and flammability properties of polypropylene/carbon nanotube nanocomposites. *Polymer* **2004**, *45*, 4227–4239. [CrossRef]
4. Bikiaris, D.N.; Vassiliou, A.; Pavlidou, E.; Karayannidis, G.P. Compatibilisation effect of PP-g-MA copolymer on iPP/SiO$_2$ nanocomposites prepared by melt mixing. *Eur. Polym. J.* **2005**, *41*, 1965–1978. [CrossRef]
5. Bikiaris, D.; Vassiliou, A.; Chrissafis, K.; Paraskevopoulos, K.M.; Jannakoudakis, A.; Docoslis, A. Effect of acid treated multi-walled carbon nanotubes on the mechanical, permeability, thermal properties and thermo-oxidative stability of isotactic polypropylene. *Polym. Degrad. Stab.* **2008**, *93*, 952–967.

6. Gao, S.L.; Mäder, E. Characterisation of interphase nanoscale property variations in glass fibre reinforced polypropylene and epoxy resin composites. *Compos. Part A Appl. Sci. Manuf.* **2002**, *33*, 559–576. [CrossRef]
7. Oever, M.; Peijs, T. Continuous-glass-fibre-reinforced polypropylene composites II. Influence of maleic-anhydride modified polypropylene on fatigue behaviour. *Compos. Part A Appl. Sci. Manuf.* **1998**, *29*, 227–239. [CrossRef]
8. Sanjay, M.; Arpitha, G.; Yogesha, B. Study on mechanical properties of natural-glass fibre reinforced polymer hybrid composites: A review. *Mater. Today Proc.* **2015**, *2*, 2959–2967. [CrossRef]
9. Kalaitzidou, K.; Fukushima, H.; Drzal, L.T. Multifunctional polypropylene composites produced by incorporation of exfoliated graphite nanoplatelets. *Carbon* **2007**, *45*, 1446–1452. [CrossRef]
10. Tarani, E.; Wurm, A.; Schick, C.; Bikiaris, D.N.; Chrissafis, K.; Vourlias, G. Effect of graphene nanoplatelets diameter on non-isothermal crystallization kinetics and melting behavior of high density polyethylene nanocomposites. *Thermochim. ACTA* **2016**, *643*, 94–103. [CrossRef]
11. Kalaitzidou, K.; Fukushima, H.; Drzal, L.T. Mechanical properties and morphological characterization of exfoliated graphite–polypropylene nanocomposites. *Compos. Part A Appl. Sci. Manuf.* **2007**, *38*, 1675–1682. [CrossRef]
12. Tarani, E.; Papageorgiou, D.G.; Valles, C.; Wurm, A.; Terzopoulou, Z.; Bikiaris, D.N.; Schick, C.; Chrissafis, K.; Vourlias, G. Insights into crystallization and melting of high density polyethylene/graphene nanocomposites studied by fast scanning calorimetry. *Polym. Test.* **2018**, *67*, 349–358. [CrossRef]
13. Pedrazzoli, D.; Pegoretti, A.; Kalaitzidou, K. Synergistic effect of graphite nanoplatelets and glass fibers in polypropylene composites. *J. Appl. Polym. Sci.* **2015**, *132*, 41682. [CrossRef]
14. Papageorgiou, D.G.; Kinloch, I.A.; Young, R.J. Hybrid multifunctional graphene/glass-fibre polypropylene composites. *Compos. Sci. Technol.* **2016**, *137*, 44–51. [CrossRef]
15. Papageorgiou, D.G.; Terzopoulou, Z.; Fina, A.; Cuttica, F.; Papageorgiou, G.Z.; Bikiaris, D.N.; Chrissafis, K.; Young, R.J.; Kinloch, I.A. Enhanced thermal and fire retardancy properties of polypropylene reinforced with a hybrid graphene/glass-fibre filler. *Compos. Sci. Technol.* **2018**, *156*, 95–102. [CrossRef]
16. Papageorgiou, D.G.; Liu, M.; Li, Z.; Vallés, C.; Young, R.J.; Kinloch, I.A. Hybrid poly(ether ether ketone) composites reinforced with a combination of carbon fibres and graphene nanoplatelets. *Compos. Sci. Technol.* **2019**, *175*, 60–68. [CrossRef]
17. Westerhout, R.W.J.; Waanders, J.; Kuipers, J.A.M.; Van Swaaij, W.P.M. Kinetics of the low-temperature pyrolysis of polyethene, polypropene, and polystyrene modeling, experimental determination, and comparison with literature models and data. *Ind. Eng. Chem. Res.* **1997**, *36*, 1955–1964. [CrossRef]
18. Gao, Z.; Kaneko, T.; Amasaki, I.; Nakada, M. A kinetic study of thermal degradation of polypropylene. *Polym. Degrad. Stab.* **2003**, *80*, 269–274. [CrossRef]
19. Chrissafis, K.; Paraskevopoulos, K.M.; Stavrev, S.Y.; Docoslis, A.; Vassiliou, A.; Bikiaris, D.N. Characterization and thermal degradation mechanism of isotactic polypropylene/carbon black nanocomposites. *Thermochim. ACTA* **2007**, *465*, 6–17. [CrossRef]
20. Papageorgiou, D.G.; Tzounis, L.; Papageorgiou, G.Z.; Bikiaris, D.N.; Chrissafis, K. β-nucleated propylene–ethylene random copolymer filled with multi-walled carbon nanotubes: Mechanical, thermal and rheological properties. *Polymer* **2014**, *55*, 3758–3769. [CrossRef]
21. Papageorgiou, D.G.; Bikiaris, D.N.; Chrissafis, K. Effect of crystalline structure of polypropylene random copolymers on Mechanical properties and thermal degradation kinetics. *Thermochim. ACTA* **2012**, *543*, 288–294. [CrossRef]
22. Hay, J.N.; Fitzgerald, P.A.; Wiles, M. Use of differential scanning calorimetry to study polymer crystallization kinetics. *Polymer* **1976**, *17*, 1015–1018. [CrossRef]
23. Avrami, M. Kinetics of phase change. I: General theory. *J. Chem. Phys.* **1939**, *7*, 1103–1112. [CrossRef]
24. Avrami, M. Kinetics of phase change. II Transformation-time relations for random distribution of nuclei. *J. Chem. Phys.* **1940**, *8*, 212–224. [CrossRef]
25. Avrami, M. Granulation, Phase Change, and Microstructure Kinetics of Phase Change. III. *J. Chem. Phys.* **1941**, *9*, 177–184. [CrossRef]
26. Supaphol, P. Application of the Avrami, Tobin, Malkin, and Urbanovici–Segal macrokinetic models to isothermal crystallization of syndiotactic polypropylene. *Thermochim. ACTA* **2001**, *370*, 37–48. [CrossRef]

27. Papageorgiou, G.Z.; Achilias, D.S.; Bikiaris, D.N. Crystallization Kinetics of Biodegradable Poly (butylene succinate) under Isothermal and Non-Isothermal Conditions. *Macromol. Chem. Phys.* **2007**, *208*, 1250–1264. [CrossRef]
28. Ferreira, C.I.; Dal Castel, C.; Oviedo, M.A.S.; Mauler, R.S. Isothermal and non-isothermal crystallization kinetics of polypropylene/exfoliated graphite nanocomposites. *Thermochim. ACTA* **2013**, *553*, 40–48. [CrossRef]
29. Papageorgiou, G.Z.; Achilias, D.S.; Bikiaris, D.N.; Karayannidis, G.P. Crystallization kinetics and nucleation activity of filler in polypropylene/surface-treated SiO_2 nanocomposites. *Thermochim. ACTA* **2005**, *427*, 117–128. [CrossRef]
30. Pérez–Camargo, R.A.; Saenz, G.; Laurichesse, S.; Casas, M.T.; Puiggalí, J.; Avérous, L.; Muller, A.J. Nucleation, Crystallization, and Thermal Fractionation of Poly (ε-Caprolactone)-Grafted-Lignin: Effects of Grafted Chains Length and Lignin Content. *J. Polym. Sci. B Polym. Phys.* **2015**, *53*, 1736–1750. [CrossRef]
31. Hoffman, J.D.; Davis, G.T.; Lauritzen, J., Jr. The rate of crystallization of linear polymers with chain folding. In *Treatise on Solid State Chemistry*; Hannay, N.B., Ed.; Springer: New York, NY, USA, 1976; Volume 3, pp. 497–614.
32. Huang, J.W. Dispersion crystallization kinetics, and parameters of Hoffman-Lauritzen theory of polypropylene and nanoscale calcium carbonate composites. *Polym. Eng. Sci.* **2009**, *49*, 1855–1864. [CrossRef]
33. Hoffman, J.D.; Miller, R.L. Kinetics of crystallization from the melt and chain folding in polyethylene fractions revisited: Theory and experiment. *Polymer* **1997**, *38*, 3151–3212. [CrossRef]
34. Gaur, U.; Wunderlich, B. Heat capacity and other thermodynamic properties of linear macromolecules. IV. Polypropylene. *J. Phys. Chem. Ref. Data* **1981**, *10*, 1051–1064. [CrossRef]
35. Xu, J.; Srinivas, S.; Marand, H.; Agarwal, P. Equilibrium melting temperature and undercooling dependence of the spherulitic growth rate of isotactic polypropylene. *Macromolecules* **1998**, *31*, 8230–8242. [CrossRef]
36. Thomas, D.; Staveley, L. A study of the supercooling of drops of some molecular liquids. *J. Chem. Soc.* **1952**, *889*, 4569–4577. [CrossRef]
37. Clark, E.J.; Hoffman, J.D. Regime III crystallization in polypropylene. *Macromolecules* **1984**, *17*, 878–885. [CrossRef]
38. Hikosaka, M.; Seto, T. Order of the molecular chains in isotactic polypropylene crystals. *Polym. J.* **1973**, *5*, 111–127. [CrossRef]
39. De Rosa, C.; Guerra, G.; Napolitano, R.; Petraccone, V.; Pirozzi, B. Conditions for the α1-α2 transition in isotactic polypropylene samples. *Eur. Polym. J.* **1984**, *20*, 937–941. [CrossRef]
40. Nakamura, K.; Shimizu, S.; Umemoto, S.; Thierry, A.; Lotz, B.; Okui, N. Temperature Dependence of Crystal Growth Rate for [alpha] and [beta] Forms of Isotactic Polypropylene. *Polym. J.* **2008**, *40*, 915–922. [CrossRef]
41. Corradini, P.; Napolitano, R.; Oliva, L.; Petraccone, V.; Pirozzi, B.; Guerra, G. A possible structural interpretation of the two DSC melting peaks of isotactic polypropylene in the α-modification. *Die Makromol. Chem. Rapid Commun.* **1982**, *3*, 753–756. [CrossRef]
42. Papageorgiou, D.G.; Papageorgiou, G.Z.; Bikiaris, D.N.; Chrissafis, K. Crystallization and melting of propylene–ethylene random copolymers. Homogeneous nucleation and β-nucleating agents. *Eur. Polym. J.* **2013**, *49*, 1577–1590. [CrossRef]
43. Naiki, M.; Kikkawa, T.; Endo, Y.; Nozaki, K.; Yamamoto, T.; Hara, T. Crystal ordering of α phase isotactic polypropylene. *Polymer* **2001**, *42*, 5471–5477. [CrossRef]
44. Dobreva, A.; Gutzow, I. Activity of substrates in the catalyzed nucleation of glass-forming melts. I. Theory. *J. Non Cryst. Solids* **1993**, *162*, 1–12. [CrossRef]
45. Weidenfeller, B.; Höfer, M.; Schilling, F.R. Thermal conductivity, thermal diffusivity, and specific heat capacity of particle filled polypropylene. *Compos. Part A Appl. Sci. Manuf.* **2004**, *35*, 423–429. [CrossRef]
46. Vyazovkin, S.; Burnham, A.K.; Criado, J.M.; Pérez-Maqueda, L.A.; Popescu, C.; Sbirrazzuoli, N. ICTAC Kinetics Committee recommendations for performing kinetic computations on thermal analysis data. *Thermochim. ACTA* **2011**, *520*, 1–19. [CrossRef]
47. Sbirrazzuoli, N. Determination of pre-exponential factors and of the mathematical functions f(α) or G(α) that describe the reaction mechanism in a model-free way. *Thermochim. ACTA* **2013**, *564*, 59–69. [CrossRef]
48. Vyazovkin, S. Isoconversional kinetics. In *Handbook of Thermal Analysis and Calorimetry*, 1st ed.; Brown, M.E., Gallagher, P.K., Eds.; Elsevier Science B.V.: Amsterdam, The Netherlands, 2008; Volume 5, pp. 503–538.

49. Chen, K.; Wilkie, C.A.; Vyazovkin, S. Nanoconfinement revealed in degradation and relaxation studies of two structurally different polystyrene-clay systems. *J. Phys. Chem. B.* **2007**, *111*, 12685–12692. [CrossRef]
50. Tarani, E.; Terzopoulou, Z.; Bikiaris, D.N.; Kyratsi, T.; Chrissafis, K.; Vourlias, G. Thermal conductivity and degradation behavior of HDPE/graphene nanocomposites. *J. Therm. Anal. Calorim.* **2017**, *129*, 1715–1726. [CrossRef]
51. Terzopoulou, Z.; Tarani, E.; Kasmi, N.; Papadopoulos, L.; Chrissafis, K.; Papageorgiou, D.G.; Papageorgiou, G.Z.; Bikiaris, D.N. Thermal decomposition kinetics and mechanism of in-situ prepared bio-based poly(propylene 2,5-furan dicarboxylate)/graphene nanocomposites. *Molecules* **2019**, *24*, 1717. [CrossRef]
52. Lorenzo, A.T.; Arnal, M.L.; Albuerne, J.; Müller, A.J. DSC isothermal polymer crystallization kinetics measurements and the use of the Avrami equation to fit the data: Guidelines to avoid common problems. *Polym. Test.* **2007**, *26*, 222–231. [CrossRef]

© 2019 by the authors. Licensee MDPI, Basel, Switzerland. This article is an open access article distributed under the terms and conditions of the Creative Commons Attribution (CC BY) license (http://creativecommons.org/licenses/by/4.0/).

Article

Thermal Decomposition Kinetics and Mechanism of In-Situ Prepared Bio-Based Poly(propylene 2,5-furan dicarboxylate)/Graphene Nanocomposites

Zoi Terzopoulou [1], Evangelia Tarani [2], Nejib Kasmi [1], Lazaros Papadopoulos [1], Konstantinos Chrissafis [2,*], Dimitrios G. Papageorgiou [3], George Z. Papageorgiou [4] and Dimitrios N. Bikiaris [1,*]

1. Laboratory of Polymer Chemistry and Technology, Department of Chemistry, Aristotle University of Thessaloniki, GR54124 Thessaloniki, Greece; terzoe@gmail.com (Z.T.); nejibkasmi@gmail.com (N.K.); lazaros.geo.papadopoulos@gmail.com (L.P.)
2. Solid State Physics Department, School of Physics, Aristotle University of Thessaloniki, GR54124 Thessaloniki, Greece; etarani@physics.auth.gr
3. School of Materials and National Graphene Institute, University of Manchester, Oxford Road, Manchester M13 9PL, UK; dimitrios.papageorgiou@manchester.ac.uk
4. Chemistry Department, University of Ioannina, P.O. Box 1186, 45110 Ioannina, Greece; gzpap@uoi.gr
* Correspondence: hrisafis@physics.auth.gr (K.C.); dbic@chem.auth.gr (D.N.B.); Tel.: +30-2310-99-8188 (K.C.); +30-2310-99-7812 (D.N.B.)

Received: 4 April 2019; Accepted: 1 May 2019; Published: 2 May 2019

Abstract: Bio-based polyesters are a new class of materials that are expected to replace their fossil-based homologues in the near future. In this work, poly(propylene 2,5-furandicarboxylate) (PPF) nanocomposites with graphene nanoplatelets were prepared via the in-situ melt polycondensation method. The chemical structure of the resulting polymers was confirmed by ^{1}H-NMR spectroscopy. Thermal stability, decomposition kinetics and the decomposition mechanism of the PPF nanocomposites were studied in detail. According to thermogravimetric analysis results, graphene nanoplatelets did not affect the thermal stability of PPF at levels of 0.5, 1.0 and 2.5 wt.%, but caused a slight increase in the activation energy values. Pyrolysis combined with gas chromatography and mass spectroscopy revealed that the decomposition mechanism of the polymer was not altered by the presence of graphene nanoplatelets but the extent of secondary homolytic degradation reactions was increased.

Keywords: poly(propylene 2,5 furandicarboxylate); graphene nanoplatelets; nanocomposites; bio-based polymers; thermal stability; decomposition mechanism

1. Introduction

In recent years, polymers have become a necessary part of modern life [1]. From clothes to housing, transportation to medicine and electronics, polymeric materials promote, foster and enable a sustainable society [2]. Environmental concerns have arisen though, mainly because the vast majority of these polymers is produced from non-renewable resources [3,4], so a growing interest towards the preparation of materials and chemicals from renewable resources is observed [2,5–8]. Economic reasons, such as the fluctuation of crude oil prices since fossil fuel resources have been diminishing, along with environmental ones, such as the limited biodegradability and the significant greenhouse gas emissions from the production of fossil-based materials, have shifted the attention towards bio-based raw materials and polymers [6,9–11]. For renewable polymers to enter the marketplace, they must outperform the traditional ones both in price and in performance. To accomplish this, new routes for biomass conversion and monomer generation must be used, to discover new, high performance materials [1]. Renewable raw materials such as cellulose, lignin, proteins, starch and vegetable oils have been intensively explored in order to develop a sustainable bio-based economy [12,13] and their use has been promoted legislatively, both in U.S.A. and Europe [6,14]. In parallel, research towards discovering new materials is constant, as shown by numerous papers where plants [15] or bacteria [16] have been used to produce materials suitable for applications as waste water treatment and drug delivery.

One of the most important biomass derived monomers is 2,5 furandicarboxylic acid (FDCA). FDCA can be formed by oxidative dehydration of glucose or by oxidation of 5-hydroxy-methylfurfural, and is considered a bio-derived homologue of terephthalic acid (TPA) [17,18], a monomer widely used for the production of commodity plastics such as poly(ethylene terephthalate) (PET), poly(propylene terephthalate) (PPT) and poly(butylene terephthalate) (PBT) that find many applications in today's society. In fact, FDCA is a monomer of such importance, that it has been included in the US Department of Energy list of top priority bio-based chemicals, which was published in 2004 [17]. Since that, polyesters containing furan moieties in their structure have been widely investigated, bearing promising results compared to their fossil-based counterparts [19–24]. The combination of FDCA with bio-based diols leads to fully bio-based polyesters that are already being produced on an industrial scale. Among them, poly(ethylene 2,5 furandicarboxylate) (PEF) has attracted the greatest interest, as it considered a viable alternative to PET for packaging applications, thus many parameters of its synthesis have been examined thoroughly in recent years [25–28].

Aside from PEF, many polyesters based on FDCA have been synthesized using various diols [21,29–31], one of them being 1,3-propanediol (PDO). Polyesters derived from PDO were not studied till recently, due to unavailability of the specific monomer in large quantities [32]. With the development of new processes of production though, the situation has changed and there has been increased interest towards this family of polyesters [33,34]. Poly(propylene terephthalate) (PPT) was the first polymer of this category that was available in the market [35]. Its applications were mainly the production of fibers, hence the odd number of methylene groups in the diol's structure provide better resilience and stress recovery properties compared with other terephthalic homologues from diols with an even number of methylene groups [36]. Poly(propylene furanoate) (PPF) is the bio-based counterpart of PPT. Industrially, PPF is a polyester that is used in packaging applications like multilayered materials, since it has good gas barrier properties [37,38].

The necessity to improve the produced materials has led to the use of various types of fillers as means to enhance the final product's performance. While traditional composite materials contain a big quantity of fillers bound to the polymer matrix, in nanocomposites small quantities of fillers can result in significant changes of the polymer properties, due to the enormous surface area per unit volume of the nanofillers, among other factors. Carbon nanofillers especially, ones like carbon nanotubes, graphene, and graphene oxide have been proposed as next-generation multifunctional nanofillers for the improvement of the thermal, electrical and mechanical properties of diverse matrices [39,40]. In recent publications, several types of carbon nanofillers were used enhance the properties of polymeric matrices. For example, carbon black was introduced to polyurethane polymers and it resulted in improved mechanical properties, while also increasing the wear resistance of the resulting materials [41]. The mechanical properties of polymeric materials are also enhanced by the addition of carbon fibers, especially resins. Due to their prominent strength-weight and stiffness-weight ratios, carbon fibers are excellent reinforcements for resin composites [42], and drawbacks such as poor interfacial interactions with the polymer matrix can be overcome by surface modification, resulting in materials with improved mechanical properties [43,44]. The addition of carbon nanofillers also results in materials with improved electrical properties, as shown in recent papers. For instance, graphene oxide was used to create supercapacitors that display excellent cycle stability even after 500 cycles [45]. Also, graphene sheets were found to enhance the conductivity of carbon-sulfur membranes and increasing their long term cycling stability [46]. Graphene, one of the more interesting substances in the fields of materials science, physics, chemistry, and nanotechnology, is a free-standing 2D crystal with one-atom thickness [47]. As an allotrope of carbon comprises layers of six-atom rings in a honeycombed network and can be wrapped to generate 0D fullerenes, rolled up to form 1D carbon nanotubes, and stacked to produce 3D graphite [48]. More recently, developments have been made towards the preparation of thinner forms of graphite, known as graphene nanoplatelets (GNPs). All types of graphitic material are covered by the definition of GNPs, from 100 nm thick platelets down to single layer graphene [49]. It is, however, the availability of single- or few-layer graphene that has caused the most excitement in recent times [50]. The addition of GNPs to polymers has been found to lead to substantial improvements in the thermal, mechanical and electrical properties at lower loadings than are needed with expanded graphite because of high intrinsic thermal conductivity, large specific surface area, and high two-dimensional sheet geometry [51–58].

Different types of fillers have been used for the improvement of the properties of furan-based polyesters, including nanoclays [23,57], nanocellulose [59–61], nanosized silica and titanium dioxide [28], graphene [62], graphene oxide and multi wall carbon nanotubes [63]. Recently, Paszkiewicz et al. [62] prepared PPF nanocomposites with low concentrations (0.1 and 0.3 wt.%) of few-layer graphene, and found that the addition of the filler did not change the glass transition temperature or the melting behavior of the polymer, but resulted in inhibition of the transport of oxygen molecules into and through the material. However, there is no information published about the effect of GNPs on the thermal degradation behavior and decomposition mechanism of PPF nanocomposites in the literature. Thermal stability of polymers and their nanocomposites is a crucial parameter that affects their thermal processing and their final applications.

In this work, neat PPF and PPF/GNP nanocomposites containing 0.5, 1, and 2.5 wt.% GNPs were synthesized in-situ by melt polycondensation. Then, the chemical structure of the resulting polymers was studied by ^{1}H-NMR spectroscopy. The effect of GNPs content on the thermal stability and thermal degradation of these polyesters along with a decomposition kinetics study was performed using thermogravimetric analysis (TGA), in order to reveal similarities and differences on their decomposition mechanism. Furthermore, pyrolysis-gas chromatography/mass spectroscopy (Py-GC/MS) was employed on the PPF/GNP nanocomposites in order to identify the individual fragments from each sample and obtain structural information concerning the decomposition mechanism.

2. Results and Discussion

2.1. Synthesis and Molecular Characterization

The PPF nanocomposites were synthesized via the two-step melt polycondensation method. The reaction procedure is presented in Scheme 1. The obtained nanocomposites were black solids.

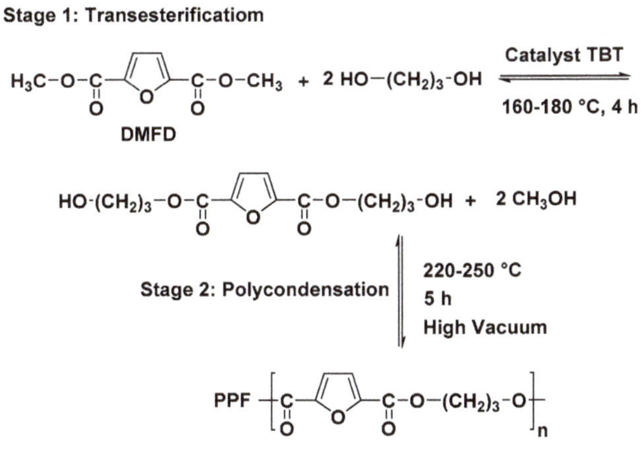

Scheme 1. Synthesis of PPF.

The intrinsic viscosity values of PPF and PPF/GNP nanocomposites with 0.5, 1, 2.5 wt.% GNPs were 0.62, 0.51, 0.5 and 0.57, respectively. The structure of the prepared materials was verified by ^{1}H-NMR spectroscopy, presented in Figure 1, along with the peak assignments. In the spectrum of neat PPF, the propylene glycol protons appear at lower values, at 4.75 ppm for the protons near the oxygen atoms (b) which are the most deprotected, and at 2.44 ppm for the (c) protons. As for the furan ring protons (a), they are the most deprotected, due to the π electron system of the ring and the carbonyl groups, and they appear at 7.45 ppm. The above results are in accordance with our previous work [23]. The same peaks were recorded for the nanocomposites thus confirming that the addition of GNPs had no effect on the molecular structure of the prepared polyesters.

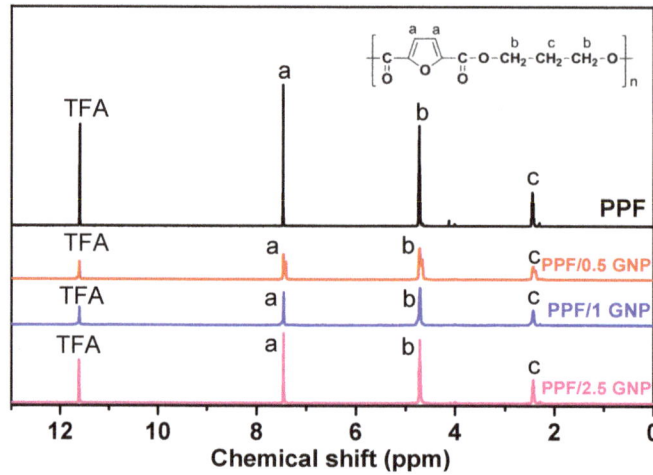

Figure 1. PPF structure, ^{1}H-NMR spectra and peak assignments of PPF and PPF/GNP nanocomposites.

The crystalline structure of the materials was assessed with WAXD analysis. As seen in the diffractograms of Figure 2, PPF and PPF/GNP nanocomposites were found to be amorphous, as they present a broad peak at 2θ = 22°. In the second diffractogram (Figure 2b) it is observed that the GNPs present a peak at 2θ = 26.5°, with d = 3.35 Å. In the nanocomposite materials, the same peak can be seen in all three polymers, and its intensity is increasing by increasing the loading of the material with the nanofiller, evidence of an intercalated polymer/GNPs structure. Similar results were shown in other studies examining graphene nanocomposites [64,65].

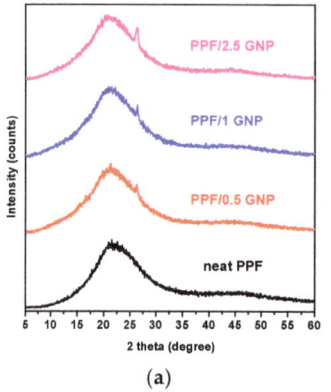

(a)

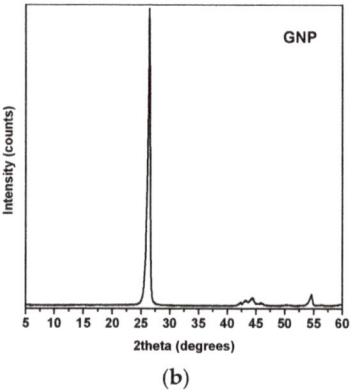

(b)

Figure 2. X-ray diffraction patterns of (a) PPF/GNP nanocomposites; (b) GNP.

2.2. Thermal Stability of PPF/GNP Nanocomposites

Thermogravimetric analysis (TGA) was performed with the objective of determining the thermal stability, as well as the influence of GNP content (0.5–2.5 wt.%) on the thermal properties of PPF. The TGA thermograms and derivative mass loss (dTG) curves of PPF/GNP nanocomposites with different filler content at a heating rate of 10 °C/min under nitrogen atmosphere are shown in Figure 3. In Figure 3a, the TGA curve of the GNPs is also presented (dashed curve) for comparative purposes. Analyzing the TGA results, it can be deduced that no remarkable mass loss has occurred until 325 °C, proving the excellent thermal stability of PPF-based materials. Mass loss curves of all studied samples are seemingly almost identical up to ~400 °C, consisting of a one-step procedure and obtaining the same curve shape. Above 400 °C the neat PPF and PPF/GNP nanocomposites decomposed. The residual amount of the nanocomposites increased with increasing the content of GNPs, indicating that thermal decomposition of PPF is retarded in the nanocomposites, due to the decomposition of GNPs that have initial degradation temperature above 600 °C (TGA thermograms of GNPs- dashed curve). From the dTG curves, it can be concluded that the degradation is carried out as a one-step process for all the studied samples as only one peak is observed. Concerning the effect of filler's content on the thermal properties of PPF, GNPs seem to have little effect on the temperature that the maximum decomposition rate takes place $T_{d,max}$, Table 1. A very small enhancement in thermal stability at the initial stage of degradation, of nearly 3 °C, and 6 °C appeared for PPF/0.5 GNP, and PPF/2.5GNP composites, respectively, using the temperature at 2% weight loss as a comparison point. Thermal stability is very important for polymeric materials as it is often the limiting factor both in processing and in end-use applications.

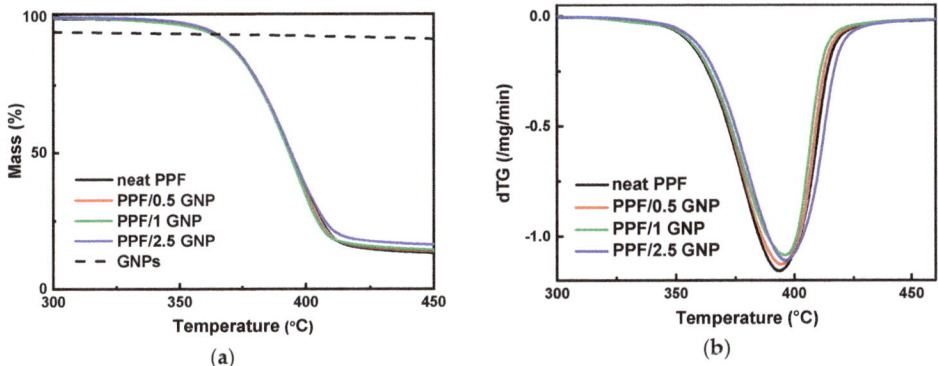

Figure 3. (a) TGA thermograms; (b) dTG curves of PPF/GNP nanocomposites with different weight fractions of GNPs recorded with the heating rate of 10 °C/min under a nitrogen atmosphere.

Table 1. TGA results of all studied samples.

Sample	$T_{d,max}$ (°C)
Neat PPF	393.2
PPF/0.5 GNP	394.2
PPF/1 GNP	395.9
PPF/2.5 GNP	396.3

2.3. Thermal Degradation Mechanism of Neat PPF and PPF/2.5GNP Nanocomposite

As was already concluded from the mass loss curves (TGA) and dTG of the PPF/GNP nanocomposites with 0.5, 1 and 2.5 wt.% of GNPs that the difference between them is very small. In order to analyze more thoroughly the degradation mechanism of PPF and PPF/2.5 GNP composite, the composite which presented the highest thermal stability enhancement, kinetic parameters (activation energy E, pre-exponential factor A) and the conversion function $f(\alpha)$ must be evaluated. So, neat PPF and PPF/2.5GNP nanocomposite was studied using a model fitting method. The relationship between kinetic parameters and extent of conversion (α) for neat PPF and PPF/2.5 GNP nanocomposite can be found using the mass curves recorded at four heating rates of 5, 10, 15, and 20 °C/min under nitrogen (Figure 4).

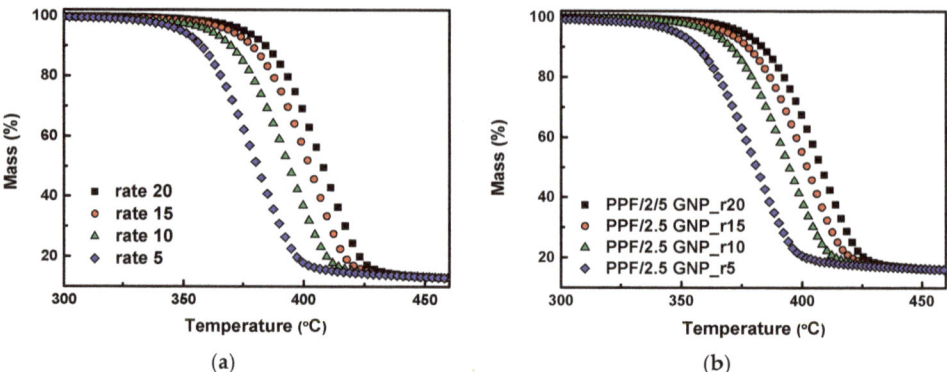

Figure 4. Mass (%) curves of (**a**) neat PPF; (**b**) PPF/2.5 GNP nanocomposite at heating rates of 5, 10, 15, and 20 °C/min under nitrogen.

For the determination of the activation energy by using multiple heating rates, two iso-conversional methods were used: (1) the integral method of Ozawa, Flynn and Wall (OFW) [66], and (2) the isoconversional method of Friedman [67]. The calculated values of activation energy versus the extent of conversion α for neat PPF and PPF/2.5 GNP nanocomposite is shown in Figure 5. The differences in the values of E calculated by the OFW and Friedman methods can be explained by a systematic error due to improper integration [68]. According to Figure 5, it seems that the calculated activation energy values of neat PPF and PPF/2.5 GNP nanocomposite slightly increase with increasing the extent of conversion "α" presenting less variability among the mean value, especially in the case of the OFW plot. The mean values of neat PPF and PPF/GNP nanocomposites were found to be 193.5 and 195.8 kJ/mol and 183.5 and 188.5 kJ/mol using Friedman and OFW methods, respectively.

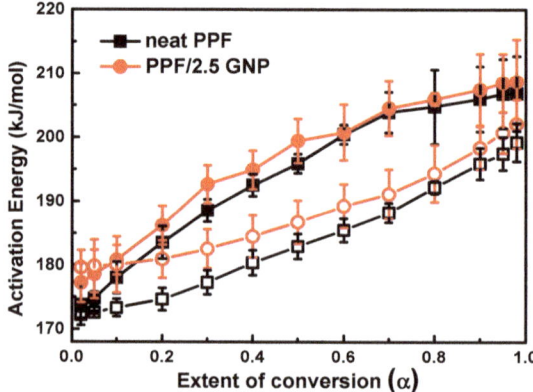

Figure 5. The dependence of activation energy (E) on the extent of conversion (α) for the thermal degradation of neat PPF and PPF/2.5GNP nanocomposite as calculated by Friedman method (filled symbols), and Ozawa Flynn Wall analysis (open symbols).

Model-fitting was complementary used in the thermal degradation studies of these materials. This method involves fitting different models to α versus temperature curves and simultaneously determining the activation energy E and the pre-exponential factor A. So, the kinetic model and the parameters for the four heating rates were determined by multivariate non-linear regression method. For this reason, 16 different reaction models were examined through the comparison of the experimental and theoretical data for the conversion range of $0 < \alpha < 1$. First, it is considered that the degradation of the samples can be described only by one mechanism, without presuming the exact mechanism. If the result of the fitting cannot be considered as accepted, then we must proceed to the fitting of the experimental data with a combination of two mechanisms.

The form of the conversion function, given by the best fitting for the neat PPF, is the mechanism of autocatalysis n-order (Cn) described by the equation $f(\alpha) = (1-\alpha)^n (1 + K_{cat} X)$, where K_{cat} is the autocatalysis rate constant and X the extent of conversion of the autocatalytic reactions. The results of the fitting can be seen for neat PPF (continuous black lines) in Figure 6 with a correlation coefficient of 0.9998; small divergences appear in the final stages of degradation. However, there isn't any further improvement in the quality of the fitting using two or more reaction mechanisms, since the differences among the regression coefficient values are rather small.

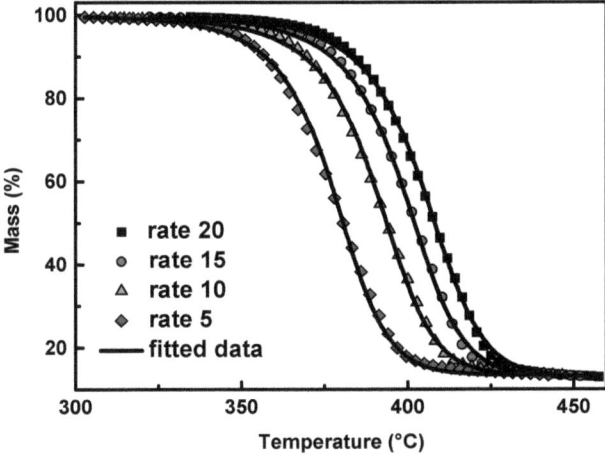

Figure 6. Mass (%) curves of neat PPF at heating rates of 5, 10, 15, and 20 °C/min in a nitrogen atmosphere (symbols) and the corresponding fitting curves with the Cn reaction model (continuous black lines).

Multivariate non-linear regression method also indicated that the Cn model, the mechanism of autocatalysis n-order, fit better to thermal decomposition of PPF/2.5 GNP nanocomposite with a correlation coefficient value of 0.9998, Figure 7. This is in consistence with the iso-conversional method (Figure 5) in which the activation energy of materials slightly increases with increasing the extent of conversion, suggesting that a single-step reaction mechanism may efficiently describe the degradation.

The calculated values of the activation energy, pre-exponential factor and the reaction order for PPF and PPF/2.5 GNP nanocomposite using the nth-order with autocatalysis (Cn) model are summarized in Table 2. It was found that the same model describes the reaction mechanism of both neat PPF and PPF/2.5 GNP nanocomposite. It is worthwhile noting that that the activation energy values are close to the ones calculated from OFW analysis, as well as Friedman's method (Figure 5). Comparing the activation energies, PPF/2.5 GNP nanocomposite has slightly higher values than neat PPF. This improvement in thermal stability of nanocomposite is associated with the 2-dimensional planar structure of GNPs. The mobility of polymer matrix is restricted, and the chemical reactivity of the corresponding chains is lower increasing activation energy and eventually retarding more the thermal degradation of the nanocomposites.

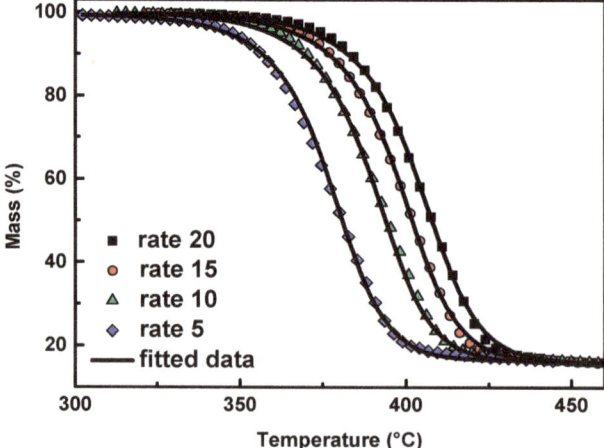

Figure 7. Mass (%) curves of PPF/2.5 GNP nanocomposite at heating rates of 5, 10, 15, and 20 °C/min in a nitrogen atmosphere (symbols) and the corresponding fitting curves with the Cn model (continuous black lines).

Table 2. Activation energy, pre-exponential factor and reaction order of neat PPF and PPF/2.5 GNP nanocomposite.

Sample	Mechanism	Activation Energy/kJmol^{-1}	Pre-exponential Factor/s^{-1}	Reaction Order/n	Log K$_{cat}$	Regression Coefficient
Neat PPF	Cn	187.0	12.2	1.36	0.81	0.9998
PPF/2.5 GNP	Cn	191.1	12.6	1.43	0.82	0.9998

The degradation mechanism was afterwards studied with Py-GC/MS measurements. The compounds are pyrolyzed in a pre-selected temperature and followed by the separation of the evolved pyrolysis products by a GC capillary column and subsequent detection via mass spectroscopy. This method enables the determination of the exact degradation routes that take place when a polymeric material is heated in high temperatures in inert atmosphere.

The degradation mechanisms of furan-based polyesters were first studied by our group, including this of PPF [23,69]. In general, they follow the degradation paths of their terephthalate homologues. Similarly to all polyesters with β-hydrogen atoms on their macromolecular chains, they degrade mainly by heterolytic scission reactions of the hydrogen in β position to the ester bond [70–73]. This degradation route leads to the evolution of vinyl-ended and carboxyl-ended products. Additionally, homolytic scission reactions of the acyl-oxygen and alkyl-oxygen bonds can occur, especially under higher pyrolysis temperatures [24,31].

Total ion chromatographs of PPF and PPF/2.5 GNP after pyrolysis at 360 °C and 400 °C are presented in Figure 8 and the corresponding compounds, identified via their mass spectra, are presented in Table 3. As expected, the main pyrolysis products are vinyl- and carboxyl- ended molecules that result from β-scission reactions (Scheme 2). Other classes of products are some hydroxyl-ended, methoxy-ended molecules as well as some aldehydes. It should be noted that since the mass spectra of the molecular ions of aldehydes are very weak, we cannot be completely confident about their identification even though they are known to be released from polyesters during heating. All the possible degradation routes have been explained in detail in our previous publications [22,24,25,31,69,74].

In both samples, the pyrolysis products identified were identical, meaning that GNP doesn't affect the degradation mechanism of PPF. The slight increase in the activation of energy calculated by TGA is caused by the ability of graphene layers to hinder gas diffusion in the polymeric matrix as they can reduce chain mobility [75,76]. When comparing the GC patterns of PPF and PPF/2.5 GNP it is observed that while they don't have any noteworthy differences at 400 °C, there is a significant increase in the intensity of some peaks, in the presence of GNP at 360 °C suggesting their larger relative amount. Those peaks in Rt = 7.00, 7.35, 11.19, 11.30, 17.76, 20.27, 21.19 and 21.58 min are highlighted in Figure 8a and the majority was identified as compounds that are produced via secondary degradation routes, including -OH ended compounds from acyl-oxygen homolysis and aldehydes from α-hydrogen bond scission. Therefore, the presence of GNP in the PPF matrix resulted in a pronounced occurrence of homolytic degradation reactions in comparison with neat PPF.

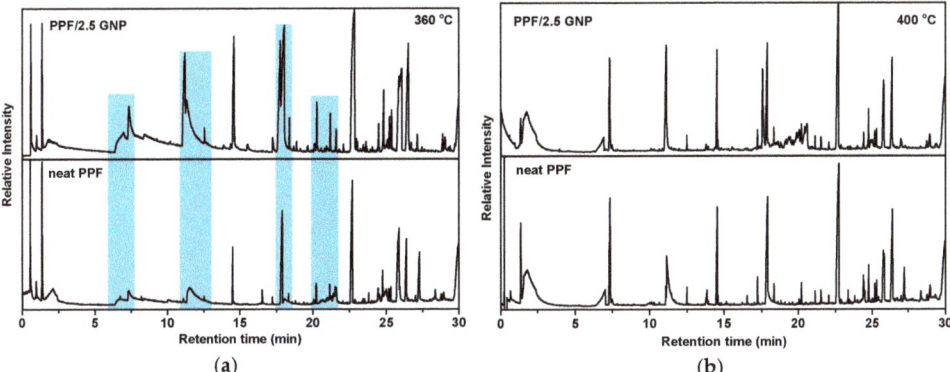

Figure 8. Total Ion Gas chromatographs of neat PPF and PPF/2.5 GNP after pyrolysis at (**a**) 360 °C; (**b**) 400 °C. The highlighted peaks correspond to compounds that are evolved via secondary degradation routes.

Table 3. Possible pyrolysis products of neat PPF and PPF/2.5 GNP nanocomposite.

PPF Rt (min)		PPF/2.5 GNP Rt (min)		Mw (amu)	Possible Product
360	400	360	400		
-	0.24			44	CO_2 or acetaldehyde
0.55	-	0.59		76	1,3-propanediol
1.36	1.36	1.37	1.37	44	CO_2 or acetaldehyde
-	1.8	1.79	1.78	68	furan
6.74	7.02	7.00	6.93	112	furan-2-carboxylic acid
7.33	7.35	7.35	7.34	152	allyl furan-2-carboxylate
11.09	11.09	11.19	11.13	162	di(furan-2-yl)methanone

Table 3. Cont.

PPF Rt (min)		PPF/2.5 GNP Rt (min)		Mw (amu)	Possible Product
360	400	360	400		
11.51	11.16	11.30	-	170	
12.50	12.51	12.53	12.51	190	
-	13.88	13.88	13.82	196	
14.52	14.57	14.61	14.56	236	
-	14.62	14.65	14.62	238	
-	-	17.76	17.58	254	
17.89	17.91	18.05	17.90	264	
20.22	-	20.27	20.23	322	
21.14	21.15	21.19	21.15	256	
21.52	21.54	21.58	21.54	336	
22.69	22.73	22.83	22.70	348	
24.42	24.50	24.47	24.43	388	
24.75	24.77	24.84	24.76	358	unidentified
25.18	25.19	25.23	25.19	366	
25.31	25.32	25.38	25.32	350	

Table 3. Cont.

PPF Rt (min)		PPF/2.5 GNP Rt (min)		Mw (amu)	Possible Product
360	400	360	400		
25.88	25.85	25.88	25.81	392	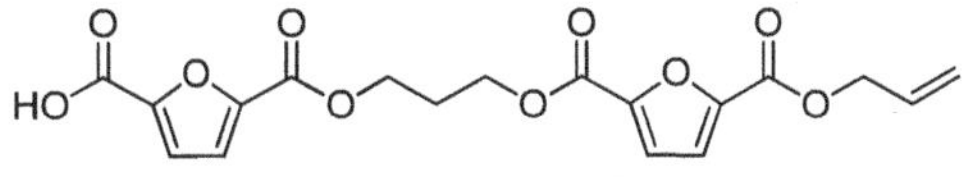
26.38	26.43	26.53	26.41	432	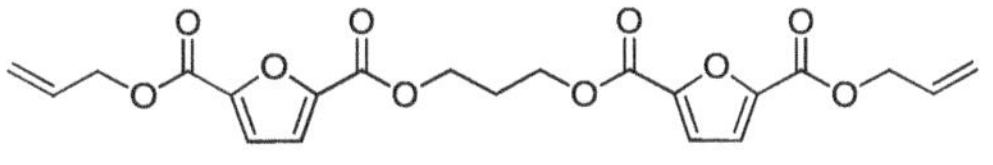
27.28	27.25	27.14	27.00	434	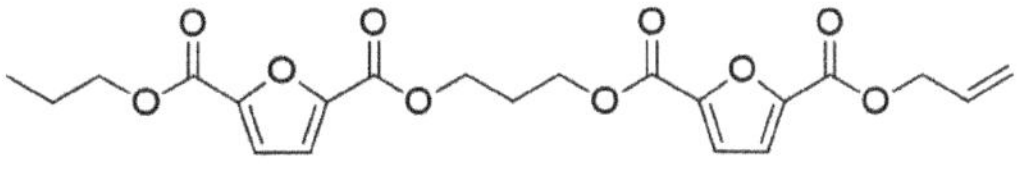

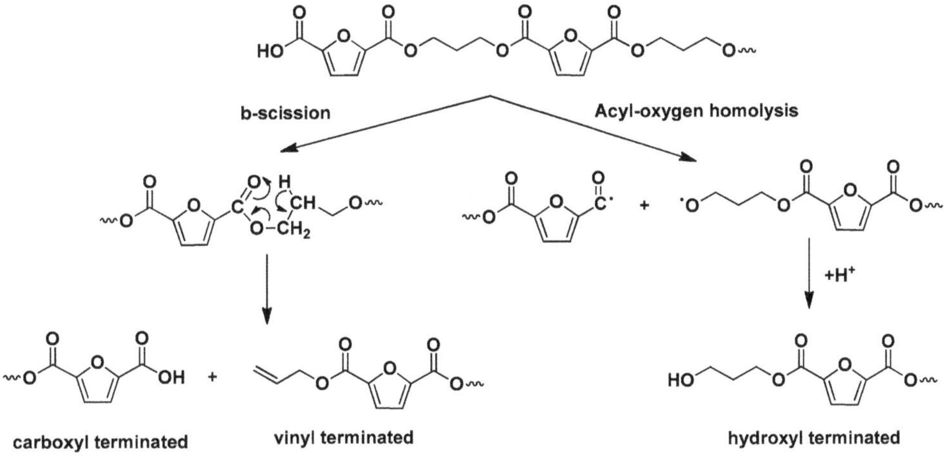

Scheme 2. Basic degradation routes of PPF.

3. Materials and Methods

3.1. Materials

2,5-Furandicarboxylic acid (purum 97%), propylene glycol anhydrous 99.6% (PG) and tetrabutyltitanate (TBT) catalyst of analytical grade were purchased from Sigma-Aldrich (Taufkirchen, Germany). GNPs with the trade name xGnP® - Grade M were supplied by XGSciences (Lansing MI, USA). According to the manufacturer, the nanoplatelets have an average thickness of 6–8 nm, surface area 120–150 m^2/g and average particle diameter of 5 μm. All other materials and solvents used were of analytical grade and were purchased from Sigma-Aldrich.

3.2. Synthesis of 2,5-Dimethylfuran-dicarboxylate (DMFD)

2,5-Furandicarboxylic acid (15.6 g), anhydrous methanol (200 mL) and concentrated sulfuric acid (2 mL) were placed in a round bottom flask (500 mL) and the mixture was refluxed for 5 h. The excess of the methanol was distilled off, and the solution was filtered through a disposable Teflon membrane filter. During filtration, dimethylfurandicarboxylate (DMFD) was precipitated as white powder and, after cooling, distilled water (100 mL) was added. The dispersion was partially neutralized by adding Na_2CO_3 5% *w/v* during stirring, while pH was measured continuously. The white powder was filtered and the solid was washed several times with distilled water and dried. The isolated white dimethyl ester was recrystallized with a mixture of 50/50 v/v methanol/water. After cooling, 2,5-DMFD was precipitated in the form of white needles. The reaction yield was calculated at 75%.

3.3. PPF and Nanocomposites Synthesis

PPF was synthesized through the two-stage melt polycondensation (esterification and polycondensation) in a glass batch reactor [21]. DMFD and propylene glycol in a molar ratio of diester/diol = 1/2.2 were charged into the reaction tube of the polyesterification apparatus with 500 ppm of TBT. The reaction mixture was heated at 160 °C under argon flow for 1.5 h, at 170 °C for additional 1.5 h and finally at 180 °C for 2 h. This first step (transesterification) is considered complete after the collection of almost all the theoretical amount of CH_3OH, which was removed from the reaction mixture by distillation and collected in a graduate cylinder. In the second step of polycondensation, vacuum (5.0 Pa) was applied slowly over a time of about 30 min to remove the excess of diol, to avoid excessive foaming and to minimize oligomer sublimation, which is a potential problem during the melt polycondensation. The temperature was gradually increased (1.5 h) to 220 °C, while stirring speed was also increased to 720 rpm. The reaction continued at this temperature for 1.5 h. Successively, the temperature was increased to 235 °C for 1.5 h and to 250 °C for additional 2 h. PPF-based GNP nanocomposites containing 0.5, 1 and 2.5 wt.% of GNPs were in-situ prepared using also the two-stage melt polycondensation method. Nanofillers were added to the propylene glycol and the dispersion was subjected to sonication for 15 min to obtain a uniform dispersion. Afterwards, the dispersion was added to the reaction tube together with DMFD and TBT catalyst. The reaction continued, as above described for the synthesis of neat PPF. After the polycondensation reaction was completed, neat PPF and PPF/GNP nanocomposites were easily removed, milled and washed with methanol.

3.4. Intrinsic Viscosity Measurements

Intrinsic viscosity $[\eta]$ measurements were performed using a Cannon Ubbelohde viscometer (State College, PA, USA) at 30 °C in a mixture of phenol/1,1,2,2-tetrachloroethane (60/40, w/w). The sample was kept in the above-mentioned mixture at 90 °C until complete dissolution was achieved. The solution was then cooled to room temperature and filtered through a Teflon disposable membrane filter.

3.5. Nuclear Magnetic Resonance (NMR)

^{1}H-NMR spectra of polyesters were obtained with a Bruker spectrometer (Billerica, MA, USA) operating at a frequency of 400 MHz. A sample concentration equal to 5% w/v in deuterated trifluoroacetic acid (d-TFA) was used. The number of scans was 10 and the sweep width was 6 kHz.

3.6. Wide Angle X-Ray Diffraction Patterns (WAXD)

X-ray diffraction measurements of the samples were performed using a MiniFlex II XRD system from Rigaku Co. (Tokyo, Japan), with CuK_α radiation (λ = 0.154 nm) in the angle 2θ range from 5° to 60°.

3.7. Thermogravimetric Analysis (TGA)

Thermogravimetric analysis of the PPF and PPF/GNP nanocomposites were carried out using a SETARAM SETSYS TG-DTA 16/18 instrument (Caluire, France) by heating the samples from 25 to 600 °C in a 50 mL/min flow of N_2 at a heating rate of 10 °C/min. For the kinetic analysis study, neat PPF and PPF/GNP nanocomposites with 2.5 wt.% filler content (PPF/2.5 GNP) were heated at four different heating rates, namely 5, 10, 15, and 20 °C/min. Samples (4.5 ± 0.5 mg) were placed in alumina crucibles, while an empty alumina crucible was used as a reference. Continuous recordings of sample temperature, sample weight, first derivative, and heat flow were taken.

3.8. Pyrolysis-Gas Chromatography–Mass Spectroscopy (Py-GC/MS)

For Py-GC/MS analysis of polyesters a very small amount of each material is "dropped" initially into the "Double-Shot" EGA/PY-3030D Pyrolyzer (Frontier Laboratories Ltd., Fukushima Japan) using a CGS-1050Ex (Kyoto, Japan) carrier gas selector. For pyrolysis analysis (flash pyrolysis) each sample was placed into the sample cup which afterwards fell free into the Pyrolyzer furnace. The pre-selected pyrolysis temperatures were 360 and 400 °C and the GC oven temperature was heated from 50 to 300 °C at 20 °C/min. Those two temperatures were selected based on the EGA pyrogram and represent the sample prior and after thermal decomposition. Sample vapors generated in the furnace were split (at a ratio of 1/50), a portion moved to the column at a flow rate of 1 mL/min, pressure 53.6 kPa and the remaining portion exited the system via the vent. The pyrolyzates were separated using temperature programmed capillary column of a Shimadzu QP-2010 Ultra Plus (Kyoto, Japan) gas chromatograph and analysed by a Shimadzu MS-QP2010SE mass spectrometer at 70 eV. Ultra ALLOY® metal capillary column from Frontier Laboratories Ltd. (Fukushima, Japan) was used containing 5% diphenyl and 95% dimethylpolysiloxane stationary phase, column length 30 m and column ID 0.25 mm. For the mass spectrometer the following conditions were used: Ion source heater 200 °C, interface temperature 300 °C, vacuum 10^{-4}–10^{0} Pa, m/z range 10–500 amu and scan speed 10.000. The chromatograph and spectra retrieved by each experiment were subjected to further interpretation through Shimadzu and Frontier post-run software (Kyoto, Japan).

4. Conclusions

In this work, PPF/GNP nanocomposites containing 0.5, 1 and 2.5 wt.% GNPs were successfully synthesized via the in-situ transesterification and polycondensation method. The addition of GNPs did not affect the intrinsic viscosity, or the chemical structure of the nanocomposites as shown by the ^{1}H-NMR spectra and the viscosity measurements. The crystallinity of the as received materials was assessed by WAXD measurements. The diffractograms showed that the materials are amorphous. Thermal properties with focus on thermal stability and degradation mechanism were evaluated. All samples had strong thermal stability, as no remarkable mass loss was observed until 325 °C and have been decomposed in similar one-step procedures as shown in the dTG curves. The study of degradation kinetics through thermogravimetry revealed a small increase in the activation energy value for the nanocomposite with 2.5 wt.% GNP in comparison with neat PPF. The nanofillers did not alter the degradation pathways of PPF, as both the β-scission and the acyl-oxygen homolysis occur. Instead, they affected the balance between the primary heterolytic scission reactions and the homolytic degradation routes. The presence of GNP in the PPF matrix resulted in a pronounced occurrence of homolytic degradation reactions in comparison with neat PPF as more -OH ended compounds were detected in the chromatographs.

Author Contributions: Conceptualization, D.N.B. and D.G.P.; methodology, L.P., Z.T. and E.T.; investigation, Z.T., E.T., L.P. and N.K.; formal analysis, Z.T. and E.T.; writing—original draft preparation, E.T., Z.T. and L.P.; writing—review and editing, Z.T., G.Z.P. and D.G.P.; supervision, D.N.B. and K.C.

Conflicts of Interest: The authors declare no conflict of interest.

References

1. Bornscheuer, U.T. Feeding on plastic. *Science* **2016**, *351*, 1154–1155. [CrossRef]
2. Zhu, Y.; Romain, C.; Williams, C.K. Sustainable polymers from renewable resources. *Nature* **2016**, *540*, 354–362. [CrossRef]
3. Schneiderman, D.K.; Hillmyer, M.A. 50th Anniversary Perspective: There Is a Great Future in Sustainable Polymers. *Macromolecules* **2017**, *50*, 3733–3749. [CrossRef]
4. Gandini, A.; Silvestre, A.J.D.; Neto, C.P.; Sousa, A.F.; Gomes, M. The furan counterpart of poly(ethylene terephthalate): An alternative material based on renewable resources. *J. Polym. Sci. Part A Polym. Chem.* **2009**, *47*, 295–298. [CrossRef]

5. Haider, T.P.; Völker, C.; Kramm, J.; Landfester, K.; Wurm, F.R. Plastics of the Future? The Impact of Biodegradable Polymers on the Environment and on Society. *Angew. Chem. Int. Ed.* **2019**, *58*, 50–62. [CrossRef]
6. Sustainability. Available online: https://ec.europa.eu/epsc/topics/sustainability_en (accessed on 28 March 2019).
7. Xie, H.; Wu, L.; Li, B.-G.; Dubois, P. Biobased Poly(ethylene-co-hexamethylene 2,5-furandicarboxylate) (PEHF) Copolyesters with Superior Tensile Properties. *Ind. Eng. Chem. Res.* **2018**, *57*, 13094–13102. [CrossRef]
8. Xie, H.; Wu, L.; Li, B.-G.; Dubois, P. Poly(ethylene 2,5-furandicarboxylate-mb-poly(tetramethylene glycol)) multiblock copolymers: From high tough thermoplastics to elastomers. *Polymer (Guildf)* **2018**, *155*, 89–98. [CrossRef]
9. Bio-Based Products. Available online: http://ec.europa.eu/growth/sectors/biotechnology/bio-based-products_is (accessed on 28 March 2019).
10. Legrand, S.; Jacquel, N.; Amedro, H.; Saint-Loup, R.; Pascault, J.-P.; Rousseau, A.; Fenouillot, F. Synthesis and properties of poly (1,4-cyclohexanedimethylene-co-isosorbide terephthalate), a biobased copolyester with high performances. *Eur. Polym. J.* **2019**, *115*, 22–29. [CrossRef]
11. Blache, H.; Méchin, F.; Rousseau, A.; Fleury, É.; Pascault, J.-P.; Alcouffe, P.; Jacquel, N.; Saint-Loup, R. New bio-based thermoplastic polyurethane elastomers from isosorbide and rapeseed oil derivatives. *Ind. Crops Prod.* **2018**, *121*, 303–312. [CrossRef]
12. Delidovich, I.; Hausoul, P.J.C.; Deng, L.; Pfützenreuter, R.; Rose, M.; Palkovits, R. Alternative monomers based on lignocellulose and their use for polymer production. *Chem. Rev.* **2015**, *116*, 1540–1599. [CrossRef]
13. Gandini, A.; Lacerda, T.M.; Carvalho, A.J.F.; Trovatti, E. Progress of Polymers from Renewable Resources: Furans, Vegetable Oils, and Polysaccharides. *Chem. Rev.* **2016**, *116*, 1637–1669. [CrossRef]
14. Agriculture, U.S.D. of WHAT IS BIOPREFERRED? Available online: https://www.biopreferred.gov/BioPreferred/faces/pages/AboutBioPreferred.xhtml (accessed on 28 March 2019).
15. Shi, Z.; Jia, C.; Wang, D.; Deng, J.; Xu, G.; Wu, C.; Dong, M.; Guo, Z. Synthesis and characterization of porous tree gum grafted copolymer derived from Prunus cerasifera gum polysaccharide. *Int. J. Biol. Macromol.* **2019**, *133*, 964–970. [CrossRef]
16. Hu, Q.; Zhou, N.; Gong, K.; Liu, H.; Liu, Q.; Sun, D.; Wang, Q.; Shao, Q.; Liu, H.; Qiu, B.; et al. Intracellular Polymer Substances Induced Conductive Polyaniline for Improved Methane Production from Anaerobic Wastewater Treatment. *ACS Sustain. Chem. Eng.* **2019**, *7*, 5912–5920. [CrossRef]
17. Werpy, T.; Petersen, G.; Aden, A.; Bozell, J.; Holladay, J.; White, J.; Manheim, A.; Eliot, D.; Lasure, L.; Jones, S. *Top Value Added Chemicals from Biomass. Volume 1-Results of Screening for Potential Candidates from Sugars and Synthesis Gas*; Department of Energy: Washington, DC, USA, 2004.
18. Pan, T.; Deng, J.; Xu, Q.; Zuo, Y.; Guo, Q.; Fu, Y. Catalytic Conversion of Furfural into a 2,5-Furandicarboxylic Acid-Based Polyester with Total Carbon Utilization. *ChemSusChem* **2013**, *6*, 47–50. [CrossRef]
19. Papageorgiou, G.Z.; Tsanaktsis, V.; Papageorgiou, D.G.; Chrissafis, K.; Exarhopoulos, S.; Bikiaris, D.N. Furan-based polyesters from renewable resources: Crystallization and thermal degradation behavior of poly(hexamethylene 2,5-furan-dicarboxylate). *Eur. Polym. J.* **2014**, *67*, 383–396. [CrossRef]
20. Papageorgiou, G.Z.; Tsanaktsis, V.; Papageorgiou, D.G.; Exarhopoulos, S.; Papageorgiou, M.; Bikiaris, D.N. Evaluation of polyesters from renewable resources as alternatives to the current fossil-based polymers. Phase transitions of poly(butylene 2,5-furan-dicarboxylate). *Polymer (United Kingdom)* **2014**, *55*, 3846–3858. [CrossRef]
21. Papageorgiou, G.Z.; Papageorgiou, D.G.; Tsanaktsis, V.; Bikiaris, D.N. Synthesis of the bio-based polyester poly(propylene 2,5-furan dicarboxylate). Comparison of thermal behavior and solid state structure with its terephthalate and naphthalate homologues. *Polymer (United Kingdom)* **2015**, *62*, 28–38. [CrossRef]
22. Tsanaktsis, V.; Terzopoulou, Z.; Exarhopoulos, S.; Bikiaris, D.N.; Achilias, D.S.; Papageorgiou, D.G.; Papageorgiou, G.Z. Sustainable, eco-friendly polyesters synthesized from renewable resources: Preparation and thermal characteristics of poly(dimethyl-propylene furanoate). *Polym. Chem.* **2015**, *6*, 8284–8296. [CrossRef]
23. Papadopoulos, L.; Terzopoulou, Z.; Bikiaris, D.N.; Patsiaoura, D.; Chrissafis, K.; Papageorgiou, D.G.; Papageorgiou, G.Z. Synthesis and Characterization of In-Situ-Prepared Nanocomposites Based on Poly(Propylene 2,5-Furan Dicarboxylate) and Aluminosilicate Clays. *Polymers (Basel)* **2018**, *10*, 937. [CrossRef]

24. Terzopoulou, Z.; Kasmi, N.; Tsanaktsis, V.; Doulakas, N.; Bikiaris, D.N.; Achilias, D.S.; Papageorgiou, G.Z. Synthesis and characterization of bio-based polyesters: Poly(2-methyl-1,3-propylene-2,5-furanoate), Poly(isosorbide-2,5-furanoate), Poly(1,4-cyclohexanedimethylene-2,5-furanoate). *Materials (Basel)* **2017**, *10*, 801. [CrossRef]
25. Terzopoulou, Z.; Karakatsianopoulou, E.; Kasmi, N.; Majdoub, M.; Papageorgiou, G.Z.; Bikiaris, D.N. Effect of catalyst type on recyclability and decomposition mechanism of poly(ethylene furanoate) biobased polyester. *J. Anal. Appl. Pyrolysis* **2017**, *126*, 357–370. [CrossRef]
26. Kasmi, N.; Papageorgiou, G.Z.; Achilias, D.S.; Bikiaris, D.N. Solid-State polymerization of poly(Ethylene Furanoate) biobased Polyester, II: An efficient and facile method to synthesize high molecular weight polyester appropriate for food packaging applications. *Polymers (Basel)* **2018**, *10*, 471. [CrossRef]
27. Terzopoulou, Z.; Karakatsianopoulou, E.; Kasmi, N.; Tsanaktsis, V.; Nikolaidis, N.; Kostoglou, M.; Papageorgiou, G.Z.; Lambropoulou, D.A.; Bikiaris, D.N. Effect of catalyst type on molecular weight increase and coloration of poly(ethylene furanoate) biobased polyester during melt polycondensation. *Polym. Chem.* **2017**, *8*, 6895–6908. [CrossRef]
28. Achilias, D.S.; Chondroyiannis, A.; Nerantzaki, M.; Adam, K.V.; Terzopoulou, Z.; Papageorgiou, G.Z.; Bikiaris, D.N. Solid State Polymerization of Poly(Ethylene Furanoate) and Its Nanocomposites with SiO_2 and TiO_2. *Macromol. Mater. Eng.* **2017**, *302*, 1–15. [CrossRef]
29. Tsanaktsis, V.; Bikiaris, D.N.; Guigo, N.; Exarhopoulos, S.; Papageorgiou, D.G.; Sbirrazzuoli, N.; Papageorgiou, G.Z. Synthesis, properties and thermal behavior of poly(decylene-2,5-furanoate): A biobased polyester from 2,5-furan dicarboxylic acid. *RSC Adv.* **2015**, *5*, 74592–74604. [CrossRef]
30. Papageorgiou, G.Z.; Terzopoulou, Z.; Tsanaktsis, V.; Achilias, D.S.; Triantafyllidis, K.; Diamanti, E.K.; Gournis, D.; Bikiaris, D.N. Effect of graphene oxide and its modification on the microstructure, thermal properties and enzymatic hydrolysis of poly(ethylene succinate) nanocomposites. *Thermochim. Acta* **2015**, *614*, 116–128. [CrossRef]
31. Terzopoulou, Z.; Tsanaktsis, V.; Nerantzaki, M.; Papageorgiou, G.Z.; Bikiaris, D.N. Decomposition mechanism of polyesters based on 2,5-furandicarboxylic acid and aliphatic diols with medium and long chain methylene groups. *Polym. Degrad. Stab.* **2016**, *132*, 127–136. [CrossRef]
32. Bikiaris, D.N.; Papageorgiou, G.Z.; Giliopoulos, D.J.; Stergiou, C.A. Correlation between Chemical and Solid-State Structures and Enzymatic Hydrolysis in Novel Biodegradable Polyesters. The Case of Poly(propylene alkanedicarboxylate)s. *Macromol. Biosci.* **2008**, *8*, 728–740. [CrossRef]
33. Haas, T.; Jaeger, B.; Weber, R.; Mitchell, S.F.; King, C.F. New diol processes: 1,3-propanediol and 1,4-butanediol. *Appl. Catal. A Gen.* **2005**, *280*, 83–88. [CrossRef]
34. Kluge, M.; Pérocheau Arnaud, S.; Robert, T. 1,3-Propanediol and its Application in Bio-Based Polyesters for Resin Applications. *Chem. Africa* **2018**. [CrossRef]
35. Wang, B.; Li, C.Y.; Hanzlicek, J.; Cheng, S.Z.D.; Geil, P.H.; Grebowicz, J.; Ho, R.-M. Poly(trimethylene terephthalate) crystal structure and morphology in different length scales. *Polymer (Guildf)* **2001**, *42*, 7171–7180. [CrossRef]
36. Ward, I.M.; Wilding, M.A.; Brody, H. The mechanical properties and structure of poly(m-methylene terephthalate) fibers. *J. Polym. Sci. Polym. Phys. Ed.* **1976**, *14*, 263–274. [CrossRef]
37. Nederberg, F.; Bell, R.L.; Torradas, J.M. Furan-based Polymeric Hydrocarbon Fuel Barrier Structures. CN105848891A, 10 August 2016.
38. Sustainable Bioplastics. Available online: http://sorona.com/ (accessed on 28 March 2019).
39. Spitalsky, Z.; Tasis, D.; Papagelis, K.; Galiotis, C. Carbon nanotube–polymer composites: Chemistry, processing, mechanical and electrical properties. *Prog. Polym. Sci.* **2010**, *35*, 357–401. [CrossRef]
40. Otaegi, I.; Aramburu, N.; Müller, A.; Guerrica-Echevarría, G. Novel Biobased Polyamide 410/Polyamide 6/CNT Nanocomposites. *Polymers (Basel)* **2018**, *10*, 986. [CrossRef] [PubMed]
41. Dong, M.; Li, Q.; Liu, H.; Liu, C.; Wujcik, E.K.; Shao, Q.; Ding, T.; Mai, X.; Shen, C.; Guo, Z. Thermoplastic polyurethane-carbon black nanocomposite coating: Fabrication and solid particle erosion resistance. *Polymer (Guildf)* **2018**, *158*, 381–390. [CrossRef]
42. Ma, L.; Zhu, Y.; Wang, M.; Yang, X.; Song, G.; Huang, Y. Enhancing interfacial strength of epoxy resin composites via evolving hyperbranched amino-terminated POSS on carbon fiber surface. *Compos. Sci. Technol.* **2019**, *170*, 148–156. [CrossRef]

43. Ma, L.; Li, N.; Wu, G.; Song, G.; Li, X.; Han, P.; Wang, G.; Huang, Y. Interfacial enhancement of carbon fiber composites by growing TiO$_2$ nanowires onto amine-based functionalized carbon fiber surface in supercritical water. *Appl. Surf. Sci.* **2018**, *433*, 560–567. [CrossRef]
44. Wu, Z.; Cui, H.; Chen, L.; Jiang, D.; Weng, L.; Ma, Y.; Li, X.; Zhang, X.; Liu, H.; Wang, N.; et al. Interfacially reinforced unsaturated polyester carbon fiber composites with a vinyl ester-carbon nanotubes sizing agent. *Compos. Sci. Technol.* **2018**, *164*, 195–203. [CrossRef]
45. Zhang, J.; Zhang, Z.; Jiao, Y.; Yang, H.; Li, Y.; Zhang, J.; Gao, P. The graphene/lanthanum oxide nanocomposites as electrode materials of supercapacitors. *J. Power Sources* **2019**, *419*, 99–105. [CrossRef]
46. Liu, M.; Meng, Q.; Yang, Z.; Zhao, X.; Liu, T. Ultra-long-term cycling stability of an integrated carbon-sulfur membrane with dual shuttle-inhibiting layers of graphene "nets" and a porous carbon skin. *Chem. Commun.* **2018**, *54*, 5090–5093. [CrossRef]
47. Wang, Y.; Li, Z.; Wang, J.; Li, J.; Lin, Y. Graphene and graphene oxide: Biofunctionalization and applications in biotechnology. *Trends Biotechnol.* **2011**, *29*, 205–212. [CrossRef]
48. Rao, C.N.R.; Sood, A.K.; Subrahmanyam, K.S.; Govindaraj, A. Graphene: The New Two-Dimensional Nanomaterial. *Angew. Chem. Int. Ed.* **2009**, *48*, 7752–7777. [CrossRef]
49. Jang, B.Z.; Zhamu, A. Processing of nanographene platelets (NGPs) and NGP nanocomposites: A review. *J. Mater. Sci.* **2008**, *43*, 5092–5101. [CrossRef]
50. Novoselov, K.S.; Geim, A.K.; Morozov, S.V.; Jiang, D.; Zhang, Y.; Dubonos, S.V.; Grigorieva, I.V.; Firsov, A.A. Electric field effect in atomically thin carbon films. *Science* **2004**, *306*, 666–669. [CrossRef]
51. Kalaitzidou, K.; Fukushima, H.; Drzal, L.T. A new compounding method for exfoliated graphite–polypropylene nanocomposites with enhanced flexural properties and lower percolation threshold. *Compos. Sci. Technol.* **2007**, *67*, 2045–2051. [CrossRef]
52. Lee, C.; Wei, X.; Kysar, J.W.; Hone, J. Measurement of the elastic properties and intrinsic strength of monolayer graphene. *Science* **2008**, *321*, 385–388. [CrossRef]
53. Young, R.J.; Liu, M.; Kinloch, I.A.; Li, S.; Zhao, X.; Vallés, C.; Papageorgiou, D.G. The mechanics of reinforcement of polymers by graphene nanoplatelets. *Compos. Sci. Technol.* **2018**, *154*, 110–116. [CrossRef]
54. Liu, M.; Papageorgiou, D.G.; Li, S.; Lin, K.; Kinloch, I.A.; Young, R.J. Micromechanics of reinforcement of a graphene-based thermoplastic elastomer nanocomposite. *Compos. Part A Appl. Sci. Manuf.* **2018**, *110*, 84–92. [CrossRef]
55. Young, R.J.; Kinloch, I.A.; Gong, L.; Novoselov, K.S. The mechanics of graphene nanocomposites: A review. *Compos. Sci. Technol.* **2012**, *72*, 1459–1476. [CrossRef]
56. Marsden, A.J.; Papageorgiou, D.G.; Vallés, C.; Liscio, A.; Palermo, V.; Bissett, M.A.; Young, R.J.; Kinloch, I.A. Electrical percolation in graphene–polymer composites. *2D Mater.* **2018**, *5*, 32003. [CrossRef]
57. Martino, L.; Guigo, N.; van Berkel, J.G.; Sbirrazzuoli, N. Influence of organically modified montmorillonite and sepiolite clays on the physical properties of bio-based poly(ethylene 2,5-furandicarboxylate). *Compos. Part B Eng.* **2017**, *110*, 96–105. [CrossRef]
58. Papageorgiou, D.G.; Kinloch, I.A.; Young, R.J. Graphene/elastomer nanocomposites. *Carbon* **2015**, *95*, 460–484. [CrossRef]
59. Codou, A.; Guigo, N.; van Berkel, J.G.; de Jong, E.; Sbirrazzuoli, N. Preparation and characterization of poly (ethylene 2,5-furandicarboxylate/nanocrystalline cellulose composites via solvent casting. *J. Polym. Eng.* **2017**, *37*, 869–878. [CrossRef]
60. Codou, A.; Guigo, N.; van Berkel, J.G.; De Jong, E.; Sbirrazzuoli, N. Preparation and crystallization behavior of poly (ethylene 2,5-furandicarboxylate)/cellulose composites by twin screw extrusion. *Carbohydr. Polym.* **2017**, *174*, 1026–1033. [CrossRef]
61. Matos, M.; Sousa, A.F.; Silva, N.H.C.S.; Freire, C.S.R.; Andrade, M.; Mendes, A.; Silvestre, A.J.D. Furanoate-based nanocomposites: A case study using poly(butylene 2,5-furanoate) and poly(butylene 2,5-furanoate)-co-(butylene diglycolate) and bacterial cellulose. *Polymers (Basel)* **2018**, *10*, 810. [CrossRef]
62. Paszkiewicz, S.; Janowska, I.; Pawlikowska, D.; Szymczyk, A.; Irska, I.; Lisiecki, S.; Stanik, R.; Gude, M.; Piesowicz, E. New functional nanocomposites based on poly(Trimethylene 2,5-furanoate) and few layer graphene prepared by in situ polymerization. *Express Polym. Lett.* **2018**, *12*, 530–542. [CrossRef]
63. Lotti, N.; Munari, A.; Gigli, M.; Gazzano, M.; Tsanaktsis, V.; Bikiaris, D.N.; Papageorgiou, G.Z. Thermal and structural response of in situ prepared biobased poly(ethylene 2,5-furan dicarboxylate) nanocomposites. *Polymer (Guildf)* **2016**, *103*, 288–298. [CrossRef]

64. Tarani, E.; Papageorgiou, D.G.; Valles, C.; Wurm, A.; Terzopoulou, Z.; Bikiaris, D.N.; Schick, C.; Chrissafis, K.; Vourlias, G. Insights into crystallization and melting of high density polyethylene/graphene nanocomposites studied by fast scanning calorimetry. *Polym. Test.* **2018**, *67*, 349–358. [CrossRef]
65. Chieng, B.W.; Ibrahim, N.A.; Yunus, W.M.Z.W.; Hussein, M.Z. Poly(lactic acid)/poly(ethylene glycol) polymer nanocomposites: Effects of graphene nanoplatelets. *Polymers (Basel)* **2014**, *6*, 93–104. [CrossRef]
66. Ozawa, T. A New Method of Analyzing Thermogravimetric Data. *Bull. Chem. Soc. Jpn.* **1965**, *38*, 1881–1886. [CrossRef]
67. Budrugeac, P.; Segal, E.; Perez-Maqueda, L.A.; Criado, J.M. The use of the IKP method for evaluating the kinetic parameters and the conversion function of the thermal dehydrochlorination of PVC from non-isothermal data. *Polym. Degrad. Stab.* **2004**, *84*, 311–320. [CrossRef]
68. Vyazovkin, S. Modification of the integral isoconversional method to account for variation in the activation energy. *J. Comput. Chem.* **2001**, *22*, 178–183. [CrossRef]
69. Tsanaktsis, V.; Vouvoudi, E.; Papageorgiou, G.Z.; Papageorgiou, D.G.; Chrissafis, K.; Bikiaris, D.N. Thermal degradation kinetics and decomposition mechanism of polyesters based on 2,5-furandicarboxylic acid and low molecular weight aliphatic diols. *J. Anal. Appl. Pyrolysis* **2015**, *112*, 369–378. [CrossRef]
70. Buxbaum, B. The Degradation of Poly (ethylene terephthalate). *Angew. Chem. Int. Ed.* **1968**, *7*, 182–190. [CrossRef]
71. Montaudo, G.; Puglisi, C.; Samperi, F. Primary thermal degradation mechanisms of PET and PBT. *Polym. Degrad. Stab.* **1993**, *42*, 13–28. [CrossRef]
72. Villain, F.; Coudane, J.; Vert, M. Thermal degradation of polyethylene terephthalate: Study of polymer stabilization. *Polym. Degrad. Stab.* **1995**, *49*, 393–397. [CrossRef]
73. Villain, F.; Coudane, J.; Vert, M. Thermal degradation of poly(ethylene terephthalate) and the estimation of volatile degradation products. *Polym. Degrad. Stab.* **1994**, *43*, 431–440. [CrossRef]
74. Konstantopoulou, M.; Terzopoulou, Z.; Nerantzaki, M.; Tsagkalias, J.; Achilias, D.S.; Bikiaris, D.N.; Exarhopoulos, S.; Papageorgiou, D.G.; Papageorgiou, G.Z. Poly(ethylene furanoate-co-ethylene terephthalate) biobased copolymers: Synthesis, thermal properties and cocrystallization behavior. *Eur. Polym. J.* **2017**, *89*, 349–366. [CrossRef]
75. Tarani, E.; Terzopoulou, Z.; Bikiaris, D.; Kyratsi, T.; Chrissafis, K.; Vourlias, G. Thermal conductivity and degradation behavior of HDPE/graphene nanocomposites. *J. Therm. Anal. Calorim.* **2017**, *129*, 1715–1726. [CrossRef]
76. Papageorgiou, D.G.; Terzopoulou, Z.; Fina, A.; Cuttica, F.; Papageorgiou, G.Z.; Bikiaris, D.N.; Chrissafis, K.; Young, R.J.; Kinloch, I.A. Enhanced thermal and fire retardancy properties of polypropylene reinforced with a hybrid graphene/glass-fibre filler. *Compos. Sci. Technol.* **2018**, *156*, 95–102. [CrossRef]

Sample Availability: Samples of the compounds are not available from the authors.

© 2019 by the authors. Licensee MDPI, Basel, Switzerland. This article is an open access article distributed under the terms and conditions of the Creative Commons Attribution (CC BY) license (http://creativecommons.org/licenses/by/4.0/).

Article

Continuous Monitoring of Shelf Lives of Materials by Application of Data Loggers with Implemented Kinetic Parameters

Bertrand Roduit [1,*], Charles Albert Luyet [1], Marco Hartmann [1], Patrick Folly [2], Alexandre Sarbach [2], Alain Dejeaifve [3], Rowan Dobson [3], Nicolas Schroeter [4], Olivier Vorlet [4], Michal Dabros [4] and Richard Baltensperger [4]

[1] AKTS SA, Technopôle 1, 3960 Sierre, Switzerland; c.luyet@akts.com (C.A.L.); m.hartmann@akts.com (M.H.)
[2] armasuisse, Science and Technology Centre, 3602 Thun, Switzerland; Patrick.Folly@ar.admin.ch (P.F.); alexandre.sarbach@ar.admin.ch (A.S.)
[3] PB Clermont EURENCO Group, Rue de Clermont, 176-4480 Engis, Belgium; a.dejeaifve@eurenco.com (A.D.); r.dobson@eurenco.com (R.D.)
[4] School of Engineering and Architecture of Fribourg, HES-SO University of Applied Sciences and Arts Western Switzerland, Bd de Pérolles 80, 1700 Fribourg, Switzerland; nicolas.schroeter@hefr.ch (N.S.); Olivier.Vorlet@hefr.ch (O.V.); Michal.Dabros@hefr.ch (M.D.); Richard.Baltensperger@hefr.ch (R.B.)
* Correspondence: b.roduit@akts.com

Received: 30 May 2019; Accepted: 11 June 2019; Published: 13 June 2019

Abstract: The evaluation of the shelf life of, for example, food, pharmaceutical materials, polymers, and energetic materials at room or daily climate fluctuation temperatures requires kinetic analysis in temperature ranges which are as similar as possible to those at which the products will be stored or transported in. A comparison of the results of the evaluation of the shelf life of a propellant and a vaccine calculated by advanced kinetics and simplified 0th and 1st order kinetic models is presented. The obtained simulations show that the application of simplified kinetics or the commonly used mean kinetic temperature approach may result in an imprecise estimation of the shelf life. The implementation of the kinetic parameters obtained from advanced kinetic analyses into programmable data loggers allows the continuous online evaluation and display on a smartphone of the current extent of the deterioration of materials. The proposed approach is universal and can be used for any goods, any methods of shelf life determination, and any type of data loggers. Presented in this study, the continuous evaluation of the shelf life of perishable goods based on the Internet of Things (IoT) paradigm helps in the optimal storage/shipment and results in a significant decrease of waste.

Keywords: shelf life; internet of things; IoT; data loggers; advanced kinetic analysis; vaccines; propellants; mean kinetic temperature

1. Introduction

One of the most important daily-life uses of kinetic investigations of the thermal behavior of materials is the possibility of the application of the computed kinetic parameters for the prediction of materials' properties at temperatures higher or lower than those used during data collection. Everybody is faced a few times each day with labels indicating the shelf life of items. Buying any daily-use products, storing chemicals, propellants, medicines, and thousands of temperature-sensitive products, one checks or carefully monitors the date of their validity expressed by common expressions, such as "best before" or "expiry date". The information depicting the period of time in which the item's properties fulfill certain criteria is of great importance, not only for individuals, but also has great economic significance. The improper handling of a batch of, e.g., expensive vaccines may result in the waste of millions of dollars or may be dangerous for the population. According to the World

Health Organization, the losses associated with temperature excursions in health care come to ca. $35 \cdot 10^9$ USD per year [1]. It is therefore obvious that enormous efforts are undertaken in order to monitor the conditions of the handling (storage) of any kind of product and to evaluate the impact of time–temperature parameters on the material properties.

The problem of the deterioration of the products' properties is extremely complicated, due to the fact that many parameters influence the rate of their deterioration. However, in general, independently of the kind of material, one of the main parameters influencing the shelf life is the temperature, and general suggestions given by producers inform users about the temperature range and duration of storage. More and more often, especially in large-scale handling, the monitoring of the temperature, as a decisive factor of the material's stability, is conducted by electronic devices (data loggers) that continuously collect the temperature–time data, which allow for the evaluation of the correctness of the storage.

The estimation of the aging extent belongs to a typical problem of the application of kinetic analyses in solving daily-life problems. The evaluation of the kinetic parameters of the main deterioration process allows the prediction of the rate of this process under any temperature fluctuations. However, in many cases, such a kinetic analysis is much more difficult to perform than during common kinetic measurements carried out in laboratories when the experimental time is limited to a few hours and when thousands of experimental data may be collected. The kinetic description of the process based on experimental points collected in the temperature (or time) domain, which is significantly different than that in which the properties of investigated material are of interest, may not be precise enough. Therefore, an additional difficulty in the application of the kinetic approach for the prediction of the shelf life of products arises from the fact that the kinetic parameters should be evaluated in the temperature ranges which are as similar as possible to those at which the products will be stored (or transported) in. This, in turn, results in very time- and effort-consuming experiments to supply the number of data points necessary, which is generally a few orders of magnitude smaller than the data collected in common kinetic experiments. These sparse data have to be elaborated by specific kinetic and statistic approaches in order to give the "best model combinations" with meaningful prediction bands which could be successfully applied during a shelf life [2]. In any case, such a kinetic analysis has to be validated by some experimental data lying in the considered time–temperature domain in which the product is stored.

In the present study, we propose the merging of the time–temperature profiles with the modified kinetic approach described in our previous paper [2], which is well suited for the estimation of the kinetics of the deterioration of any temperature-sensitive products. Our proposal is illustrated by the determination of the shelf lives of products which are relatively temperature resistant, such as propellants, and those which are very sensitive to temperature excursions, such as vaccines. These two classes of compounds can be used as examples representing the boundary conditions occurring during storage and the procedures applied for the evaluation of their aging kinetics can be used for almost all kinds of products.

The knowledge of temperature and its fluctuations is a very important factor for the estimation of the shelf life of materials, therefore, its continuous monitoring during storage or shipping has become more common. Information collected by data loggers, among other methods, monitor storage (shipping) temperatures to ensure their quality by checking whether the temperature of sensitive products is in accordance with approved temperature specifications.

The correct estimation of the material's shelf life can be performed only in the case when both the correct kinetic parameters and temperature data are known. Therefore, it is of great importance to combine the temperature–time data, which are collected and stored in the data loggers, with a kinetic analysis of the deterioration process. The application of the data loggers to the monitoring of only the temperature may be significantly extended this way, by a much more precise evaluation of the deterioration extent of products.

This study will address the advantages of implementing kinetic parameters into data loggers, allowing continuous online monitoring of the change of material properties. Additionally, it will be illustrated that simplified applications of the Arrhenius equation, such as in the form of the mean kinetic temperature approach or predictions based on the 0th or 1st order kinetic models, may result, in certain cases, in imprecise results.

2. Results and Discussion

2.1. Experimental

Stability tests of materials based on different experimental techniques monitoring their properties are performed by artificial aging either at (i) one single temperature or (ii) at a number of different temperatures (multi-temperature aging procedure). The single-temperature aging procedure is cheaper, however, as it is generally based on simplified assumptions, although it cannot precisely assess the safe storage life at an arbitrarily chosen temperature. The multi-temperature aging [3,4] procedure is more time- and effort-consuming, however, it enables a more precise prediction of a material's behavior for a wide range of temperature profiles after the calculation of the kinetic parameters and the selection of the best kinetic model, which can be verified by using, for example, the Akaike and Bayesian information criteria [2,5,6]. It may give the answer if a material is sufficiently stable to be stored for a given period of time, for example, 10 years at a specific temperature, such as ambient storage conditions, e.g., 25 °C ($t_{25} \geq 10$ years), or for more specific temperature and time profiles. Based on the bootstrap sets of estimated kinetic parameters, the prediction can be enhanced by the estimation of the prediction band (PB) in the form of, e.g., the upper and lower 95 percentiles (PB 95% confidence). The simulations presented in this study were done with the AKTS-Thermokinetics Software [5].

2.2. Kinetic Analysis

2.2.1. Determination of the Reaction Rate and Kinetic Triplets

The reaction progress can be defined as follows:

$$\alpha = \frac{Y - Y_0}{Y_{end} - Y_0}, \qquad (1)$$

where Y, Y_0, and Y_{end} represent the value characterizing the certain material property at time t, $t = 0$, and $t = t_{end}$, respectively.

The residual sum of squares (RSS) can be used to compute the parameters used for simulation:

$$RSS = \sum_{i=1}^{N} \left(Y_{i,exp} - Y_{i,cal}\right)^2, \qquad (2)$$

where the indices represent the value of an experimental point and its calculated value and N is the total number of points collected discontinuously. The conversion rate is expressed as:

$$\frac{d\alpha}{dt} = A \cdot \exp\left(-\frac{E}{R} \cdot \frac{1}{T}\right) f(\alpha), \qquad (3)$$

where t is the time, T is the temperature, R is the gas constant, E is the activation energy, A is the pre-exponential factor, α is the reaction extent, and $f(\alpha)$ is a differential form of the conversion function depending on the reaction model.

Although there is a significant number of various reaction models, $f(\alpha)$, they all can be reduced to three major types when considering the dependence of the reaction progress on the time in isothermal conditions: Accelerating, decelerating, and S-shaped (logistic or sigmoidal function). Each of these types has a characteristic "reaction profile" or "kinetic curve", the terms frequently used to describe a dependence of α or $d\alpha/dt$ on t for a given T.

In the present study, we applied the S-shaped model:

$$f(\alpha) = (1-\alpha)^n \alpha^m. \tag{4}$$

Under isothermal conditions, such sigmoidal reaction models may be considered as accelerating at the beginning (when α is close to 0) and decelerating at the reaction end (when α is close to 1) so that the process rate reaches its maximum at some intermediate values of the extent of the conversion. The sigmoidal reaction model turns into the nth order model (decelerating type under isothermal conditions) for $m = 0$; and to the 0th or 1st order models if $m = 0$ and $n = 0$ or 1, respectively. This means that the quality of fit of multi-temperature aging data by the sigmoidal reaction model is always equally good (if $m = 0$) or better (if $m \neq 0$) than those obtained with the nth order model. Sigmoidal reaction models may also be used to describe accelerating type reactions if $n = 0$ and $m > 0$. One should note that the solution of the reaction rate (Equation (4)) using the 'sigmoidal reaction model' if both n and $m \neq 0$ implies the presence of a very small amount of reaction progress, α_0, at time $t = 0$. In this study, we assumed that α_0 amounts to 1×10^{-10}.

Despite the applied model in kinetic analysis, the number of data points should be larger than the number of fitted parameters applied. The application of too many parameters which are fitted to a small number of data points leads to overfitting, which is manifested by the nonsensical values of the calculated parameters and the reduced predictive performance of the model. Methods based on information theory, such as the Akaike and Bayesian information criteria (AIC and BIC) [2,6,7] and, recently, [8], can be used to assess the statistical relevance of the fitted parameters, n and/or m, and find the optimum number of parameters.

2.2.2. Propellants: Application of Kinetic Analyses for Shelf Life Predictions

Using nitrocellulose-based propellants as an example, it is necessary to consider that the decomposition products may influence their chemical stability. In order to prevent these undesired processes, small amounts of stabilizing compounds are added to the propellants in order to react with the decomposition products, therefore preventing their reactions with the parent material. Surface modification of the propellant during aging may also change the ballistic properties and the shelf life of propellants. The basic information concerning the assessment of the stability of propellants and safe lifetime are presented in the study of de Klerk [9].

The optimal procedure for the investigation of a propellant's aging should be based on the application of more than one experimental technique monitoring changes of the material during storage. The main reason for analyzing different properties is the fact that certain material properties are inherently less stable than others and can vary differently during temperature excursions.

The determination of the kinetics of the investigated phenomena is difficult due to the fact that the number of experimental points in rationally limited periods of time is relatively small. To perform a meaningful kinetic analysis, having few experimental points collected at only two or three temperatures, we propose the significant optimization of the experimental procedure required for the correct kinetic description of the investigated process. The optimization of the experimental procedure is based on decreasing temperature–time domains which, in turn, allows avoidance of the necessity of collecting experimental points during a few months or years. Using the proposed method, it is possible to verify the selection of the best kinetic model and computed kinetic parameters by the experimental points collected after several days or even years by checking if they are lying inside the prediction bands. After successful validation of the kinetic analysis with experimental data, it was possible to uncover the differences of the reaction course for the various propellant properties in different climates and storage.

Aging can give rise to many phenomena, which may modify the thermal behavior of composite propellants. The aim of our study was to compare the results obtained during the investigation of the artificial aging performed by different analytical techniques in which different physico-chemical phenomena occur in the material. We applied our kinetic and statistical approach to the results obtained by four different methods in which specific material behaviors were monitored, namely: The pressure firing (PF), gas evolution (VST), stabilizer depletion (UPLC) [10], and the heat evolution (HFC) [11].

The results of the kinetic analysis for all applied testing methods are depicted below in Figures 1–4. The plots display:

- Top section: Fit of experimental data at three temperatures (solid circles) by the best model chosen according to Akaike (AIC) [2,6] and Bayesian (BIC) [2,7] criteria and by commonly applied 0th and 1st order kinetic models (curves are marked as «best», «0», and «1», respectively).
- Middle section: Long-term prediction of the reaction course according to the best model containing prediction bands with 95% confidence. The empty circles indicate the results of the additional experiments not used during the kinetic analysis which were applied for the verification of the simulations. The plot additionally contains the simulated course of the reaction at a lower temperature (50 °C) with one experimental point.
- Bottom section: Comparison of the prediction of the reaction course at 20 °C over 10 years and for climatic category A1 (diurnal seasonal storage according to [10] using the best, 0th, and 1st order models.

The results of the kinetic analysis based on AIC and BIC criteria are displayed in Table 1 for four analytical methods: PF, VST, UPLC, and HFC, respectively.

Table 1. The statistical AIC and BIC weights, sum of residual squares RSS, number of data and parameters used in simulations, initial and final values of measured quantities (Y_{init} and Y_{end}), and the evaluated kinetic parameters (activation energy, E; pre-exponential factor, A; reaction order exponents, n and m) calculated for the testing methods, PF, VST, UPLC, and HFC, respectively. For each method, the statistic and kinetic parameters were calculated for the fixed integer reaction order exponents, $n = 0, 1, 2$, and 3, and for adjustable fitted n and m values. The results for the best models according to AIC and BIC criteria are displayed in bold. Only in one case was the reaction order model with a fixed integer n value (PF, $n = 3$) the best from the statistical point of view. Interpretation of the AIC and BIC criteria are explained in detail in [2].

	wAIC (%)	wBIC (%)	No. of param.	No. of data	RSS	E (kJ·mol^{-1})	Ln(A*s) (-)	n (-)	m (-)	Y_{init}	Y_{end}
PF	**78.59**	**82.22**	**2**	**15**	**6.66 × 10^4**	**206.5**	**58.30**	**3**	**0**	**3599**	**5000**
	12.31	8.25	3	15	6.61 × 10^4	202.8	56.95	2.85	0	3599	5000
	9.09	9.51	2	15	8.88 × 10^4	179.5	48.52	2	0	3599	5000
	~0	~0	2	15	2.15 × 10^5	147.4	36.96	1	0	3599	5000
	~0	~0	2	15	9.81 × 10^{13}	129.0	30.19	0	0	3599	5000
VST	**59.1**	**56.46**	**3**	**14**	**1.37 × 10^{-1}**	**143.1**	**35.30**	**0**	**0.20**	**0.34**	**5**
	25.67	24.53	3	14	1.54 × 10^{-1}	141.6	34.10	−1.33	0	0.34	5
	9.66	12.42	2	14	2.36 × 10^{-1}	146.9	36.16	0	0	0.34	5
	5.15	6.06	4	14	1.35 × 10^{-1}	144.0	35.88	0.57	0.27	0.34	5
	~0	~0	2	14	3.75 × 10^{-1}	150.9	37.72	1	0	0.34	5
	~0	~0	2	14	5.47 × 10^{-1}	154.7	39.18	2	0	0.34	5
	~0	~0	2	14	7.33 × 10^{-1}	158.4	40.61	3	0	0.34	5
UPLC	**55.51**	**48.10**	**3**	**14**	**9.62 × 10^{-3}**	**145.8**	**37.13**	**0.62**	**0**	**0.61**	**0**
	44.47	51.88	2	14	1.32 × 10^{-2}	148.2	38.16	1	0	0.61	0
	~0	~0	2	14	4.05 × 10^{-2}	157.6	41.89	2	0	0.61	0
	~0	~0	2	14	7.37 × 10^{-2}	169.1	46.31	3	0	0.61	0
	~0	~0	2	14	6.71 × 10^{-1}	150.0	38.27	0	0	0.61	0
HFC	**99.59**	**98.68**	**4**	**28**	**6.71 × 10^2**	**138.6**	**30.50**	**−4.79**	**0.25**	**0**	**4000**
	0.40	1.31	3	28	1.10 × 10^3	137.6	30.80	0	0.37	0	4000
	~0	~0	3	28	1.83 × 10^3	143.3	30.84	−12.40	0	0	4000
	~0	~0	2	28	1.92 × 10^4	164.8	38.43	0	0	0	4000
	~0	~0	2	28	2.13 × 10^4	166.5	38.03	1	0	0	4000
	~0	~0	2	28	2.37 × 10^4	168.2	39.63	2	0	0	4000
	~0	~0	2	28	2.58 × 10^4	169.9	40.23	3	0	0	4000

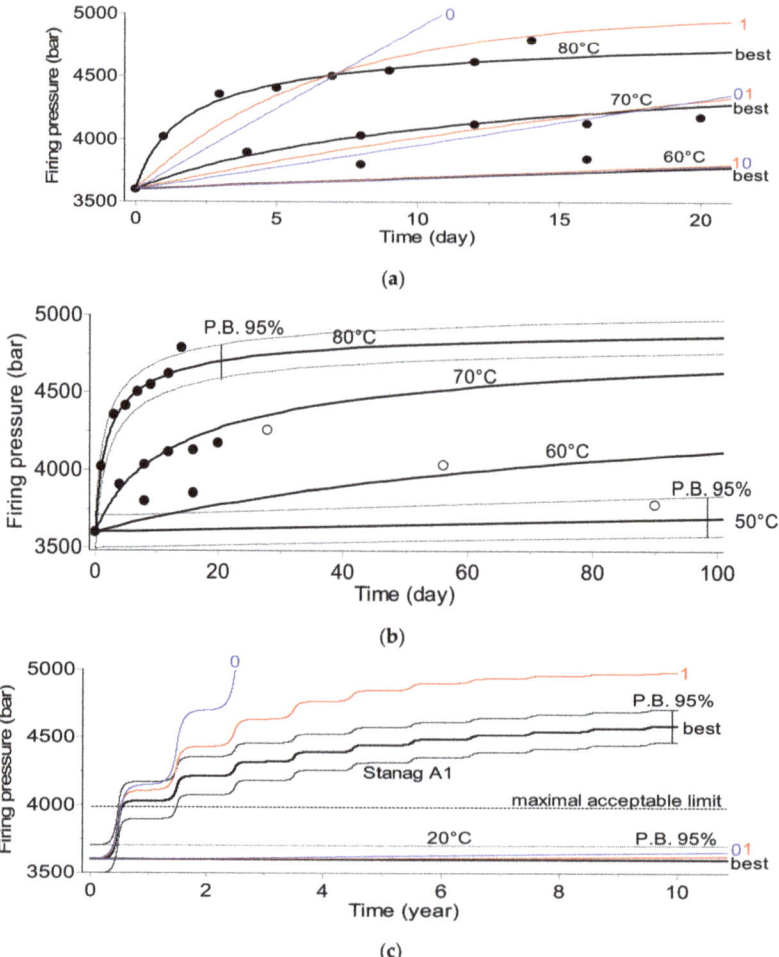

Figure 1. Pressure firing (PF) test. (**a**) Predictions of pressure firing peak values based on 15 experimental points (solid symbols) collected during 20 days in the temperature range of 60–80 °C. Predictions based on the best model are marked in bold, numbers 0 and 1 placed on the solid lines depict the predictions based on the 0th and 1st reaction order, respectively. (**b**) Prediction curves at 50, 60, and 70 °C were verified by the experimental points marked by the open circles. The prediction bands (dotted lines) were determined by the bootstrap method. (**c**) Comparison of the prediction using the best, 0th, and 1st order models at 20 °C over 10 years and for climatic category A1 (diurnal seasonal storage) according to STANAG 2895. The arbitrarily chosen acceptable limit of each measured quantity is marked by a dashed line.

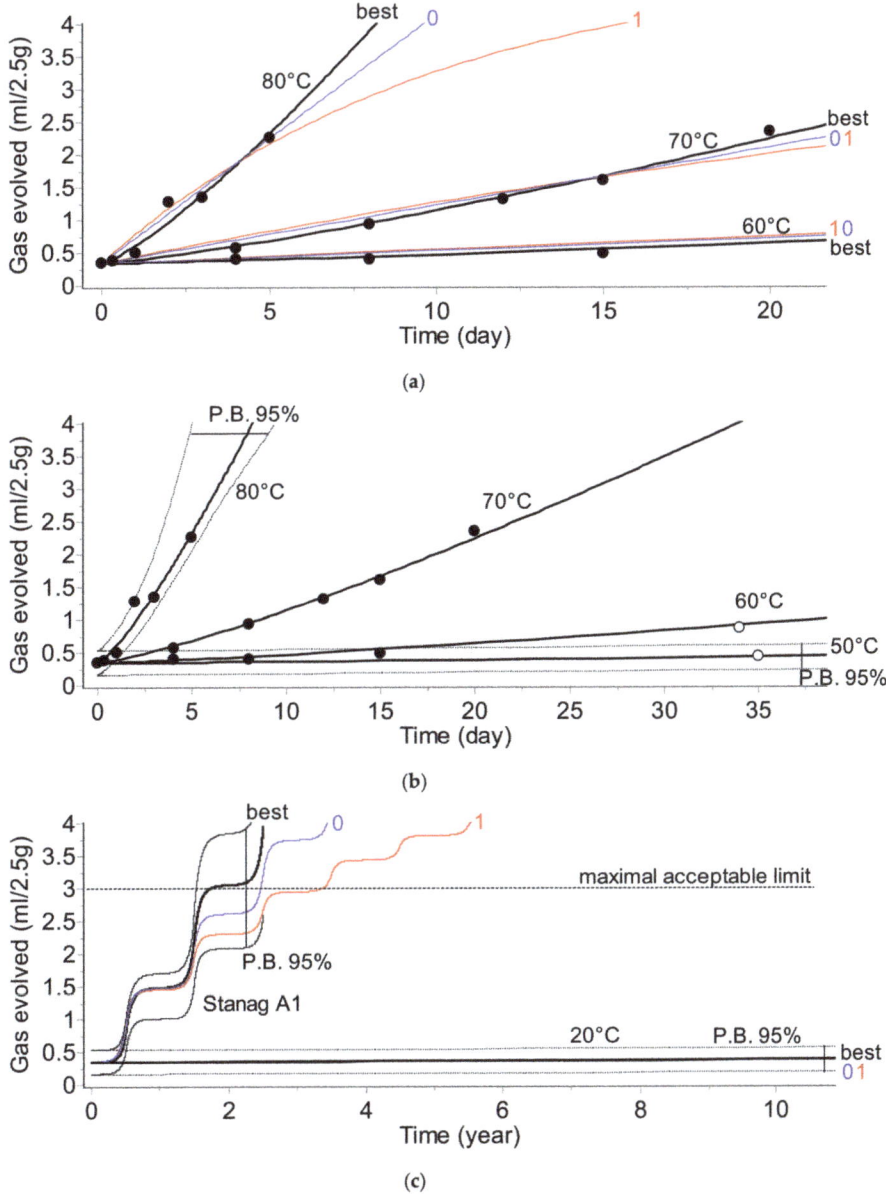

Figure 2. Gas evolution (VST) test. The plots present: (**a**) The fit of experimental data by the best (bold), 0th, and 1st order models. (**b**) Prediction of the long-term reaction course for temperatures of 80, 70, 60, and 50 °C (predictions are verified by experimental data marked as empty circles). (**c**) 10-year predictions for 20 °C and climate category STANAG A1.

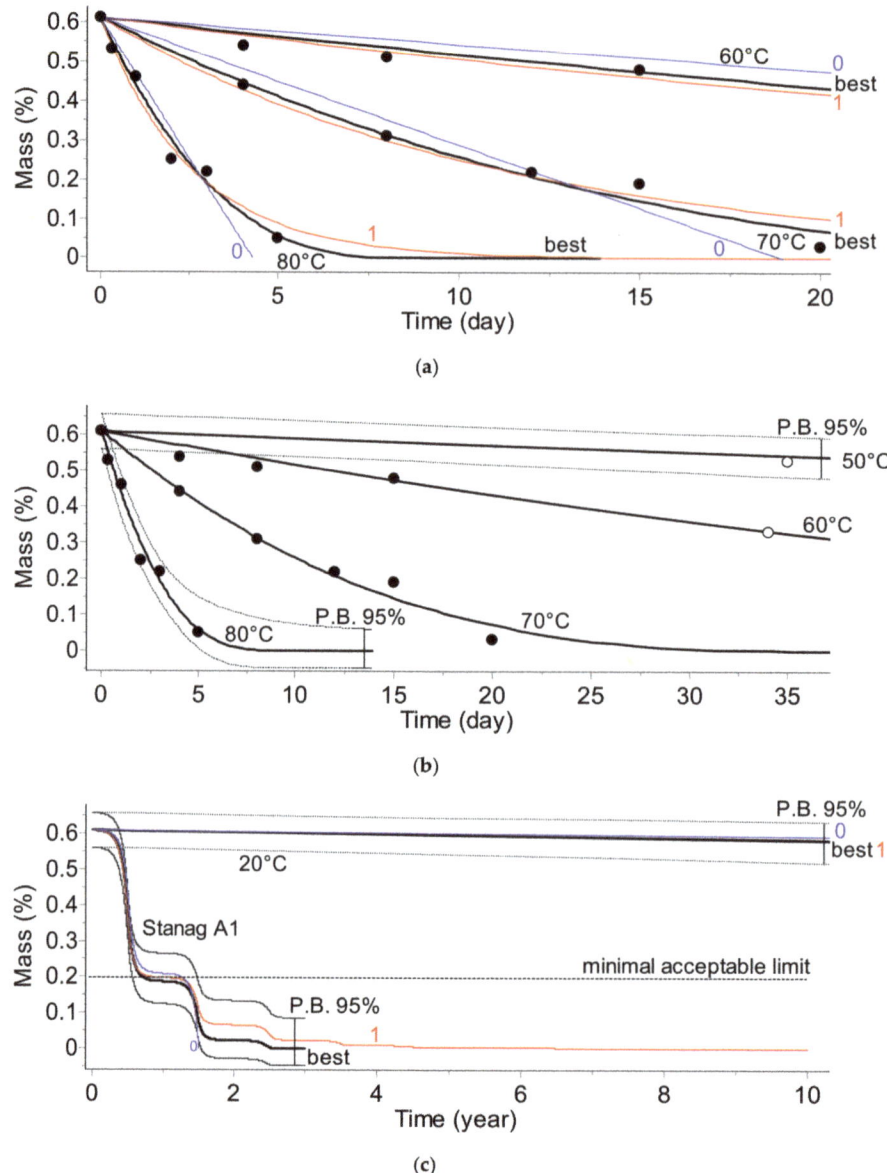

Figure 3. Stabilizer depletion (UPLC) test. (**a**) The fit of experimental data by the best (bold), 0th, and 1st order models. (**b**) Prediction of the long-term reaction course for temperatures of 80, 70, 60, and 50 °C (predictions are verified by experimental data marked as empty circles). (**c**) 10-year predictions for 20 °C and climate category STANAG A1.

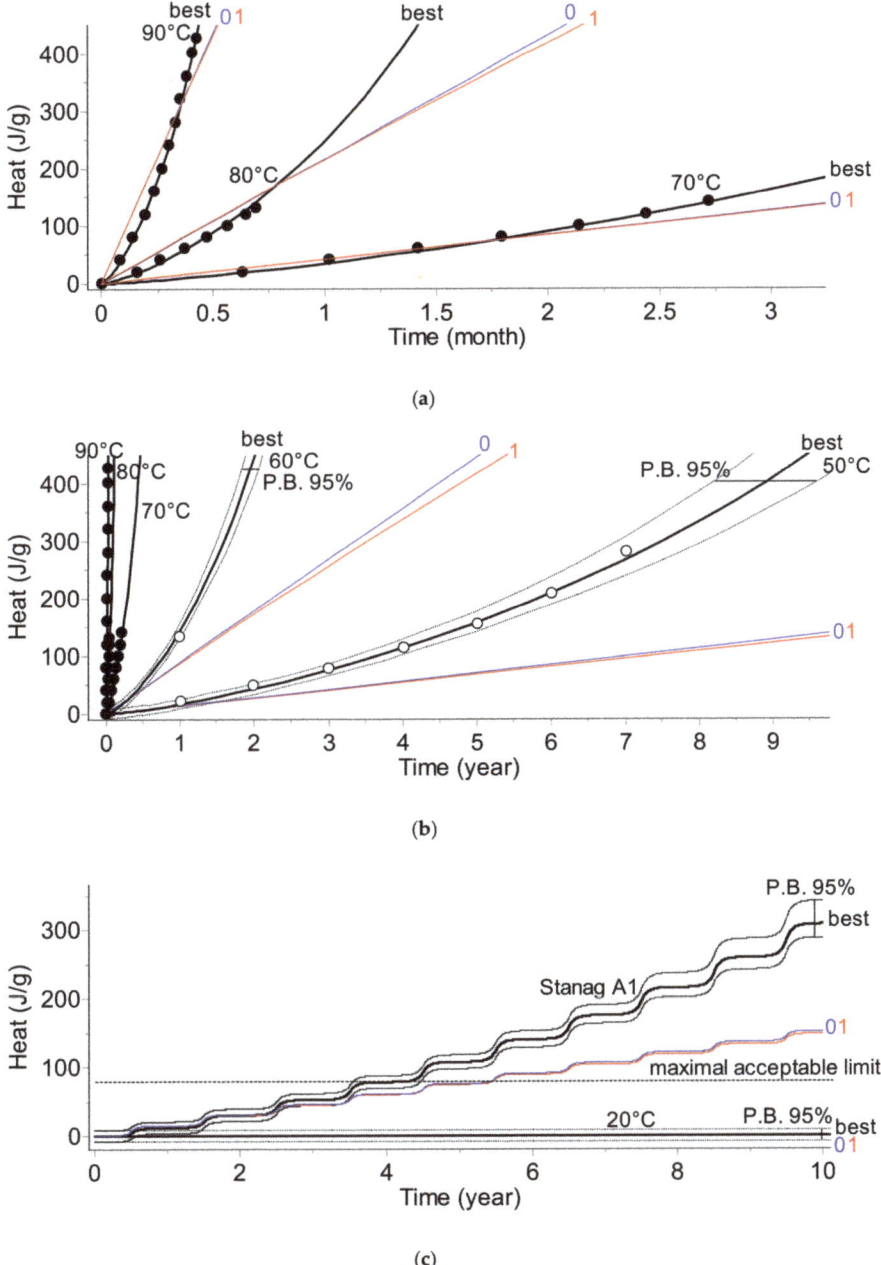

Figure 4. Heat evolution (HFC) test. (**a**) The fit of experimental data collected at 90, 80, and 70 °C by the best (bold), 0th, and 1st order models. (**b**) Prediction of the long-term reaction course for temperatures in the range of 90–50 °C. The predictions for 60 and 50 °C are verified by experimental data marked as empty circles collected once a year. (**c**) 10-year predictions for 20 °C and climate category STANAG A1.

Results depicted in Figures 1–4 clearly indicate that the application of the simplified assumptions concerning the form of the kinetic function, f(α), may influence the predictions of the shelf life of the propellants. In the two analytical methods, namely gas evolution monitoring and stabilizer depletion, this difference is not large; however, the pressure firing and, especially the heat evolution test recommended by STANAG 4582, the predictions based on simplified kinetics (assuming 0th or 1st order reaction) differ significantly for the results obtained by advanced kinetic analysis. For example, for the climate category, STANAG A1, the shelf life evaluated for the advanced kinetics amounts to circa 4 years whereas the predictions based on first order kinetics result in a shelf life of ca. 5.2 years (Figure 4c). Depending on the applied method chosen for monitoring the propellant properties (PF, VST, UPLC, or HFC), the E values were in the range of 138.6 to 206.5 kJ·mol^{-1}.

The presented results clearly show that the kinetic parameters which are going to be implemented into the data loggers should be evaluated by an advanced kinetic analysis. The combination of the time–temperature data with simplified kinetics may result in an imprecise evaluation of the aging extent.

2.2.3. Pharmaceuticals

The Peculiarities of the Application of Kinetics for the Evaluation of the Shelf Life

Evaluation of the thermal stability of pharmaceuticals is a difficult task due to the influence of numerous factors on the rate of their deterioration. These factors include the thermal stability of the active component; interaction between active ingredients and excipients; methods of packing; and temperature, light, and moisture conditions encountered during storage, shipment, and handling. Pharmaceuticals should at all times be stored under the conditions recommended by the manufacturer to prevent deterioration which can result in a loss of potency and efficacy. Certain medications, such as vaccines and biological medicines, need to be stored under refrigeration in order to maintain the stated potency and ensure safety of the product until its expiry date.

Over the past several decades, numerous scientific papers and books have addressed the problem of the evaluation of the shelf life of pharmaceuticals, see, for example, the book edited by Carstensen and Rhodes [12], comprehensive reviews of Kartoglu and Milstien [13], the book of Waterman [14] and his papers [15,16], or recent publications of Fan et al. [17], Fu et al. [18], Almalik et al. [19], Faya et al. [20], Khan et al. [21], Clénet et al. [22,23], or Clancy et al. [24].

The determination of the shelf life of medicines is an important task of the pharma-industry and the World Health Organization (WHO), whose primary role is to direct international health within the United Nation's system [25,26]. It is also important for institutions attempting to supply medicines to populations in countries where transport and storage facilities do not fulfill the criteria required for the preservation of products requiring cold chain management. The term "cold chain" refers to the transportation and storage of drug products, such as vaccines, insulin, and biological medicines requiring stable refrigerated conditions. The proteins present in these products often have vulnerable structures and their unfolding at higher temperatures significantly decreases the medicine's activity.

An evaluation of the change in the medicine's potency during storage and transportation is generally based on recording their temperature which should follow the approved profiles. The list of WHO-recommended temperature-monitoring devices for storage and transportation [27] contains electronic shipping indicators, vaccine vial monitors, and user-programmable temperature data loggers.

The "controlled temperature chain" (CTC) is an innovative approach to vaccine management, allowing vaccines to be kept at temperatures outside of the traditional cold chain of +2 to +8 °C for a limited period of time under monitored and controlled conditions, as appropriate to the stability of the antigen. A CTC typically involves a single excursion of the vaccine into ambient temperatures not exceeding +40 °C for a period of time not shorter than 3 days [28].

In order to be sure that vaccines have not been exposed to temperatures higher than +40 °C, a "peak threshold indicator" must accompany the vaccines at all times when, in a CTC, the temperature exposure of the vaccines is monitored. This indicator is a card with a sticker, which changes color from light grey to black as soon as the temperature exposure has exceeded +40 °C. If this is the case, all vaccines in that vaccine carrier must be discarded, following an appropriate investigation and documentation of the event. Additionally, temperature monitoring's impact on the vaccines' potency retention is carried out by means of vaccine vial monitors (VVMs). A VVM is a label containing a heat-sensitive material which is placed on a vaccine vial to register cumulative heat exposure over time. The combined effects of time and temperature cause the inner square of the VVM to darken, gradually and irreversibly. A direct relationship exists between the rate of color change and temperature. Vaccine vial monitors and peak temperature threshold indicators protect potency and quality by monitoring cumulative and peak exposure to heat.

Shelf Life Evaluation Criteria Derived from the Arrhenius Equation

In a more advanced procedure of monitoring the thermal behavior of vaccines, the mean kinetic temperature (MKT) is used. This method was introduced by Haynes et al. [29] who addressed the fact that climate-based temperature variation in uncontrolled pharmaceutical storage makes it difficult to select a single temperature for use in product expiry testing. A detailed description of the MKT concept can be found in [30], its modification in [31], and limitations in [32], where one finds the recommended caution in using the MKT to evaluate temperature excursions.

The MKT is defined as a single calculated temperature at which the total amount of degradation over a particular period is equal to the sum of the individual degradations that would occur at various temperatures. The MKT may be considered as an isothermal storage temperature that simulates the non-isothermal effects of storage temperature variation, i.e., that corresponds to the same kinetic effects of a time–temperature distribution. The calculation gives increased weighting to higher temperature excursions than normal arithmetic methods, recognizing the accelerated rate of thermal degradation of materials at higher temperatures. The commonly used formula for MKT calculation was introduced by Haynes and is given by:

$$T_{MKT} = \frac{\Delta H/R}{-\ln\left(\frac{e^{\frac{-\Delta H}{RT_1}} + e^{\frac{-\Delta H}{RT_2}} + \ldots + e^{\frac{-\Delta H}{RT_n}}}{n}\right)}, \quad (5)$$

where T_{MKT} is the mean kinetic temperature in degrees Kelvin, ΔH is the activation energy in kJ·mol^{-1}, R is the gas constant in J·mol^{-1}·K^{-1}, T_1 to T_n are the temperatures at each of the sample points in degrees Kelvin.

Continuous evaluation of the MKT seems to better characterize the impact of time–temperature parameters on the rate of deterioration of products rather than, for example, the mean temperature recorded during long-term storage (shipment). However, the application of the concept of the MKT has significant drawbacks from the point of view of advanced kinetics. Restrictions, such as the necessity of collecting data in the same time-intervals, the assumption that the activation energy amounts to 83.144 kJ·mol^{-1}, or that the reaction of the deterioration is a one-step first order reaction, indicate that the MKT approach is roughly linked with advanced kinetic analysis. A default value of 83.144 kJ·mol^{-1} is typically used because it is supposed to be an acceptable approximation for most pharmaceutical compounds. According to Seevers et al. [30], it is an average value of activation energy for breaking most covalent bonds. However, according to [32], for a wide range of pharmaceuticals, ΔH in Equation (5) may be in the range of 42 to 125 kJ·mol^{-1}. Furthermore, the application of the 1st order model as a rule may result in an incorrect evaluation of the aging extent for most products (see, e.g., Figure 4).

The information received from data loggers, vaccine vial monitors, peak temperature threshold indicators, and other devices recommended by the WHO [27] are the only indicators as to whether the storage fulfills a specific refrigeration temperature criterion. Therefore, the implementation of continuous evaluation of the MKT into data loggers does not change the situation in which it is impossible to continuously monitor the actual degree of the deterioration of samples. The MKT is essentially just another way to express the impact of temperature during sample exposure which does not bring information of whether the permissible reaction progress limit is reached. This issue can be solved only if the kinetics of the deterioration or inactivation process can be merged with the time/temperature profiles recorded by data loggers. Application of the MKT concept for more complicated temperature profiles is described in the papers of Okeke at al. [33,34].

Although a comprehensive discussion of the problem of the kinetics of medicine inactivation is beyond the scope of this study, numerous scientific papers have appeared to better address this issue, see, e.g., Kumru at al. [35] or WHO guidelines [25]. However, general remarks concerning the specific treatment of sparse data collected during stability studies of medicines should be addressed because an evaluation of the kinetic parameters of medicines' deterioration differs from commonly applied kinetic workflows (see [2,22,23] and the references cited herein). One of the main reasons for this situation is the fact that an evaluation of the kinetics of the degradation of vaccines can be done only in time- and effort-consuming experiments and therefore the number of data points which could be applied in kinetic analysis is relatively small, often in the range of 20 to 30. This, in turn, requires modification of the kinetic and statistical approaches applied during standard kinetic analysis [2].

We are aware that a variety of methods of vaccine stability testing (such as biological assays or chemical and physical studies) may be applied for evaluating vaccine immunogenicity or efficacy changes. We treat this issue from the kinetics point of view by considering the change of the chosen parameter which is used for the evaluation of vaccine potency or activity after its normalization in the range of 0 to 1 (or 0–100%), which is commonly applied in typical kinetic studies. Kinetic analysis applied for the evaluation of vaccine degradation rates is generally carried out by the accelerated degradation test in which the investigated products are exposed to temperatures greater than those recommended for vaccine storage (typically 5, 25, or 37 °C). During the kinetic analysis of the data, often, the simplified kinetics models, such as 0th or 1st order kinetic functions, are applied. Such models fail to correctly describe the complicated course of decomposition of biological materials, which frequently show complex and multi-step degradation behavior (see [23] and the references cited herein). The rate constant derived from simplified models is often of little value during an advanced kinetic workflow; the prediction of half-lives of vaccines only from the Arrhenius plot depicting the rate of material degradation [36] is not precise enough because it is based on only one of three required kinetic parameters, namely the activation energy. The two other equally important kinetic parameters, i.e., the pre-exponential factor in the Arrhenius equation, A, and the form of the kinetic function, $f(\alpha)$, are not considered in simplified kinetic approaches. A more precise kinetic description of the decomposition of biological compounds [37,38] was obtained with kinetic parameters of two-step models, which better mimic the complicated decomposition of the investigated samples. The application of the autocatalytic kinetic model in the evaluation of the shelf life of pharmaceuticals was presented in the review by Brown and Glass [39].

For illustration purposes, we used the pharmaceutical product studied by an advanced kinetic approach for which the kinetic parameters are known. A freeze-dried measles vaccine was investigated in our former study [2] for which the criteria for the discrimination of the best kinetic models were done using the AKTS-Thermokinetics Software [5] and based on the information theory introduced by Akaike [6] and its Bayesian counterpart [7]. The deterioration rate of this pharmaceutical product, called throughout our study the "model vaccine", is characterized by the following kinetic parameters (see Table 5 in [2]) used in all our simulations:

$$\frac{d\alpha}{dt} = 6.95 \cdot 10^{19} \cdot exp\left(-\frac{156.26 \cdot 10^3}{RT}\right)(1-\alpha)^2 + 2.28 \cdot 10^{12} \cdot exp\left(-\frac{121.13 \cdot 10^3}{RT}\right). \tag{6}$$

The kinetic simulations for the course of deterioration of the model vaccine kept in the temperatures of 2.1 and 7.9 °C, i.e., those lying in the commonly applied temperature range characteristic for the "cold chain", are displayed in Figure 5. The presented results show that the expression "cold chain" is very imprecise from a kinetics point of view. For the boundary temperatures characteristic of the cold chain (2 °C < T < 8 °C), the shelf lives of a model vaccine may significantly vary from 880 to 3365 days.

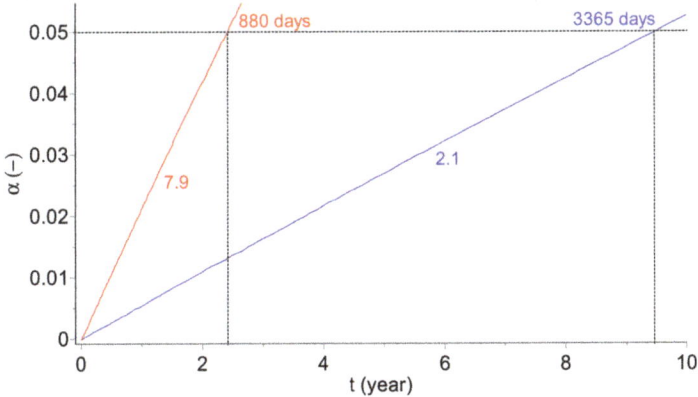

Figure 5. Model vaccine: Simulation of the long-term prediction of the change of the viral titer degree at temperatures of 2.1 and 7.9 °C (marked on the curves) lying in the range of 2 to 8 °C applied during the cold chain storage. Note the severe change of the vaccine shelf life (set arbitrarily as a time when 5% of the viral titer loss is reached) from 880 to 3365 days for 7.9 and 2.1 °C, respectively, despite fulfilling the cold chain criterion in both cases.

To illustrate the possible pitfalls resulting from the application of a simplified MKT approach, we present the simulations of the deterioration of samples of the "model vaccine" kept at 5 °C for 30 days followed by a rapid temperature excursion to 22 °C (Figure 6) and to 10, 20, 30, and 40 °C (Figure 7). A comparison of the real reaction progress with those based on the temperatures obtained by the computed MKT indicates the severe differences in time to reach the same extent of vaccine decomposition (see Figure 6). For a viral titer, the read-out is generally expressed in Log(pfu/vial). To be in line with the commonly applied expression of the extent of the material deterioration in the kinetic studies, instead of the viral titer in Log(pfu/vial), the reaction progress, α, is used in this study together with an arbitrarily chosen critical value of 5% for the permissible limit of degradation following a temperature excursion out of, e.g., the "cold chain". The application of the real kinetics for time out of refrigeration (TOR), which is very important in pharma logistics, enables precise quantification of the impact of a temperature excursion on the reaction progress, which amounts to 5% (arbitrarily set limit for the shelf life considerations) after 67.5 days. However, the predicted times to reach the same level of decomposition are considerably overestimated and amount to ca. 110, 121, and 138 days when including into Equation (6) the MKT for E = 42, 83.1, and 125 kJ·mol^{-1}, respectively.

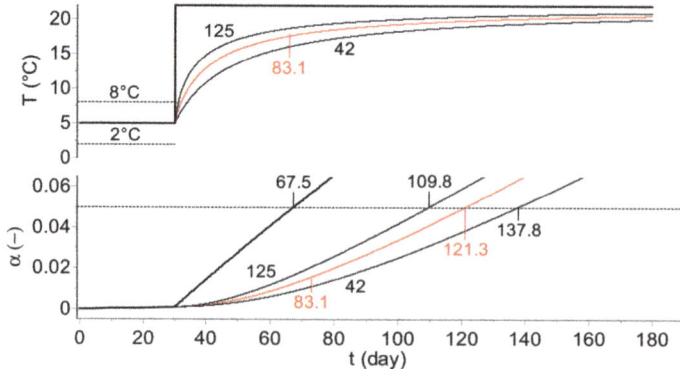

Figure 6. Model vaccine: Simulation of the long-term prediction of the change of virus infectivity (infectious titer) during the temperature excursion from 5 to 22 °C (TOR procedure). Top plot: time dependence of the real (bold) and MKT temperatures, bottom plot: the degree of infectious titer for the real (bold) and MKT temperatures. The mean kinetic temperatures were calculated for three activation energy values (marked on the curves in kJ·mol^{-1}). Note that application of the MKT leads to the overestimation of the shelf life: the value of 5% of degradation is reached after 67.5 days if following the real temperature, whereas according to the MKT, depending on the assumed activation energy, this loss of infectious titer is reached after ca. 110, 121, and 138 days for E = 42, 83.1, and 125 kJ·mol^{-1}, respectively.

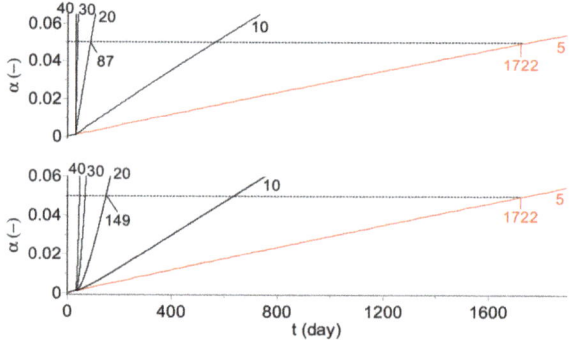

Figure 7. Model vaccine: The change (α) of virus infectivity during 30 days at 5 °C followed by the temperature excursions (10, 20, 30, and 40 °C marked in °C on the curves), for the real temperature course (top) and for MKT temperatures calculated by the assumption that E = 83.144 kJ·mol^{-1} (bottom). Bold curves show the vaccine deterioration progress at 5 °C, the dependences temperature vs. time for the real temperature and the MKT are identical at a constant temperature. The predicted limit of the shelf life, set arbitrarily as a time when a 5% loss of the infectious titer is reached, amounts at 5 °C to ca. 1722 days.

Figure 7 displays the extent of the infectious titer of the model vaccine occurring during 30 days of storage in the cold chain at 5 °C, followed by the temperature excursions to 10, 20, 30, and 40 °C. The top part of the plot depicts the reaction extent (calculated according to Equation 6) for the real temperature whereas in the bottom plot the MKT (calculated for E = 83.144 kJ·mol^{-1}) has been used for the reaction course estimation. For a temperature excursion of 20 °C, as presented in Figure 7, the shelf life amounts to 87 or 149 days at the real temperature and MKT, respectively. The results shown in Figures 6 and 7 indicate that not only the arbitrarily assumed value of the activation energy used for MKT calculation but also the difference between the real temperature and evaluated MKT influence the determination of the correct shelf life value.

2.3. Continuous Shelf-Life Estimation by Using Data Logger

The AKTS-Thermokinetics Software [5] allows the determination of the reaction extent, α (in our case the degree of the infectious titer change), in any temperature mode. Figure 8 shows the dependence of α on the time during temperature variations during a 3-year long storage in a cold chain (2 °C < T < 8 °C), followed by storage at ambient conditions with daily temperature fluctuations corresponding to the month of March in New Delhi (India). After three years of storage in the cold chain, the reaction extent amounts to ca. 3.6%. After removal from the cold chain and exposure of the vaccine to ambient temperature fluctuations, the shelf life (=5%) is reached after 15 days (see Figure 8b).

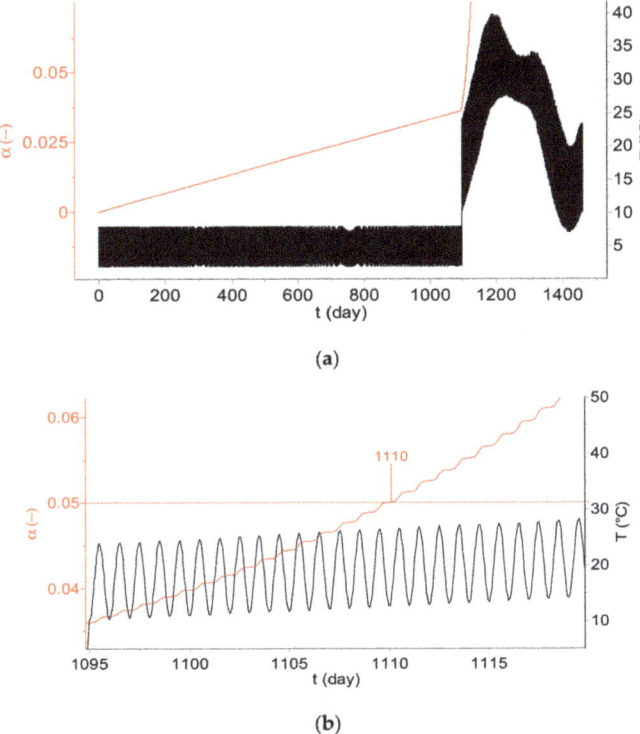

Figure 8. (**a**) Degree of the loss of infectious titer, α, after 3 years of storage in the cold chain (temperature variations between 2 and 8 °C), followed by storage at a real atmospheric temperature profile corresponding to the month of March in New Delhi (India). (**b**) The storage period between 1095 and 1120 days. The shelf life is reached 15 days after removal from the cold chain.

Online determination of the reaction extent (degree of the infectious titer change) based on the kinetic parameters and temperature–time data is of great importance because it allows immediate evaluation of the remaining shelf life as a function of the excursion temperature. This can be done using the TTT (transformation–time–temperature) plot presented in Figure 9, in which the position of the oblique bold line calculated for the 5% reaction extent allows immediate determination of the time at which, for an arbitrarily chosen temperature (30 °C on the plot), the loss of infectious titer reaches the set value of 5%.

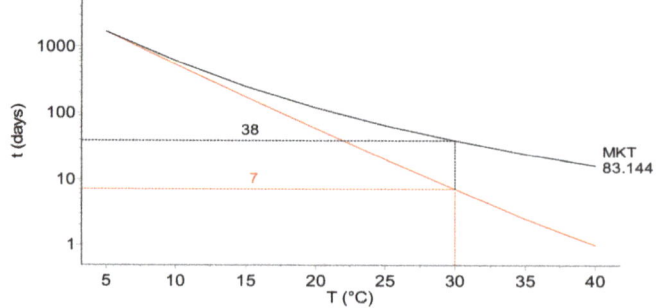

Figure 9. TTT (transformation–time–temperature) diagram presenting the shelf life as a function of the value of the excursion temperature after 30 days of storage in the cold chain (5 °C). The simulations show the dependences for the real temperature (bold) and the MKT calculated by the assumption of the activation energy as $E = 83.144$ kJ·mol^{-1}. At the excursion temperature of 30 °C, the infectious titer loss, set arbitrarily to 5%, is reached after ca. 7 days. Note that the application of the MKT leads to severe overestimation of the shelf life. Depending on the assumed values of the activation energy, the same infectivity loss is reached after ca. 51 days for $E = 42$, 38 days for 83.1, and 29 days for 125 kJ·mol^{-1}, respectively.

Such a TTT diagram allows quick evaluation of how many days at a specific constant temperature (in the depicted case, in a range of 5 to 40 °C) the vaccine removed from the cold chain will fulfil the required criterion of usability. The duration of the shelf life of the model vaccine kept 30 days at 5 °C after exposition to 30 °C, evaluated from the presented plot, amounts to 7 days for the real temperature and to 38 days for the MKT. Also, in this case, the arbitrary choice of the E value required for the MKT evaluation changes the predicted value of the shelf life: The virus infectivity loss at 30 °C after one-month storage at 5 °C estimated for the E values of 42, 83.1, and 125 kJ·mol^{-1} will amount to 51, 38, and 29 days, respectively. All these values are significantly overestimated when compared to the 7 days predicted for the real temperature course. A similar but simplified concept of the TTT diagram, which was based on the 1st order kinetics only, was presented by Ammann [40].

The TTT diagram depicted in Figure 9 can be applied for quick evaluation of the duration of the shelf life only at a constant temperature. The procedure proposed in this study allows for its application under any temperature mode. In our approach, the reaction extent (whatever parameter or approach is used for the evaluation of this parameter) is calculated continuously online at any customized period of time. Knowing at any time the reaction extent and the actual temperature, it is possible to continuously evaluate the remaining time until the sample reaches the shelf life value. The scheme of our approach is presented in Figure 10. For the sake of clarity, the concept is illustrated by isothermal temperature variations in an arbitrarily chosen range of 15 to 25 °C in relatively long periods of time. During real data logger applications, the presented long isothermal steps are replaced by the short time periods set by the user, allowing for the application of the TTT approach at any temperature mode.

Figure 10a presents the set of temperature segments recorded by the data logger. In segment no.1, the temperature of the sample amounts to 20 °C and at its end, the reaction extent (degree of the deterioration) reaches the value of α_1 (see Figure 10b). The sample aging progress displayed in Figure 10c occurs along the horizontal line at a temperature of 20 °C, which at the end of segment no.1 crosses the isoconversional line for α_1 at the point I′.

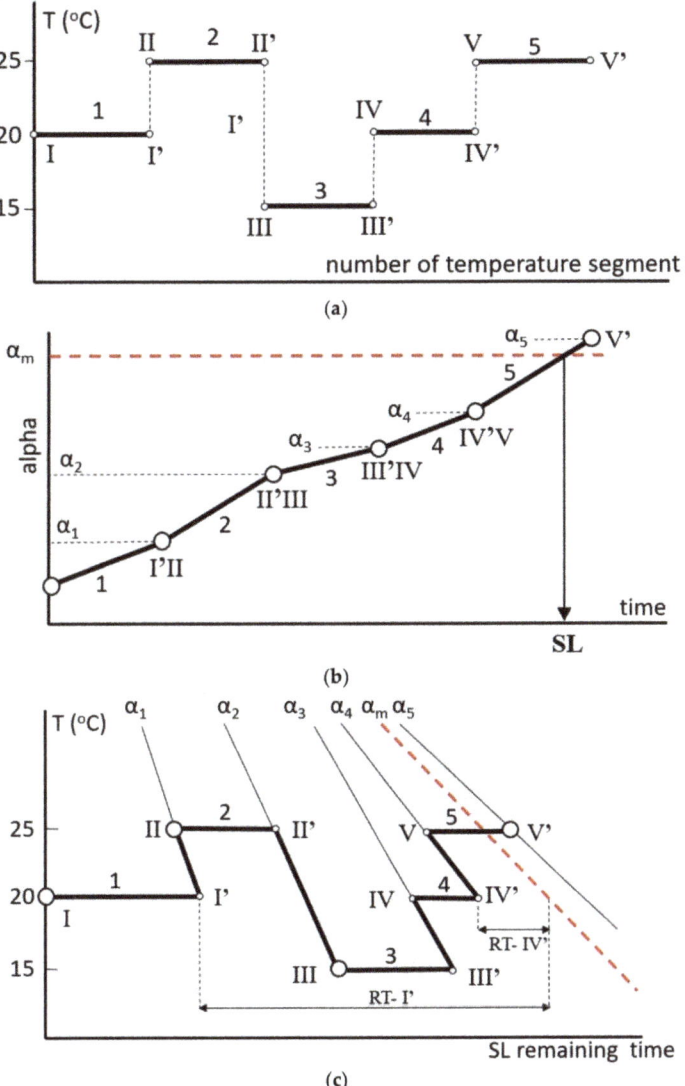

Figure 10. Scheme of the estimation of the remaining shelf life of the model vaccine during temperature variations. (a) Time–temperature dependence during the storage; (b) change of the reaction extent, α, as a function of the time and temperature profile shown above, the dashed line presents the maximal acceptable degradation extent which determines the shelf life value; (c) TTT plot presenting the mutual dependence of the time, temperature, and reaction progress. The dashed line (plot 10b) depicts the maximal acceptable degradation degree and their intersection with the actual degradation extent indicates the time in which the shelf life (SL) is reached. The determination of the remaining shelf life during storage displayed in Figure 10c is explained in the text.

Immediately after the end of segment no.1 begins segment no. 2, which occurs at 25 °C. The change in temperature from 20 to 25 °C results in a shift from I′ to II, which depicts the point at which the sample has the decomposition extent, α_1 at a temperature of 25 °C. The sample aging in segment 2 proceeds along the horizontal line II–II′ at 25 °C and at the end of segment (point II′), the sample aging progress reaches the value of α_2. In segment no. 3 on the TTT diagram, point II' moves to the position marked by III, which presents the intersection of the isoconversional line of α_2 with the horizontal line at the temperature of 15 °C. The further stages of the sample aging occur according to the presented scenario. The determination of the remaining time (RT) to reach the shelf life value is explained (for the sake of clarity) for segments 1 and 4 only, which proceed at a temperature of 20 °C. The length of the horizontal line drawn at the temperature of the respective stage from point I' at the end of segment 1 till the intersection with the isoconversional line drawn for the maximal allowed the reaction extent (marked by α_m) amounts to RT-I' and at the end of segment 4 (point no. IV') to RT-IV'. The shelf life is reached before the end of segment no.5.

The simplified example presented in Figure 10 is used only for the basic explanation of the concept of the remaining shelf life calculations. In real applications, these values (characterized by the lengths of the arrows, RT-I' and RT-IV') are calculated continuously at the end of each segment collected by the data logger. The frequency of the data collection is arbitrarily set by the user.

The typical information displayed from the data logger concerning the online evaluation of the remaining shelf life is shown in Figure 11.

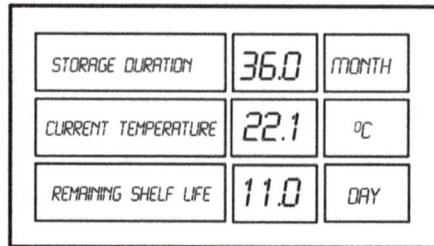

Figure 11. Online information concerning the remaining shelf life received from the data logger.

The implementation into the data logger of the TTT diagram based on the kinetic parameters also allows online prediction of the dependence remaining shelf life–isothermal temperature for the arbitrarily chosen variable. Figure 12 illustrates the basic concept of how the remaining time is evaluated at any temperature for the sample at a temperature of 20 °C and having the deterioration progress of $\alpha = 3.6\%$ (point A).

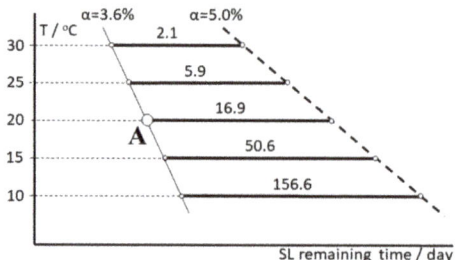

Figure 12. Dependence of the shelf life (SL) remaining time on the isothermal temperature which allows the prediction of the usability of the sample stored at 20 °C with a deterioration progress of 3.6% (point A). The remaining values of the shelf life at temperatures between 10 and 30 °C are displayed in days on the thick horizontal lines.

After increasing the temperature from 20 to 30 °C, the remaining time decreases from 16.9 to 2.1 days, and decreasing the temperature from 20 to 10 °C leads to increasing the shelf life remaining to 156.6 days. It is also possible to arbitrarily set the value of the SL-remaining time (e.g., 5.0 days) and determine the value of the maximal temperature at which this period of time is not exceeded and the vaccine remains within the approved shelf life specifications. The examples of data displayed by the data logger for both scenarios are shown in Figure 13.

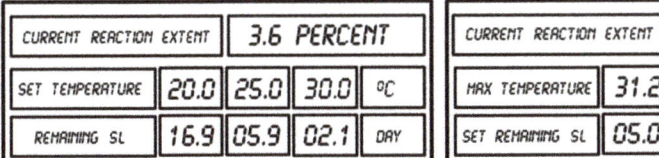

Figure 13. Display of the shelf life remaining for the sample with a deterioration progress of 3.6% at the user set temperature (left) or the maximal isothermal temperature for the shelf life remaining time (right) set by the user.

In summary, we would like to underline that the chosen methods of estimating the shelf life of pharmaceuticals presented in this study are used for illustration purposes only. The proposed procedure of implementing and merging kinetic analysis with continuously recorded time–temperature data is independent of the chosen criteria of the shelf life definition, storage and/or shipment conditions, applied kinetic approach, procedures used during accelerated stability studies, etc. We are aware that dozens of institutions, organizations, and scientific groups have already published an enormous number of papers, documents, and regulations concerning the above issues. It is not our goal to participate in discussions about the definition of the shelf life, such as those presented in the paper of Capen et al. [41]. Presenting our simulation, we keep in mind that the temperature is only one out of a few parameters influencing the deterioration rate of pharmaceuticals and the real rate of deterioration may depend on the relative humidity or other parameters. However, this fact does not change the general conclusion arising from our study, which is the following: The best available comprehensive kinetic analysis implemented into data loggers together with continuously monitoring of time–temperature data allows for continuous monitoring of the remaining shelf life of medicine products. The application of the kinetics based on a simplified 0th or 1st order kinetic approach may lead to imprecise predictions. Our study illustrates that it is advantageously possible to determine the exact time during which a batch of the product is expected to remain within the approved shelf life specifications for any temperature fluctuations.

2.4. Basic Technical Information about the Application of Data Loggers with Implemented Kinetic Data

This study describes the implementation of advanced kinetic parameters into data loggers which collect time–temperature data belonging to the "Internet of Things (IoT)" approach introduced in 1999. The term, IoT, refers to an evolution in computer technology and communication aiming to connect objects together via the Internet through either wired, wireless, or hybrid systems. The IoT enables anytime, anywhere, anything, and any media communications, for example, information can be sensed about the environment, such as temperature, humidity, localization, light exposure, etc. The interaction and communication among "smart things/objects" is reliable and occurs in real-time. A comprehensive review concerning IoT is presented in [42], and its general application is described in [43]. A general application of IoT for monitoring perishable goods is given for cold chain management [44,45] and specific applications in the area of healthcare are shown in [46,47] and in [48,49] for agriculture. Due to the rapidly increasing number of applications of IoT, the McKinsey's Global Institute predicts that IoT will have an economic impact of between 4 and 11 × 10^{12} USD by 2025 [50].

According to [51], monitoring systems in the current market involve the recording of only raw environmental data, e.g., temperature and/or humidity. Only rarely are these used for controlling environment parameters throughout the storage and transportation process. However, perishable goods (such as, e.g., the vaccines used by us for illustration purposes) are degraded during transportation to customers due to longer delivery routes and improper handling methods or the necessity of removal from the cold chain environment. An enormous amount of data is collected in the monitoring system. There is, however, little attention given towards further investigation of transforming the data to predict the current online quality of goods, such as, e.g., the deterioration extent or the actual shelf life value.

In our study, we presented a solution in which not only the environmental data (we use temperature as an example) but also the on-time degradation extent is monitored. Technical details of the application of data loggers are beyond the scope of this study, therefore only the main information, which can be interesting for non-specialists, were given.

The SensorTag developed by Texas Instruments was applied in this study, however, the described concept is not limited to any specific data logger supplier or this type of electronic device. The data logger reads various experimental conditions at regular time intervals, such as temperature and/or relative humidity. After some adjustments, customization and optimization of the original firmware by AKTS for SL prediction purposes, all readings can be stored in the memory.

Communication with a SensorTag and its programming is done by a smartphone (or similar device) and the AKTS mobile app. All key information, such as the kinetic parameters, the permissible limits of the goods' properties, the duration of the intervals between two successive readings, etc., are transmitted wirelessly via Bluetooth using a customized application and stored in the data logger. The concept is not restricted to Bluetooth communication.

We applied Bluetooth technology essentially because it has become a standard for (i) wireless transmission of data, (ii) availability and accessibility, (iii) ease of use, and (iv) energy efficiency. Once launched and the data logger starts recording required experimental conditions, it is possible to store ca. 10,000 data in its memory. To save battery, customized firmware switches to sleep mode automatically between two successive readings. It is also possible to put the device in sleep mode for a specific period of time so that it can be transported without restrictions by airplanes. Due to merging time–temperature data with the kinetic parameters implemented in the memory, the extent of aging can be computed and displayed continuously for the thermal history provided by the electronic sensor. Information about the aging extent may be displayed in different modes:

(i) Through small LED lights, whose displays may be set up to vary depending on the reaction progress computed online.

(ii) Option (i) can be enhanced by using a smartphone which, in proximity of the SensorTag and after identification through a QR code, can access wirelessly the stored temperature readings and display graphically the evolution of the temperature profile and determined reaction progress of the investigated property of the goods.

(iii) Tracking information can be sent to a central computer where the AKTS-Datalogger Manager Software continuously collects all readings submitted by smartphones or similar transmitters (after data encryption and QR code identification) along with geo-location information.

All authorized senders and receivers have instant access to collected information, which allows the temperature and deterioration extent of stored (shipped) products to be maintained within approved temperature specifications.

3. Materials and Methods

3.1. Propellant

As a model sample, the nitrocellulose-based propellant was used. All analyses were carried out in compliance with the NATO operational procedures.

3.1.1. Analytical Methods

Pressure Firing (PF)

The PF testing method is based on the measurement of the maximum pressure during firing. In general, the measuring system employs a piezoelectric transducer flush mounted in the chamber of the test barrel. The pressure developed by the gases from the burning propellant exerts a force on the transducer through the cartridge case wall, causing the transducer to deflect, and creating a measurable electric charge which, after calibration, is converted into the pressure. The propellant aging increases the maximum pressure value.

The measurements were performed with materials aged at 60, 70, and 80 °C using a home-made experimental arrangement. The data obtained for propellant aged at 50 °C were used for verification of the simulation procedure. The pressure was measured by a piezo-resistive pressure sensor while speed measurements were done by the photoelectric sensor placed in the light barrier.

Gas Evolution (VST)

VST is based on the measurement of the volume of gas evolved on heating under specified conditions. The sample was heated in an evacuated tube and the volume of gas evolved was determined by either the mercury manometric method or by using a pressure transducer.

Experiments were carried out using a tube Vacuum Stabilizer Tester (OZM Research, Hrochův Týnec, Czech Republic) equipped with a Pressure Transmitter DMP 331 (BD Sensors GmbH, Thierstein, Germany). The measurements were performed with materials aged at 60, 70, and 80 °C. The data obtained for propellant aged at 50 °C were used for verification of the simulation procedure.

Ultra-Performance Liquid Chromatography (UPLC)

UPLC was used for stability testing by chromatographic monitoring of the stabilizer concentration which in fresh propellant amounts to ca. 1% and decreases during propellant aging. This method allows monitoring of the amount of stabilizer and allows detection if its concentration in the propellant does not drop below a safety level. The testing procedure fulfills the requirements AOP-48 Ed.2 [10]. UPLC analyses were carried out on a UPLC-MS Acquity chromatograph (Waters, Milford, MA, USA) with the following experimental settings: Column Acquity UPLC BEH C18 1.7 µm 2.1 × 50 mm, isocratic mode, column temperature of 45 °C, flow rate of 0.7 mL·min^{-1}, injection volume of 0.5 µL, and measuring time of 1.2 min. The measurements were performed with materials aged at 60, 70, and 80 °C. The data obtained for propellant aged at 50 °C were used for verification of the simulation procedure.

Heat Flow Calorimetry (HFC)

HFC monitors the evolution of heat during the decomposition of the propellants, allowing detection of the heat generation in the µW range. With such a high sensitivity, it is possible to investigate the early stages of decomposition, i.e., for the reaction progresses, α, in the range from 0 to 0.05. Due to the fact that we have the results of very long experiments at a relatively low temperature (50 °C, 7 years), the results obtained by means of HFC were used for the validation of our kinetic and statistical approaches. In the kinetic analysis depicted in this study, the duration of the experimental domain was restricted to 3 months and the maximal heat evolved was lower than 450 J·g^{-1}. Kinetic analysis was performed with arbitrarily chosen discontinuous data in the range of 70 to 90 °C. The testing procedure based on HFC is described in STANAG 4582 [11]. The measurements were performed using a TAM IV microcalorimeter (TAInstruments, New Castle, DE, USA). Data of the samples aged at 70, 80, and 90 °C were used for kinetic analysis. Additional experiments carried out once a year at 50 and 60 °C were used for verification of the simulation results.

3.2. Vaccine

A freeze-dried live-attenuated virus measles vaccine was used in our study as a model substance.

3.2.1. Stability Monitoring

The kinetics of the deterioration of the vaccine were studied by the plaque assay method by determining the infectious level of virus preparations. The virus titers are expressed as log10 of plague forming units per vial. The experimental results presented by Allison at al. [52] obtained at 31, 37, and 45 °C were simulated by both, one-, and two steps models [2].

4. Conclusions

The prediction of the shelf life of nitrocellulose-based propellants and a model vaccine was performed by an advanced kinetic analysis based on discontinuously collected experimental points. It was achieved by modifications of the often applied kinetic and model selection approaches for the phenomenological description of the reaction course versus time. A specific model selection tool was required because the available data were in the form of sparse experimental points. The criteria for comparing models were based on Akaike and Bayesian information theory. The presented results confirmed that the applied procedure helped in balancing between the goodness of fit, the number of parameters, and the models to be used. The results of the simulation of the shelf lives of the considered materials clearly indicate that the application of simplified kinetic models (it is often assumed that the investigated reaction course can be sufficiently well described by 0th or 1st order models) do not result in the correct prediction of their long-term stability. Even worse, depending on the applied test method, these simplified models result in an underestimation (Figure 1) or overestimation (Figures 2 and 4 for the propellant and Figure 7 for the vaccine) of the shelf life. The correct kinetic evaluation of a material's behavior is therefore of great importance from both safety and financial points of view. Our results show that the prediction of shelf life can be based on the results collected by any analytical technique: For the propellant, the simulations were conducted using data from pressure firing, gas evolution, stabilizer concentration, and heat flow calorimetry techniques. In the kinetic analysis of the vaccine, determination of the rate of the loss of virus infectivity (infectious titer) was used.

We implemented the kinetic parameter evaluated by the advanced kinetic approach into a programmable data logger, i.e., an electronic device that records chosen specific parameters (such as temperature or humidity) over time. Collection of the temperature–time dependences with a chosen frequency (once an hour, day or week) introduced into the kinetic simulation procedure led to continuous estimation of the aging degree. Due to the fact that such a procedure continuously supplies the current degree of deterioration, it has major advantages compared with the common case, in which data loggers record only the temperature–time dependence.

Knowledge of the real rate (not determined by simplified kinetics) of the aging process of any goods allows the prediction of the time–reaction extent dependence for any arbitrarily chosen temperature fluctuation. More generally, the presented procedure enables the estimation of the long-term behavior of any material whose property changes may be expressed by both the kinetic analysis and temperature–time dependences provided by data loggers. The described approach is universal and can be applied for any goods, any methods of shelf life determination, and any type of data loggers.

Author Contributions: Conceptualization, B.R.; methodology, B.R and R.B.; software, B.R., C.A.L., and N.S.; resources, P.F., A.S., A.D., R.D., N.S., O.V., and M.D.; formal analysis, B.R. and M.H.; data curation, P.F., A.S., A.D., and R.D.; writing—original draft preparation, B.R.; writing—review and editing, B.R., R.D., and R.B.; supervision, B.R.; funding acquisition, B.R., P.F., A.S., A.D., R.D., N.S., O.V., and M.D.

Funding: This research received no external funding.

Conflicts of Interest: The authors declare no conflict of interest.

References

1. Cargosense Homepage. Available online: http://www.cargosense.com/supply-chain-intelligence-the-internet-of-things-iot-and-its-impact-on-healthcare-logistics/ (accessed on 24 April 2019).
2. Roduit, B.; Hartmann, M.; Folly, P.; Sarbach, A.; Baltensperger, R. Prediction of thermal stability of materials by modified kinetic and model selection approaches based on limited amount of experimental points. *Thermochim. Acta* **2014**, *579*, 31–39. [CrossRef]
3. Brown, M.E.; Maciejewski, M.; Vyazovkin, S.; Nomen, R.; Sempere, J.; Burnham, A.; Opfermann, J.; Strey, R.; Anderson, H.L.; Kemmler, A.; et al. Computational aspects of kinetic analysis. Part A: The ICTAC kinetics project-data, methods and results. *Thermochim. Acta* **2000**, *355*, 125–143. [CrossRef]
4. Vyazovkin, S.; Burnham, A.K.; Criado, J.M.; Pérez-Maqueda, L.A.; Popescu, C.; Sbirrazzuoli, N. ICTAC kinetics committee recommendations for performing kinetic computations on thermal analysis data. *Thermochim. Acta* **2011**, *520*, 1–19. [CrossRef]
5. AKTS-Thermokinetics Software Version 5.1. Available online: http://www.akts.com (accessed on 13 January 2019).
6. Akaike, H. A new look at the statistical model identification. *IEEE Trans. Autom. Control* **1974**, *19*, 716–723. [CrossRef]
7. Burnham, K.P.; Anderson, D.R. *Model Selection and Multimodel Inference: A Practical Information-Theoretic Approach*, 2nd ed.; Springer: New York, NY, USA, 2002.
8. Bohn, M.A. Assessment of description quality of models by information theoretical criteria based on Akaike and Schwarz-Bayes applied with stability data of energetic materials. In Proceedings of the 46th International Annual Conference of ICT on "Energetic Materials—Performance, Safety and System Applications", Karlsruhe, Germany, 23–26 June 2015; pp. 6.1–6.23.
9. De Klerk, W.P.C. Assessment of Stability of propellants and safe lifetimes. *Propellants Explos. Pyrotech.* **2015**, *40*, 388–393. [CrossRef]
10. NATO AOP-48. *Explosives, Nitrocellulose Based Propellants—Stability Test Procedures and Requirements Using Stabilizer Depletion*, 2nd ed.; North Atlantic Treaty Organization: Brussels, Belgium; Military Agency for Standardization: Brussels, Belgium, 2008.
11. NATO STANAG 4582. *Explosives, Nitrocellulose-Based Propellants, Stability Test Procedure and Requirements Using Heat Flow Calorimetry*; Military Agency for Standardization: Brussels, Belgium; North Atlantic Treaty Organization: Brussels, Belgium, 2004.
12. Carstensen, J.T.; Rhodes, C.T. *Drug Stability, Principles and Practices*, 3rd ed.; CT Informa Healthcare: New York, NY, USA, 2007.
13. Kartoglu, U.; Milstein, J. Tools and approaches to ensure of vaccines throughout the cold chain. *Expert Rev. Vaccines* **2014**, *13*, 843–854. [CrossRef]
14. Waterman, K.C. Understanding and predicting pharmaceutical product shelf life. In *Handbook of Stability Testing in Pharmaceutical Development*; Huynh-Ba, K., Ed.; Springer: New York, NY, USA, 2000; pp. 115–135.
15. Waterman, K.C.; Adami, R.C. Accelerated aging: Prediction of chemical stability of pharmaceuticals. *Int. J. Pharm.* **2005**, *293*, 101–125. [CrossRef] [PubMed]
16. Waterman, K.C.; Carella, A.J.; Gumkowski, M.J.; Lukulay, P.; MacDonald, B.C.; Roy, M.C.; Shamblin, S.L. Improved protocol and data analysis for accelerated shelf life estimation of solid dosage forms. *Pharm. Res.* **2007**, *24*, 780–790. [CrossRef]
17. Fan, Z.; Zhang, L. One- and two-stage Arrhenius model for pharmaceutical shelf life prediction. *J. Biopharm. Stat.* **2015**, *25*, 307–316. [CrossRef]
18. Fu, M.; Perlman, M.; Lu, Q.; Varga, C. Pharmaceutical solid-state kinetic stability investigation by using moisture modified Arrhenius equation and JMP Statistical software. *J. Pharm. Biomed. Anal.* **2015**, *107*, 370–377. [CrossRef]
19. Almalik, O.; Nijhuis, M.B.; van den Heuvel, E.R. Combined statistical analyses for long-term stability data with multiple storage conditions: A simulation study. *J. Biopharm. Stat.* **2014**, *24*, 493–506. [CrossRef] [PubMed]
20. Faya, P.; Seaman, J.W., Jr.; Stamey, J.D. Using accelerated drug stability results to inform long-term studies in shelf life determination. *Stat. Med.* **2018**, *37*, 2599–2615. [CrossRef] [PubMed]

21. Khan, S.R.; Kona, R.; Faustino, P.J.; Gupta, A.; Taylor, J.S.; Porter, D.A.; Khan, M. United States Food and Drug Administration and Department of Defense shelf life extension program of pharmaceutical products: Progress and promise. *J. Pharm. Sci.* **2014**, *103*, 1331–1336. [CrossRef] [PubMed]
22. Clénet, D.; Imbert, F.; Probeck, P.; Rahman, N.; Ausar, A.F. Advanced kinetic analysis as a tool for formulation development and prediction of vaccine stability. *J. Pharm. Sci.* **2014**, *103*, 3055–3062. [CrossRef] [PubMed]
23. Clénet, D. Accurate prediction of vaccine stability under real storage conditions and during temperature excursions. *Eur. J. Pharm. Biopharm.* **2018**, *125*, 76–84. [CrossRef] [PubMed]
24. Clancy, D.; Hodnett, N.; Orr, R.; Owen, M.; Peterson, J. Kinetic model development for accelerated stability studies. *AAPS PharmSciTech* **2017**, *18*, 1158–1176. [CrossRef] [PubMed]
25. World Health Organization. Guidelines on Stability Evaluation of Vaccines. WHO/BS/06.249-Final. 2006. Available online: https://www.who.int/biologicals/publications/trs/areas/vaccines/stability/en/ (accessed on 11 January 2019).
26. Controlled Temperature Chain Working Group. *Controlled Temperature Chain: Strategic Roadmap for Priority Vaccines 2017–2020*; World Health Organization: Geneva, Switzerland, 2017. Available online: https://www.who.int/immunization/programmes_systems/supply_chain/ctc_strategic_roadmap_priority_vaccines.pdf (accessed on 13 December 2018).
27. World Health Organization. WHO PQS Prequalified Devices and Equipment, E006 Temperature Monitoring Devices. Available online: http://www.who.int/immunization/documents/financing/who_ivb_15.04/en/ (accessed on 11 January 2019).
28. WHO. *Controlled Temperature Chain (CTC)*; World Health Organization (WHO): Geneva, Switzerland, 2016. Available online: http://www.who.int/immunization/programmes_systems/supply_chain/ctc/en/ (accessed on 19 December 2018).
29. Haynes, J.D. Worldwide virtual temperatures for product stability testing. *J. Pharm. Sci.* **1971**, *60*, 927–929. [CrossRef]
30. Seevers, R.H.; Hoffer, J.; Harber, P.; Ulrich, D.A.; Bishara, R. The use of Mean Kinetic temperature (MKT) in the handling, storage and distribution of temperature sensitive pharmaceuticals. *Pharm. Outsourcing* **2009**, *10*, 12–17. Available online: https://studylib.net/doc/8854836/the-use-of-mean-kinetic-temperature--mkt-, (accessed on 15 December 2018).
31. Tong, C.; Lock, A. A computational procedure for Mean Kinetic Temperature using unequally spaced data. In Proceedings of the Joint Statistical Meeting 2015, Biopharmaceutical Section, Seattle, WA, USA, 8–13 August 2015; pp. 2065–2070.
32. Health Products Regulatory Authority (HPRA). Guide to Control and Monitoring of Storage and Transportation Temperature Conditions for Medicinal Products and Active Substances IA-G0011-2.2017. Available online: https://www.hpra.ie/docs/default-source/publications-forms/guidance-documents/ia-g0011-guide-to-control-and-monitoring-of-storage-and-transportation-conditions-v2.pdf (accessed on 15 December 2018).
33. Okeke, C.C.; Bailey, L.C.; Medwick, T.; Grady, L.T. Temperature fluctuations during mail order shipments of pharmaceutical articles using Mean Kinetic temperature approach. *Pharm. Forum* **1997**, *23*, 4155–4182.
34. Okeke, C.C.; Bailey, L.C.; Lindauer, R.F.; Medwick, T.; Grady, L.T. Evaluation of the physical and chemical stability of some drugs when exposed to temperature fluctuations during shipment. *Pharm. Forum* **1998**, *24*, 7064–7073.
35. Kumru, O.S.; Joshi, S.B.; Smith, D.E.; Middaugh, C.R.; Prusik, T.; Volkin, D.B. Vaccine instability in the cold chain: Mechanisms, analysis and formulation strategies. *Biologicals* **2014**, *42*, 237–250. [CrossRef] [PubMed]
36. Zhang, J.; Pritchard, E.; Hu, X.; Valentin, T.; Panilaitis, B.; Omenetto, F.G.; Kaplan, D.L. Stabilization of vaccines and antibiotics in silk and eliminating the cold chain. *Proc. Natl. Acad. Sci. USA* **2012**, *109*, 11981–11986. [CrossRef] [PubMed]
37. Watzky, M.A.; Morris, A.M.; Ross, E.D.; Finke, R.G. Fitting yeast and mammalianprion aggregation kinetic data with the Finke-Watzky two-step model of nucleation and autocatalytic growth. *Biochemistry* **2008**, *47*, 10790–10800. [CrossRef] [PubMed]
38. Morris, A.M.; Watzky, M.A.; Finke, R.G. Protein aggregation kinetics, mechanism and curve-fitting: A review of the literature. *Biochim. Biophys. Acta* **2009**, *1794*, 375–397. [CrossRef] [PubMed]
39. Brown, M.E.; Glass, B.D. Pharmaceutical application of the Prout-Tompkins rate equation. *Int J. Pharm.* **1999**, *190*, 129–137. [CrossRef]

40. Ammann, C. A mathematical approach to assessing temperature excursions in temperature-controlled chains. *Eur. J. Parenter. Pharm. Sci.* **2008**, *13*, 57–59.
41. Capen, R.; Christopher, D.; Forenzo, P.; Ireland, C.; Liu, O.; Lyapustina, S.; O'Neill, J.; Patterson, N.; Quinlan, M.; Sandell, D.; et al. On the shelf life of pharmaceutical products. *AAPS PharmSciTech* **2012**, *13*, 911–918. [CrossRef]
42. Madakam, S.; Ramaswamy, R.; Tripathi, S. Internet of Things (IoT): A literature review. *J. Comp. Commun.* **2015**, *3*, 164–173. [CrossRef]
43. Proceedings of the GloTs 2017: Global Internet of Things Summit, Geneva, Switzerland, 6–9 June 2017. Available online: https://ieeexplore.ieee.org/xpl/mostRecentIssue.jsp?punumber=8011434 (accessed on 13 March 2019).
44. Tsang, Y.P.; Choy, K.L.; Wu, C.H.; Ho, G.T.S.; Lam, H.Y.; Koo, P.S. An IoT-based cargo monitoring system for enhancing operational effectiveness under a cold chain environment. *Int. J. Eng. Bus. Manag.* **2017**, *9*, 1–13. [CrossRef]
45. Salunkhe, P.G.; Nerkar, R. IoT driven smart system for best cold chain application. In Proceedings of the 2016 International Conference on Global Trends in Signal Processing, Information Computing and Communication (ICGTSPICC), Jalgaon, Maharashtra, India, 22–24 December 2016; pp. 64–67. Available online: https://ieeexplore.ieee.org/document/7955270 (accessed on 20 January 2019).
46. Monteleone, S.; Sampalo, M.; Maia, R.F. A novel deployment of smart Cold Chain system using 2g-RFID-Sys temperature monitoring in medicine Cold Chain based on Internet of Things. In Proceedings of the 2017 IEE International Conference on Service Operations and Logistics, and Informatics (SOLI), Bari, Italy, 18–20 September 2017; pp. 205–210. Available online: https://ieeexplore.ieee.org/document/8120995 (accessed on 20 January 2019).
47. Mohsin, A.; Yellampalli, S.S. IoT based cold chain logistics monitoring. In Proceedings of the International Conference on Power, Control, Signals and Instrumentation Engineering (ICPCSI), Chennai, India, 21–22 September 2017; pp. 1971–1974. Available online: https://ieeexplore.ieee.org/document/8392059 (accessed on 15 January 2019).
48. Ping, H.; Wang, J.; Ma, Z.; Du, Y. Mini-review of application of IoT technology in monitoring agricultural products quality and safety. *Int. J. Agric. Biol. Eng.* **2018**, *11*, 35–45. [CrossRef]
49. Corradini, M.G. Shelf Life of Food products: From open Labeling to real-time measurements. *Annu. Rev. Food Sci. Technol.* **2018**, *9*, 251–269. [CrossRef] [PubMed]
50. Available online: https://www.mckinsey.com/industries/high-tech/our-insights/the-internet-of-things (accessed on 15 January 2019).
51. Yuen, J.S.M.; Choy, K.I.; Lam, H.Y.; Tsang, Y.P. An Intelligent-Internet of Things (IoT) outbound Logistic knowledge management system for handling temperature sensitive products. *Int. J. Knowl. Syst. Sci.* **2018**, *9*, 23–40. [CrossRef]
52. Allison, L.M.C.; Mann, G.F.; Perkins, F.T.; Zuckerman, A.J. An accelerated stability test procedure for lyophilized measles vaccines. *J. Biol. Stand.* **1981**, *9*, 185–194. [CrossRef]

Sample Availability: Samples of the compounds are available from the authors.

© 2019 by the authors. Licensee MDPI, Basel, Switzerland. This article is an open access article distributed under the terms and conditions of the Creative Commons Attribution (CC BY) license (http://creativecommons.org/licenses/by/4.0/).

Article

Non-Isothermal Sublimation Kinetics of 2,4,6-Trinitrotoluene (TNT) Nanofilms

Walid M. Hikal [1,2,3,*] **and Brandon L. Weeks** [1]

1. Department of Chemical Engineering, Texas Tech University, Lubbock, TX 79409, USA; brandon.weeks@ttu.edu
2. Department of Mathematics, Australian College of Kuwait, Safat 13015, Kuwait
3. Department of Physics, Faculty of science, Assiut University, Assiut 71516, Egypt
* Correspondence: w.hikal@ack.edu.kw

Academic Editors: Sergey Vyazovkin and Derek J. McPhee
Received: 27 January 2019; Accepted: 20 March 2019; Published: 23 March 2019

Abstract: Non-isothermal sublimation kinetics of low-volatile materials is more favorable over isothermal data when time is a crucial factor to be considered, especially in the subject of detecting explosives. In this article, we report on the in-situ measurements of the sublimation activation energy for 2,4,6-trinitrotoluene (TNT) continuous nanofilms in air using rising-temperature UV-Vis absorbance spectroscopy at different heating rates. The TNT films were prepared by the spin coating deposition technique. For the first time, the most widely used procedure to determine sublimation rates using thermogravimetry analysis (TGA) and differential scanning calorimetry (DSC) was followed in this work using UV-Vis absorbance spectroscopy. The sublimation kinetics were analyzed using three well-established calculating techniques. The non-isothermal based activation energy values using the Ozawa, Flynn–Wall, and Kissinger models were 105.9 ± 1.4 kJ mol^{-1}, 102.1 ± 2.7 kJ mol^{-1}, and 105.8 ± 1.6 kJ mol^{-1}, respectively. The calculated activation energy agreed well with our previously reported isothermally-measured value for TNT nanofilms using UV-Vis absorbance spectroscopy. The results show that the well-established non-isothermal analytical techniques can be successfully applied at a nanoscale to determine sublimation kinetics using absorbance spectroscopy.

Keywords: 2,4,6-trinitrotoluene (TNT); sublimation; activation energy; UV spectroscopy; spin coating; explosives detection

1. Introduction

Explosives detection keeps rising as a critical issue due to the global rise in terrorist activity and needs more intensive investigation. Sublimation kinetics of explosives at a submicron scale or even nanoscale is crucial for their trace detection [1]. Measuring the sublimation kinetics of an explosive in the nanometer scale could be used to determine its persistence as well as its lifetime. 2,4,6-trinitrotoluene (TNT) is among the most widely used secondary explosives in both military and industry applications due to its low sensitivity and safe handling. However, there is a large discrepancy in the activation energy values reported for TNT's sublimation value in literature (90–141 kJ/mol).

Both isothermal and non-isothermal sublimation kinetics are often measured using thermogravimetry analysis (TGA), differential scanning calorimetry (DSC), and differential thermal analysis (DTA). Sublimation kinetics is usually measured using TGA where a flow of gas through both the balance chamber and the sample is often needed to prevent temperature and pressure build-up [2,3]. However, the mass change detectable by TGA (few nanograms) limits usefulness to samples larger than a few milligrams, depending on the sensitivity of the balance. This makes TGA an unreliable technique to study thermodynamic properties of nanofilms. Quartz crystal microbalance (QCM) [4,5] and atomic force microscopy (AFM) [6,7] have been used to isothermally determine sublimation kinetics of explosives' microparticles. However, both techniques cannot be operated neither in-situ nor non-isothermally. AFM and QCM also require long times in data collection. This is expected to introduce errors to the measured sublimation kinetics.

Recently, we reported a new in-situ methodology to isothermally determine sublimation rates, activation energies of sublimation, and vapor pressures of continuous nanofilms of low volatile materials using UV-absorbance spectroscopy [8–10]. The determined sublimation rates were shown to be more accurate than those obtained using both AFM and QCM. Using UV-Vis absorbance spectroscopy in determining sublimation kinetics has the advantage of eliminating the surface area from the rate equation. In addition, the relatively small scanned area ensures accurate measurements even with the existence of surface roughness, dislocations, and cracks. The technique was shown to be accurate when operated isothermally. It was also shown that when operated non-isothermally using one set of data at a single heating rate, values within the reported data were achieved.

In this article we illustrate the successful use of UV-Vis absorbance spectroscopy in non-isothermal measurement of TNT sublimation kinetics, using different heating rates and three different famous calculation techniques. TNT sublimation kinetics were only determined non-isothermally once before using UV-Vis spectroscopy with a crude temperature integral approximation in the kinetics equation used [11]. No other techniques have been reported to measure the sublimation kinetics of secondary explosives non-isothermally due their low melting point (80 °C for TNT). In addition, the common procedure used to determine sublimation kinetics non-isothermally in famous techniques uses different heating rates. However, this procedure is not suitable in these techniques for low volatile explosives such as TNT, 1,3,5-Trinitro-1,3,5-triazinane (RDX), and Pentaerythritol tetranitrate (PETN). For the first time, we report the use of this procedure to non-isothermally determine the activation energy of TNT sublimation using absorbance spectroscopy.

2. Experimental Section

2,4,6-trinitrotoluene (TNT) was provided by Austin explosives and was purified by a crystallization technique using the evaporation methodology from acetone solution. The resulting rod-like pure crystallites were then dissolved in acetone at room temperature. A stock solution of 0.2 M TNT was used for preparing continuous TNT films on quartz substrates by spin coating (single wafer spin processor, Laurell technologies corp., North Wales, PA, USA) of 20 μL TNT solutions at 4500 rpm for one minute. The quartz substrates were cleaned using acetone and de-ionized water before the films were deposited.

The films were characterized by a PSIA XE AFM (Santa Clara, CA, USA) in contact mode with a silicon cantilever (Nanosensor pointprobes, Nominal spring constant 5.0 N/m). Absorbance of TNT was recorded in situ using a Lambda 1050 UV/Vis/NIR (Perkin-Elmer, UK) spectrometer at 1.0 nm resolution. The spectrometer is equipped with a temperature controller, with an accuracy of 0.05 °C, allowing for in situ temperature-dependent absorbance measurements for both sample and reference. Time-driven measurements of TNT absorbance at the peak with highest absorbance were collected at different heating rates, at different temperatures below the melting point (80 °C).

3. Theory

The dependence of kinetic process rates on temperature is usually represented by the temperature-dependent rate constant, $k(T)$, and the dependence on the extent of conversion by the reaction model, $f(\alpha)$. Sublimation kinetics of a solid volatile material is usually studied using thermogravimetric analysis (TGA) and is often described by the following basic equation:

$$-\frac{dm}{dt} = k(T)f(\alpha) \tag{1}$$

where m is the mass, α is the extent of conversion which is defined as $\alpha = \frac{(w_i - w_T)}{(w_i - w_f)}$ (where w_i and w_f the initial and final mass, respectively, while w_T is the mass loss at temperature T), $f(\alpha)$ is the model function, which assumes different mathematical forms depending on the reaction mechanism [12], and $k(T)$ is the specific rate constant in g/s, whose temperature dependence is commonly described by the Arrhenius equation [13]:

$$ln\, k(T) = ln\, A - E_a/RT \quad (2)$$

where E_a is the activation energy, A is the pre-exponential/frequency factor, R the gas constant and T the absolute temperature. The plot of $ln(k)$ versus $1/T$ is linear and from the slope, the activation energy (E_a) for the sublimation is calculated. The proposed kinetic analysis, based on dynamic model-free methods using data obtained at different fixed heating rates, $\beta = \frac{dT}{dt}$ seems to be the most reliable approach. The thermal sublimation kinetics is examined using the onset temperature and the Ozawa equation given by [14]:

$$ln(\beta) = const - 1.052 \left(\frac{E_a}{RT_\alpha} \right) \quad (3)$$

Kissinger's Technique was also used to determine the activation energy of sublimation using the equation [15–17]:

$$ln\, \beta / T_m^2 = ln\left(\frac{AR}{E_a} \left[n(1-\alpha)_m^{n-1} \right] \right) - \frac{E_a}{RT_m} \quad (4)$$

where T_m is the 50% mass loss temperature at a given β value, A the pre-exponential factor, E_a the activation energy and R the gas constant. The E_a value can be calculated from the slope of $ln(\beta/T_m^2)$ as a function of $1/T_m$. Kinetic study of sublimation steps was also performed using the isoconversional method of Flynn–Wall Technique [18]. The isoconversional expression is,

$$ln(\beta) = ln\left(\frac{AE_a}{R} \right) - ln\, F(\alpha) - E_a/RT \quad (5)$$

where $F(\alpha)$ is the integral form of $f(\alpha)$. The activation energy, E_a can be calculated from a plot of $ln\, \beta$ versus $1/T$ at a fixed weight loss since the slope of such a line is given by E_a/R (gas constant 8.314 J mol^{-1} K^{-1}). The $ln\, A$ is calculated from the intercept value of the line and the derived E_a value.

In using absorbance spectroscopy to determine the sublimation kinetics, the sublimation kinetics can be determined by relating the mass loss to the decrease in the thickness of a nanofilm of the material, assuming a homogenous sublimation from the films' surface as follows: for a nanofilm of volume V, surface area S, and thickness l, we have $\frac{dm}{Sdt} = \frac{\rho dV}{Sdt} = \frac{\rho dl}{dt}$, where ρ is the density of the material. The optical absorbance (A) of a nanofilm of thickness l and absorbance coefficient α is given by the Lambert law and can be written as [19,20] $A = \alpha\, l$, thus the mass loss can be written as $\frac{\rho}{\alpha} \frac{dA}{dt}$. Hence the rate of the absorbance decrease can be used as property to monitor the mass loss and can be used to determine the sublimation kinetics.

4. Results and Discussion

TNT nanofilms used in this study have been shown to be continuous by using optical microscopy. In addition, the continuity of the films has been confirmed by atomic force microscopy. The thickness of the films was measured using AFM operated in contact mode at room temperature by removing a part of the film using a tape. The films thicknesses are ~500–600 nm with an RMS surface roughness of ~30 nm [8].

TNT nanofilms exhibit two prominent absorbance peaks centered at 264 nm and 230 nm as shown in Figure 1. However, the two peaks are broad and overlap. The location of the absorbance peaks was determined by the second derivatives of the absorbance spectra using UV Winlab Data Processor & Viewer software. The locations of the two prominent peaks do not change over the temperature range used in this work.

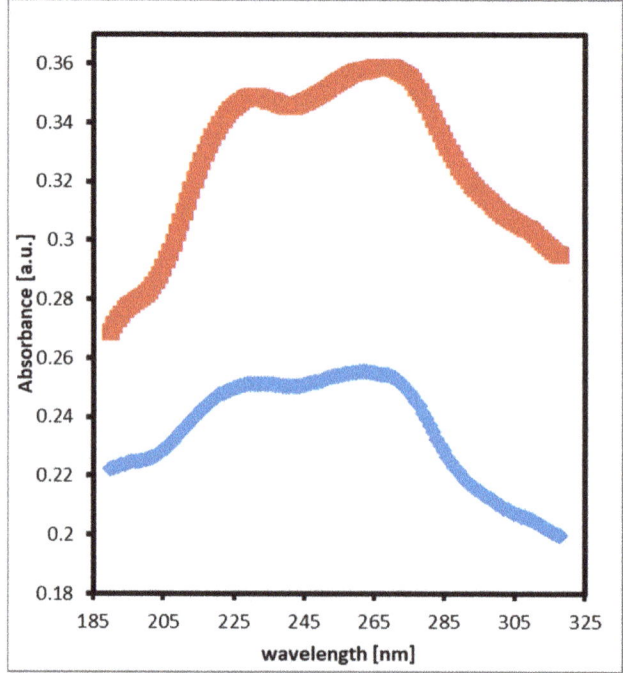

Figure 1. Absorbance spectrum of TNT nanofilms at 50 °C (bottom) and 70 °C (top) showing no shift in the absorbance peaks upon temperature change.

Figure 2 shows the normalized non-isothermal absorbance change in TNT monitored at the 264 nm peak upon heating at four different heating rates from 0.25 K/min to 1.5 K/min. The heating effect on the non-isotherms of TNT is clearly observed and consistent. The sublimation process is a single step process in the defined temperatures ranges (30–80 °C). For a specific weight loss, the sublimation temperature rises by increasing the heating rate. The non-isotherms show that the change in absorbance is insignificant at the beginning of the heating process until the temperature reaches a specific temperature known as the onset temperature (T_o). The onset temperature (T_o) can be determined as the temperature at which the line of best fit for the linear part of each isotherm intersects with the zero mass loss line (zero absorbance change line in Figure 2). Each isotherm shows a different onset temperature (T_o) that increases with increasing the heating rate.

Figure 3 shows a plot of $ln(\beta)$ versus the inverse of the absolute onset temperatures according to the Ozawa method. The plot is linear and the activation energy is calculated to be 105.9 ± 1.4 KJ/mol.

Figure 4 shows a plot of $ln(\beta/T^2)$ versus the inverse of the absolute temperatures corresponding to specific weight losses ranging from 5 to 50% for three heating rates, according to the Flynn–Wall method. The plots are linear and result in very close activation energies of sublimation values. However, for clarity, only the linear fit (black line) for the average values is shown. The activation energy was calculated to be 102.1 ± 2.7 KJ/mol.

Figure 5 represents a plot of $ln(\beta/T^2)$ versus the inverse of the absolute onset temperatures according to the Kissinger method. The plot is linear and the activation energy was calculated to be 105.8 ± 1.6 KJ/mol.

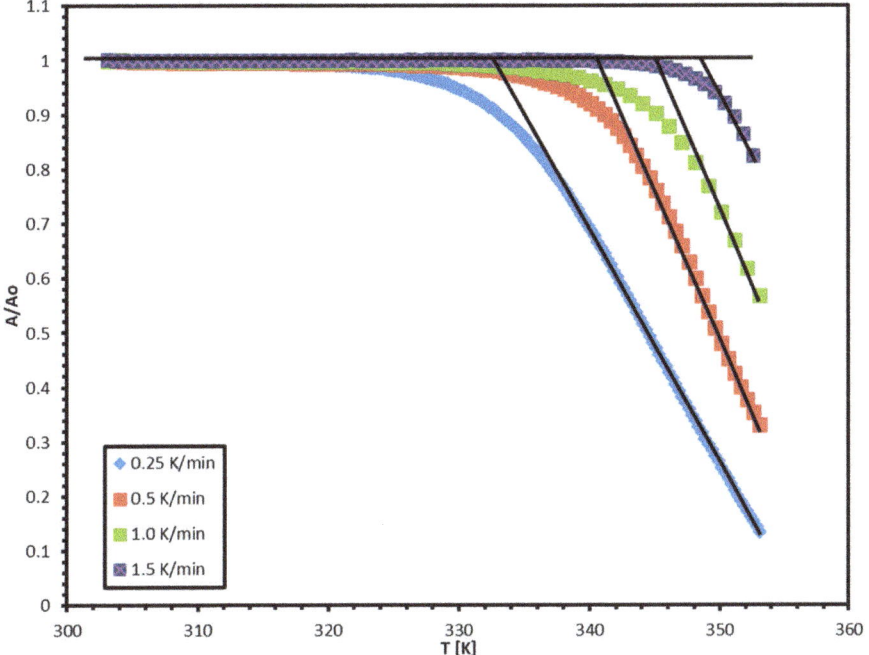

Figure 2. The effect of heating rate on the absorbance non-isotherms of TNT.

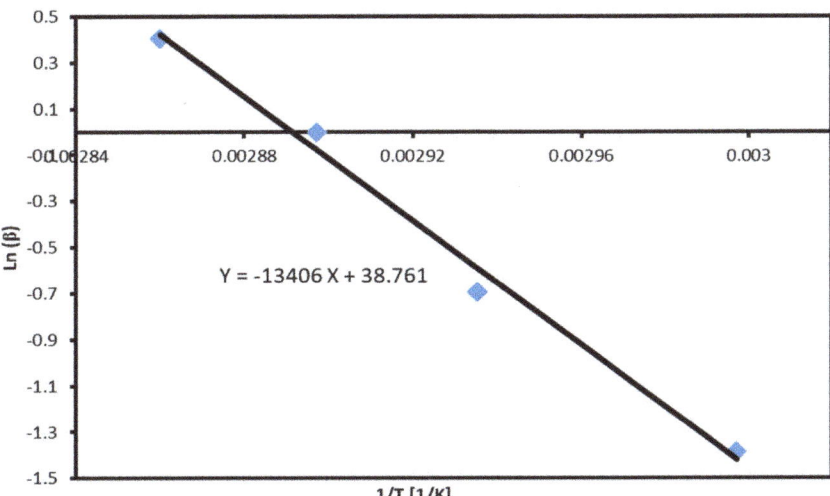

Figure 3. Plot of $ln(\beta)$ versus the inverse of the absolute onset temperatures according to the Ozawa method.

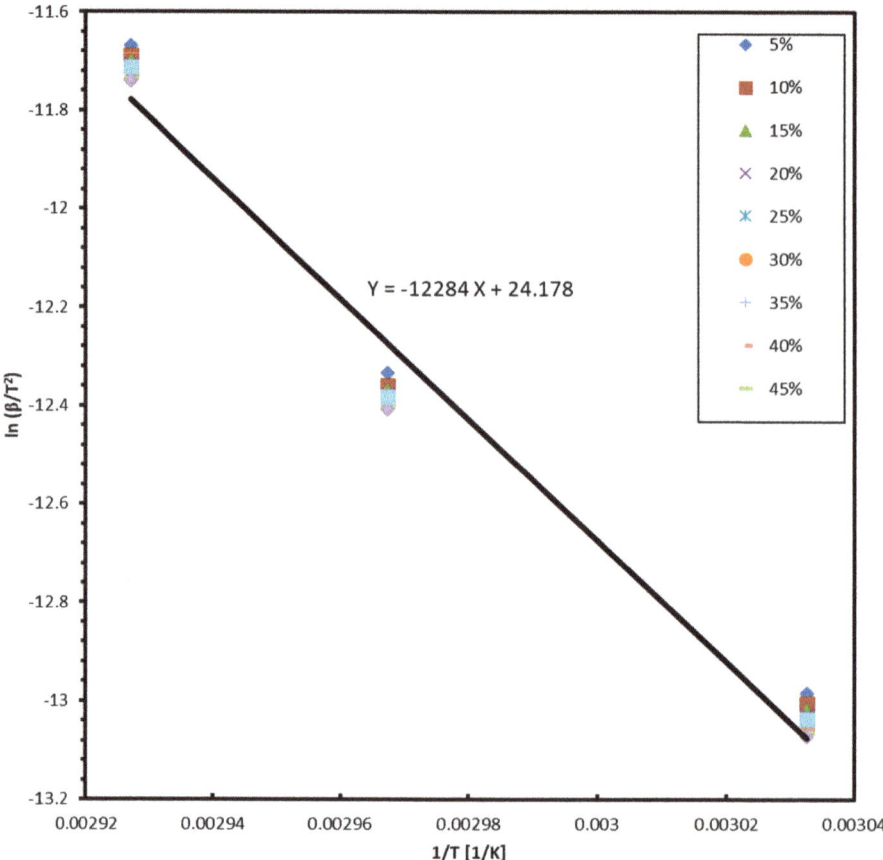

Figure 4. Plot of $ln(\beta/T^2)$ versus the inverse of the absolute temperature with weight loss from 0.05 to 0.5 in steps of 0.05 according to Flynn–Wall method (β = 0.25, 0.5, 1.0 K/min).

The results indicate good agreement between the values calculated using the Ozawa method, Flynn-wall method, and Kissinger method.

There is a large discrepancy (90–141 kJ/mol) in the activation energies described for TNT sublimation in literature [4,5,21–27]. The values measured here are in very good agreement with the values determined by the most sensitive techniques used in measuring sublimation kinetics of TNT: QCM and Knudsen effusion method 97 ± 7 and 103 kJ/mole, respectively [4,23]. In addition, the value reported here is in good agreement with the value determined isothermally using UV-Vis spectroscopy, 99.6 ± 5 kJ/mole [8]. The results illustrates that the UV-Vis absorbance spectroscopy can be used in the same manner as TGA and DSC to non-isothermally measure the activation energy of sublimation for low volatile materials, aided by the most common analytical techniques.

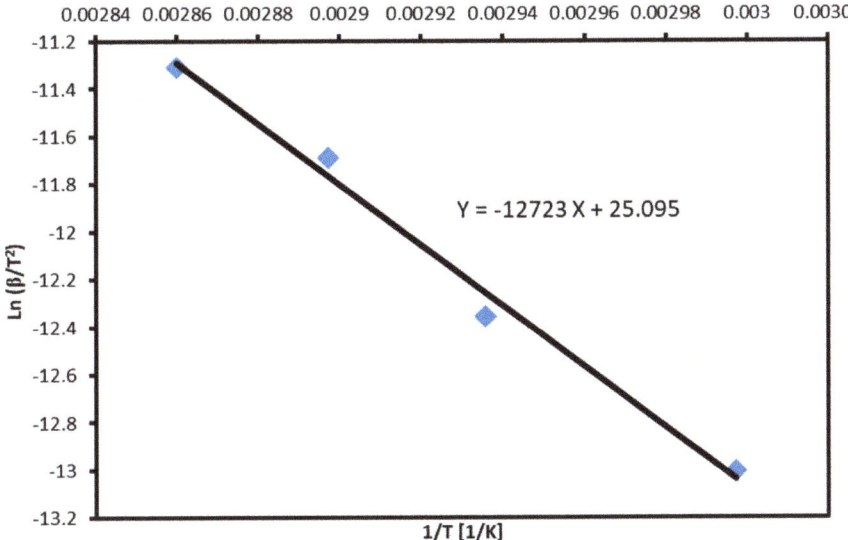

Figure 5. Plot of $ln\beta/T_o^2$ versus the inverse of the absolute temperature according to Kissinger method.

5. Conclusions

TNT nanofilms' sublimation kinetics was investigated using UV-Vis absorbance spectroscopy in the UV range. Non-isothermal heating of the nanofilms at different heating rates was used in this study to determine the activation energy for TNT sublimation. Increasing the heating rate increases the onset temperature of sublimation observed by the technique as well as the temperature corresponding to a specific mass loss. The common data collection procedure followed in common thermal analysis techniques was used in this work with absorbance spectroscopy. Three different non-isothermal analysis models were used and the activation energy for TNT sublimation was calculated. The values are in excellent agreement, and agree well with the values reported using QCM and isothermal UV-Vis absorbance spectroscopy. This method, with the validated new data collection procedure, is expected to have a high impact on the popularity of the new technique and on the field of volatile hazardous materials' detection.

Author Contributions: Conceptualization, W.M.H. and B.L.W.; methodology, W.M.H.; formal analysis, W.M.H.; investigation, W.M.H.; resources, B.L.W.; data curation, W.M.H.; writing—original draft preparation, W.M.H.; writing—review and editing, W.M.H.; supervision, B.L.W.; project administration, B.L.W.; funding acquisition, B.L.W.

Funding: This work was supported by NSF CAREER (CBET-0644832).

Conflicts of Interest: The authors declare no conflict of interest.

References

1. Czarnecki, J.; Sestak, J. Practical Thermogravimetry. *J. Therm. Anal. Calorim.* **2000**, *60*, 759–778. [CrossRef]
2. Hajimirsadeghis, S.S.; Teimouri, M.R.; Rahimi-Nasrabadi, M.; Dehghanpour, S. Non-isothermal kinetic study of the thermal decomposition of N-{bis[benzyl(methyl)amino]phosphoryl}-2,2-dichloroacetamide and N-{bis[dibenzylamino]phosphoryl}-2,2-dichloroacetamide. *J. Therm. Anal. Calorim.* **2009**, *98*, 463–468. [CrossRef]
3. Hobbs, M.L.; Nakos, J.T.; Brady, P.D. Response of a glass/phenolic composite to high temperatures. *J. Therm. Anal. Calorim.* **2011**, *103*, 543–553. [CrossRef]
4. Gershanik, A.P.; Zeiri, Y. Sublimation Rate of TNT Microcrystals in Air. *J. Phys. Chem. A* **2010**, *114*, 12403–12410. [CrossRef] [PubMed]

5. Mu, R.; Ueda, A.; Liu, Y.C.; Wu, M.; Henderson, D.O.; Lareau, R.T.; Chamberlain, R.T. Effects of interfacial interaction potential on the sublimation rates of TNT films on a silica surface examined by QCM and AFM techniques. *Surf. Sci.* **2003**, *530*, L293–L296. [CrossRef]
6. Pitchimani, R.; Burnham, A.K.; Weeks, B.L. Quantitative Thermodynamic Analysis of Sublimation Rates Using an Atomic Force Microscope. *J. Phys. Chem. B* **2007**, *11*, 9182–9185. [CrossRef] [PubMed]
7. Burnham, A.K.; Qiu, S.R.; Pitchimani, R.; Weeks, B.L. Comparison of kinetic paprameters of single crystal pentaerythritol tetranitrate using atomic force microscopy and thermogravimetric analysis: Implications on coarsening mechanisms. *J. Appl. Phys.* **2009**, *105*, 104312. [CrossRef]
8. Hikal, W.M.; Weeks, B.L. Determination of sublimation rate of 2,4,6-trinitrotoluene (TNT) nano thin films using UV-absorbance spectroscopy. *J. Therm. Anal. Calorim.* **2010**, *110*, 955–960. [CrossRef]
9. Hikal, W.M.; Paden, J.T.; Weeks, B.L. Simple method for determining the vapor pressure of materials using UV-absorbance spectroscopy. *J. Phys. Chem. B* **2011**, *115*, 13287–13291. [CrossRef] [PubMed]
10. Hikal, W.M.; Paden, J.T.; Weeks, B.L. Thermo-optical determination of vapor pressures of TNT and RDX nanofilms. *Talanta* **2011**, *87*, 290–294. [CrossRef] [PubMed]
11. Hikal, W.M.; Paden, J.T.; Weeks, B.L. Rapid estimation of the thermodynamic parameters and vapor pressures of volatile materials at the nanoscale. *ChemPhysChem* **2012**, *13*, 2729–2733. [CrossRef] [PubMed]
12. Antony Premkumar, P.; Nagaraja, K.S.; Pankajavalli, R.; Mallika, C.; Sreedharan, O.M. DTA studies on the liquidus temperatures of Cr complex with the addition of an anhydrous Ni complex. *Mater. Lett.* **2004**, *58*, 474–477. [CrossRef]
13. Gao, X.; Chen, D.; Dollimore, D. The correlation between the value of α at the maximum reaction rate and the reaction mechanisms: A theoretical study. *Thermochim. Acta* **1993**, *223*, 75–82. [CrossRef]
14. Burnham, L.; Dollimore, D.; Alexander, K. Calculation of the vapor pressure–temperature relationship using thermogravimetry for the drug allopurinol. *Thermochim. Acta* **2001**, *367–368*, 15–22. [CrossRef]
15. Kissinger, H.E. Reaction kinetics in differential thermal analysis. *Anal. Chem.* **1957**, *29*, 1702–1706. [CrossRef]
16. Budrugeac, P.; Segal, E. Applicability of the Kissinger equation in thermal analysis. *J. Therm. Anal. Calorim.* **2007**, *88*, 703–707. [CrossRef]
17. Jankovic, B.; Mantus, S. Model-fitting and model-free analysis of thermal decomposition of palladium acetylacetonate [Pd(acac)2]. *J. Therm. Anal. Calorim.* **2008**, *94*, 395–403. [CrossRef]
18. Flynn, J.H.; Wall, L.A. A quick, direct method for the determination of activation energy from thermogravimetric data. *J. Polym. Sci. Part B* **1966**, *4*, 323–328. [CrossRef]
19. Ingle, J.D.J.; Crouch, S.R. *Spectrochemical Analysis*; Prentice Hall PTR: Upper Saddle River, NJ, USA, 1988.
20. Schubert, E.F. *Light-Emitting Diodes*, 2nd ed.; Cambridge University Press: Cambridge, UK, 2006.
21. Pella, P.E. Measurement of the vapor pressures of TNT, 2,4-DNT, 2,6-DNT, and EGDN. *J. Chem. Thermodyn.* **1977**, *9*, 301–305. [CrossRef]
22. Leggett, D.C. Vapor pressure of 2,4,6-trinitrotoluene by a gas chromatographic headspace Technique. *J. Chromatogr. A* **1977**, *133*, 83–90. [CrossRef]
23. Cundall, R.B.; Palmer, T.F.; Wood, C.E.C. Vapour pressure measurements on some organic high explosives. *J. Chem. Soc. Faraday Trans. 1* **1978**, *74*, 1339–1345. [CrossRef]
24. Oxley, J.C.; Smith, J.L.; Shinde, K.; Moran, J. Determination of the Vapor Density of Triacetone Triperoxide (TATP) Using a Gas Chromatography Headspace Technique. *Propell. Explos. Pyrot.* **2005**, *30*, 127–130. [CrossRef]
25. Lenchitz, C.; Velicky, R.W. Vapor pressure and heat of sublimation of three nitrotoluenes. *J. Chem. Eng. Data* **1970**, *15*, 401–403. [CrossRef]
26. Edwards, G.T. The vapor pressure of 2,4,6-trinitrotoluene. *Trans. Faraday Soc.* **1950**, *46*, 423–427. [CrossRef]
27. Phelan, J.M.; Patton, R.T. *Sublimation Rates of Explosive Materials-Method Development and Initial Results, Sandia Report SAND2004-4525*; Sandia national laboratories: Albuquerque, CA, USA, 2004.

Sample Availability: Samples of the compounds are available from the authors.

© 2019 by the authors. Licensee MDPI, Basel, Switzerland. This article is an open access article distributed under the terms and conditions of the Creative Commons Attribution (CC BY) license (http://creativecommons.org/licenses/by/4.0/).

Article

Investigation of Size-Dependent Sublimation Kinetics of 2,4,6-Trinitrotoluene (TNT) Micro-Islands Using In Situ Atomic Force Microscopy

Yong Joon Lee and Brandon L. Weeks *

Department of Chemical Engineering, Texas Tech University, Lubbock, TX 79409, USA; yong-joon.lee@ttu.edu
* Correspondence: brandon.weeks@ttu.edu; Tel.: +1-806-834-7450

Academic Editor: Sergey Vyazovkin
Received: 18 April 2019; Accepted: 14 May 2019; Published: 17 May 2019

Abstract: Kinetic thermal analysis was conducted using in situ atomic force microscopy (AFM) at a temperature range of 15–25 °C to calculate the activation energy of the sublimation of 2,4,6-trinitrotoluene (TNT) islands. The decay of different diameter ranges (600–1600 nm) of TNT islands was imaged at various temperatures isothermally such that an activation energy could be obtained. The activation energy of the sublimation of TNT increases as the diameter of islands increases. It was found that the coarsening and the sublimation rate of TNT islands can be determined by the local environment of the TNT surface. This result demonstrates that a diffusion model cannot be simply applied to "real world" systems for explaining the sublimation behavior and for estimating the coarsening of TNT.

Keywords: AFM; TNT; explosives; coarsening; sublimation; activation energy

1. Introduction

Explosives have relatively low vapor pressures, making it difficult to detect traces of explosive in the air through vapor phase sampling. Other methods of detection have relied on collection of micro- or nanometer sized particles which can be subsequently heated, increasing the total explosive in the vapor phase [1–3]. Therefore, understanding the sublimation of explosives at given environmental conditions is crucial for the optimization of sampling and for estimating the aging/coarsening process of explosives [4]. The activation energy of sublimation of energetic materials is regarded as a very important thermodynamic parameter related to detection and aging and needs to be investigated if one is to predict the thermodynamics and aging behavior of explosives. To minimize safety risks, several studies have been recently developed using a variety of nanoscale methodologies to determine the activation energy for sublimation, including TGA, quartz crystal microbalance (QCM), UV spectroscopy, and atomic force microscopy (AFM) [5–13].

TGA is a well-known general method for determining the thermodynamic properties of materials. The activation energy of sublimation, and even the diffusion coefficient, of typical secondary explosives, 2,4,6-trinitrotoluene (TNT), Pentaerythritol tetranitrate (PETN), and hexahydro-1,3,5-trinitro-1,3,5-triazine (RDX), have been reported by measurement of mass loss during isothermal heating [5,6]. QCM is another tool which can be used to calculate the mass change of samples quantitatively by detecting the change in frequency of a quartz crystal resonance. It has been used to describe the sublimation process of explosives, the sublimation rate, and activation energy of sublimation for energetic materials [7–9]. QCM is a more accurate tool for the measurement of mass loss than TGA, but it requires high adhesion between samples and the QCM surface to achieve accurate results with high sensitivity [8]. UV absorbance spectroscopy has also been used to find the heat of sublimation of TNT, PETN, and RDX [10–12]. Only transparent films within a certain range of wavelengths can be used. Samples preparation is important in both QCM and UV absorbance spectroscopy analysis where thin films are normally prepared.

AFM, on the other hand, has the advantage that it does not require a specific sample preparation. Samples can be prepared as thin films (continuous or non-continuous) or even as single crystals, depending upon the purpose of study. Moreover, it has the ability to conduct in situ imaging of surface phenomena at the nanoscale. To calculate the sublimation rate and activation energy of evaporation for PETN quantitatively, ex-situ AFM thermal analysis was first demonstrated by imaging non-continuous PETN films [13]. In other work, the activation energy of the sublimation and nucleation of PETN was obtained using in situ AFM thermal analysis with PETN single crystals [6]. While AFM thermal analysis has been performed on various explosives, surprisingly, the activation energy of TNT sublimation has not been reported. Preparation of nano-TNT is difficult because its vapor pressure is relatively higher than those of other commonly used types of secondary explosives [14].

In this report, we measured the sublimation rate and activation energy for the sublimation of various nano-sized TNT islands (600–1600 nm), using in situ AFM thermal analysis within a low temperature range (15–25 °C). Large variances in the activation energy of TNT (90–141 kJ mol^{-1}) and vapor pressure have been previously reported [5,7,8,10,11,15–19]. This indicates that the experimentally-obtained thermodynamic parameters of TNT might vary with the condition of the system. The main goals of this study were to understand how TNT islands size affects the sublimation rate and the activation energy of sublimation. The measured sublimation rates of the TNT islands were compared to those calculated using a diffusion model [8] and values previously reported using other methods [5,10,11].

2. Results and Discussion

Figure 1 shows a series of five plots which represent the change of normalized volume for TNT islands in the various dimeter sizes (600–1600 nm) as a function of time, during annealing isothermally (15–25 °C). It was noted that the coarsening of islands at early times is typically observed at all temperatures. The decay of islands in the larger islands groups generally show non-linear shrinkage over time, whereas the volume of the smaller islands (600–800 nm) decreases nearly linearly during annealing. This might be explained by the Ostwald ripening phenomena that describes the growth of larger particles at the expense of smaller particles. Smaller particles have a higher curvature than that of a flat surface or larger particles which leads to an increase in the equilibrium vapor pressure of smaller particles [20,21]. Owing to the deviation in equilibrium vapor pressure between different size of particles, mass transfer from smaller islands to bigger islands occurs [20,21]. Therefore, the size-dependent shrinkage behavior of TNT islands on the Si-substrate can be understood as combination of two competing mechanisms, sublimation (detachment from islands) and coarsening. In the case of smaller islands, (600–800 nm) which shrink linearly, decay behavior is dominated by sublimation. Although coarsening of TNT islands is observed at low temperatures (15–25 °C), the coarsening effect decreases as temperature rises. It is reasonable to expect that the shrinkage behavior of TNT islands become dominated by the sublimation at elevated temperature ranges.

To calculate the activation energy (E_a) of sublimation of TNT islands, the Arrhenius relation is used:

$$-\frac{\rho dV}{S dt} = A \exp\left(-\frac{E_a}{kT}\right), \tag{1}$$

where ρ is the density of the bulk TNT (1.654 g/cm^3), V is the volume of the islands, S is the initial surface area of the islands before annealing, t is the measured time, A is pre-exponential factor, k is Boltzmann's constant, and T is the absolute temperature (K). Each volumetric sublimation rate of various diameter size of TNT island at five different temperature (15, 17.5, 20.6, 23.7, 25 °C) is obtained by the measurement of slope of plots after the initial coarsening is terminated. Calculated volumetric sublimation rates in each size range represents average values of 1–3 different islands in the same scanned area.

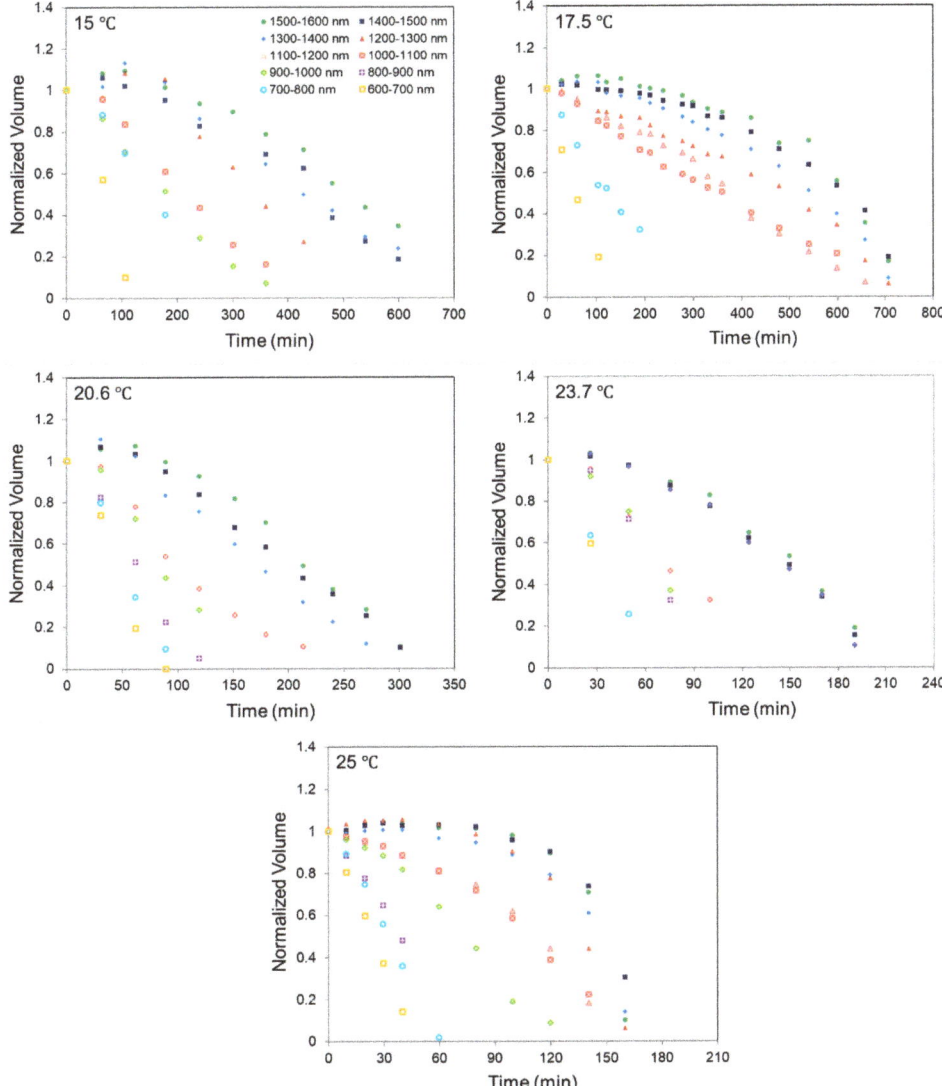

Figure 1. The normalized volume of various diameter range of 2,4,6-trinitrotoluene (TNT) islands as a function of time at different temperature (15, 17.5, 20.6, 23.7, 25 °C).

Figure 2 shows the Arrhenius plot for different size TNT islands over the temperature range of 15–25 °C. It was observed that the activation energy of sublimation increases as the diameter of TNT islands increases. For the groups of small islands in the diameter range of 600–800 nm, the values of activation energy were calculated to be 104.4 ± 2.4 kJ mol^{-1} which corresponded to the values in the lower end of range among those reported in the previous literature (90–141 kJ mol^{-1}) [5,7,8,10,11,15–19]. Within the middle size range (900–1100 nm), higher values of the activation energy were obtained (120.5 ± 6.1 kJ mol^{-1}). In the groups of large diameter size over the 1300 nm, 195.5 ± 0.05 kJ mol^{-1} were calculated, which was considerably higher than reported value, but no trend was observed between this diameter range of islands (1300–1600 nm). One might argue that interfacial interaction between native silicon oxide (silica) on the Si-wafer and TNT islands might contribute to the discrepancy of activation energy depending upon the size of islands [7]. However, this argument might be not reasonable since the island height is greater than a few hundred nanometers.

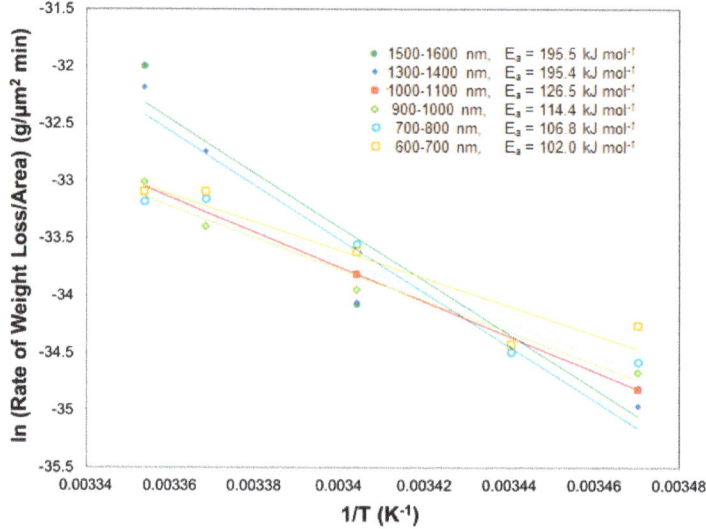

Figure 2. Arrhenius plot for TNT islands in the groups of various diameter size (600–1600 nm). The activation energy (E_a) of sublimation of TNT islands is calculated from the slope of each Arrhenius fit.

To investigate how the distance between islands affects the sublimation rate and the activation energy, decay behavior of two similar diameter size of TNT islands (~1 µm) with different neighbor distances during annealing at 15 °C were compared. In Figure 3a, island 1 is surrounded by smaller particles which might act as sources of coarsening, whereas island 2 is relatively isolated with larger neighboring particles. The plots of normalized area versus time for two similar diameter sizes of particles are shown in Figure 3b. The coarsening was observed only in island 1 at early periods of time. Afterward, the discrepancy of shrinkage rate of area between two particles slightly increases until the time (178.2 min) when the smaller neighbor islands nearly disappear, and then the sublimation rate seems to remain constant. The same observation is also found at 25 °C, as seen in Figure 3c,d. At the early time, coarsening is commonly observed in all three similar diameter size islands (1200–1300 nm). However, the difference of area shrinkage rate between three islands associated with local environment gradually increases with time. After 39.8 min, the coarsening of island 3 and 4 was still observed, while the area shrinkage rate of island 5 starts increasing because some neighbor smaller islands have been already consumed. Subsequently, shrinkage of the three islands becomes nearly equivalent after coarsening is terminated.

Figure 3d shows that the area shrinking of three islands increases, as a function of time, after coarsening is eliminated. We postulated that an increase in the sublimation rate is related to a decrease in local density of TNT islands. To validate our hypothesis, the relation between volumetric sublimation rate of total TNT islands in the scanned area, and average percentage covered by TNT islands, was analyzed at different temperatures (17.5, 20.6, 23.7, 25 °C) (Figure 4a). This was based on the assumption that local vapor density of TNT islands is associated with the surface area coverage of TNT islands on the silicon substrate. Figure 4a shows that the reduction rate in area covered by TNT islands with time, increases as temperature rises, and the sublimation rate is directly affected by the local density at higher temperature. Consequently, the activation energy of sublimation of TNT islands increases as surface area of TNT islands decreases, as observed in Figure 4b. This corresponds well with the deviation observed in activation energy of sublimation depending on diameter size of TNT islands. The increase in activation energy of sublimation with increasing size of TNT islands can be understood as a result of decrease in local density of TNT islands with time. Since small islands with diameter range of 600–800 nm are evaporated and disappeared quickly, their sublimation behavior seems to be independent on the changes of island local density during heating.

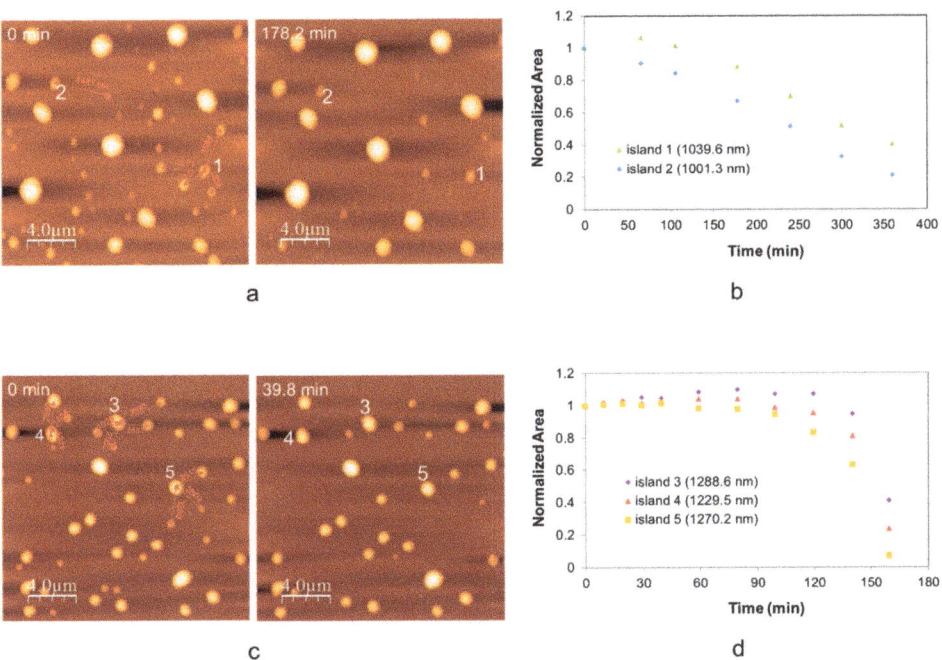

Figure 3. (**a**) Atomic force microscopy (AFM) height images (20 μm × 20 μm) of TNT islands at 0 min and 178.2 min during annealing at 15 °C, (**b**) the plot of normalized area versus time for island 1 and 2 at 15 °C, (**c**) AFM height images of TNT islands (20 μm × 20 μm) at 0 min and 39.8 min during annealing at 25 °C, and (**d**) the plot of normalized area versus time for island 3, 4, and 5 at 25 °C.

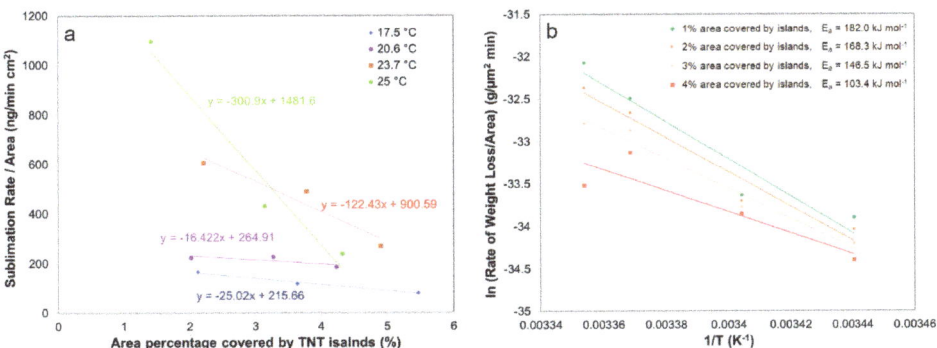

Figure 4. (**a**) Volumetric sublimation rate vs. area percentage covered by TNT islands (17.5, 20.6, 23.7, 25 °C) and (**b**) Arrhenius plot for TNT islands depending on total area percentage covered by TNT area in scanned area.

The sublimation process of hemispherical shaped islands in an open environment can be described with a previously described diffusion model [8,9]. This model assumes the quasi stationary steady state and ideal gas state where the mass sublimation rate per area of TNT islands is given by:

$$-\frac{dm}{sdt} = \frac{2\pi M D P_{sat} r_0}{A_0 k T} \quad (2)$$

where $\frac{dm}{sdt}$ is the rate of mass loss per area (kg m^{-2} s^{-1}), M is the molecular mass of TNT, D is the diffusion coefficient, P_{sat} is the vapor pressure, r_0 is the radius of islands, A_0 is the initial surface area of the islands before annealing, k is Boltzmann's constant, and T is the absolute temperature (K) [8,9]. To calculate the sublimation rate, two unknown parameters, vapor pressure and the diffusion coefficient, are required. Table 1 shows the reported vapor pressure data for the temperature range of 15–25 °C, calculated from the Clausius–Clapeyron equation [15,16,18,19]. The average value of data reported by Lenchitz et al. and Pella et al. is selected. For the diffusion coefficient, theoretical value, $D = 5.59 \times 10^{-6}$ m^{2}s^{-1}, is determined from literature, and the temperature variance is ignored [5,6,22,23].

Table 1. Comparison of reported vapor pressure data of TNT at different temperature.

	Heat of Vaporization (kJ mol^{-1})	Vapor Pressure at 15 °C (Pa)	Vapor Pressure at 20.6 °C (Pa)	Vapor Pressure at 25 °C (Pa)
Lenchitz et al. [15]	103.4	2.89×10^{-4}	6.57×10^{-4}	1.23×10^{-3}
Pella et al. [18]	99.2	2.99×10^{-4}	6.57×10^{-4}	1.19×10^{-3}
Leggett et al. [16]	141.1	6.25×10^{-5}	1.92×10^{-4}	4.51×10^{-4}
Oxley et al. [19]	137	6.37×10^{-5}	1.89×10^{-4}	4.32×10^{-4}

Table 2 shows the comparison of the experimentally measured mass sublimation rate per area of TNT islands, those calculated from diffusion model and extrapolated values for the temperature range of 15–25 °C from other literature. As seen in Table 2, measured values are 1–2 orders of magnitude smaller than those calculated by the diffusion model. The discrepancy of two values decreases as diameter size of islands increases, and the values tend to converge at higher temperature. In addition, measured mass sublimation rate per area of TNT islands are 2–3 orders of magnitude larger than those reported by Walid et al. using other techniques, UV spectroscopy and TGA [5,10]. It is noted that in their experiment, continuous nanofilms and powders were used as samples, which are expected to have a higher local vapor density of TNT. This corresponds to our argument, as discussed earlier, that sublimation rate of TNT increases as local vapor density decreases.

Since there are limitations on collecting valid data for the sublimation rate normalized to surface area for TNT, a comparison was made to PETN, where more data is available, in Table 3, to validate our argument [6,12,13]. To be compared, all values are extrapolated for the temperature range of 30 and 40 °C. It is shown that the magnitude of mass loss rate per area are divided in to two group. In the lower sublimation rate group, non-continuous islands are used as a samples, whereas continuous film and single crystals are used as a sample in the higher mass loss rate group. Similar to TNT, the same tendency shows a correlation between sublimation rate per unit area and the local vapor density is found.

In the previous work by Burnham et al., there are two orders of magnitude difference in the evaporation rate between those measured by TGA and AFM [6]. The authors explain the discrepancy occurs due to surface migration of PETN molecules or surface quality but state that the absolute mechanism is unclear [6]. Our result could be evidence that illustrates the significant difference in sublimation rate, and the activation energy for explosives is dependent on the experimental method and types of sample used.

Table 2. Comparison of mass sublimation rate per unit area for TNT at different temperature. TGA = thermogravimetric analysis.

	Sample Type	Experimental Method	Sublimation Rate at 15 °C (kg s^{-1} m^{-2})	Sublimation Rate at 20.6 °C (kg s^{-1} m^{-2})	Sublimation Rate at 25 °C (kg s^{-1} m^{-2})
Measured	Non-continuous micro islands	AFM	1.27×10^{-8} (1500–1600 nm) 1.26×10^{-8} (1000–1100 nm) 2.21×10^{-8} (600–700 nm)	2.67×10^{-8} (1500–1600 nm) 3.45×10^{-8} (1000–1100 nm) 4.20×10^{-8} (600–700 nm)	2.12×10^{-7} (1500–1600 nm) 7.52×10^{-8} (1000–1100 nm) 7.12×10^{-8} (600–700 nm)
Calculated by Diffusion theory	-	-	1.23×10^{-6} (1500–1600 nm) 1.92×10^{-6} (1000–1100 nm) 2.99×10^{-6} (600–700 nm)	2.79×10^{-6} (1500–1600 nm) 4.18×10^{-6} (1000–1100 nm) 6.29×10^{-6} (600–700 nm)	4.96×10^{-6} (1500–1600 nm) 7.32×10^{-6} (1000–1100 nm) 1.19×10^{-5} (600–700 nm)
Walid et al. [10]	Continuous nanofilm (500–600 nm)	UV spectroscopy	1.87×10^{-10}	4.11×10^{-10}	7.48×10^{-10}
Walid et al. [5]	powder	TGA	1.48×10^{-11}	3.18×10^{-11}	5.68×10^{-11}

Table 3. Comparison of mass sublimation rate per unit area for Pentaerythritol tetranitrate (PETN) at different temperature.

	Sample Type	Experimental Method	Sublimation Rate at 30 °C (kg s^{-1} m^{-2})	Sublimation Rate at 40 °C (kg s^{-1} m^{-2})
Pitchimani at al. [13]	Non-continuous micro islands	AFM	5.31×10^{-9}	4.39×10^{-8}
Burnham at al. [6]	Nano-islands on single crystal	AFM	3.00×10^{-10}	1.83×10^{-9}
Walid et al. [12]	Continuous nano-film (~100 nm)	UV spectroscopy	4.15×10^{-13}	2.28×10^{-12}
Burnham at al. [6]	Single crystal	TGA	9.29×10^{-13}	5.21×10^{-12}

3. Experimental Section

TNT powder, provided by Austin explosive, was dissolved in acetone at room temperature to prepare 0.01 M TNT solution. The solution was filtered with 0.45 μm syringe filter, and the purity of TNT was measured to be 99.67% using HPLC (BDS hypersil C-18, column size: 150 × 2.1 mm). 1 cm × 1 cm silicon wafer was cleaned by the general Piranha solution method (cleaning agent with a ratio of 3:1 sulfuric acid to hydrogen peroxide) and then, 20 μL of 0.01 M TNT solution was deposited on the Si wafer at 4000 rpm for 1 min using spin-coater (single wafer spin processor, Laurell Technologies corp., North Wales, PA, USA). At these conditions, the morphology of the TNT film typically show non-continuous hemispherical islands, and the size of particles varied with the diameter range of 0.2 μm to 1.8 μm (Figure 5). The prepared film was directly moved to the temperature stage on the AFM (XE 100, PSIA, Santa Clara, CA, USA). A conventional Peltier controller was used for in situ isothermal AFM analysis at ambient pressure open to the environment. AFM images during isothermal heating at five different temperature (15, 17.5, 20.6, 23.7, 25 °C) were collected as a function of time. In order to prevent the tip impacting the morphology of TNT islands, each image was scanned in non-contact mode with a silicon cantilever with a drive frequency of ~240 kHz imaging a 20 μm × 20 μm area.

Collected images were analyzed with the Scanning probe image processor (SPIP) v. 6.7.0 (Image Metrology) and WSXM 5.0 v. 8.4 software [24].

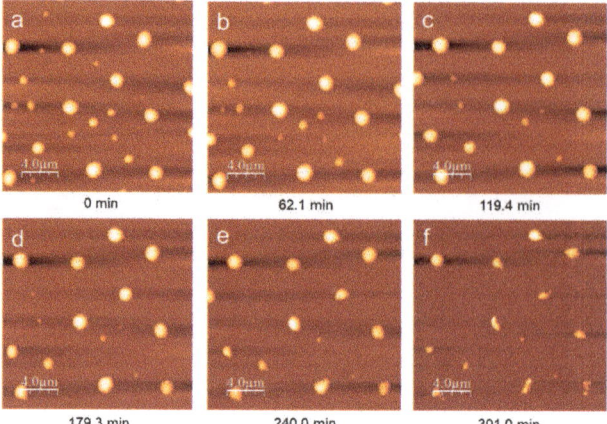

Figure 5. Time series of AFM height images for TNT islands during annealing at 20.6 °C. All images are 20 µm × 20 µm.

4. Conclusions

In this study, we used in situ AFM thermal analysis to observe the decay behavior of nanoscale TNT islands. We observed, for the first time, that coarsening of nanoscale TNT islands occurs in the temperature range of 15–25 °C, and we showed that this coarsening effect can be affected by the local distribution of TNT islands. The activation energies calculated for the sublimation of small TNT islands, which are 600–1100 nm in diameter agree well with the values reported in the literature. Due to the significant change in local vapor density when TNT is annealed, however, the activation energies of large diameter TNT islands (1300–1600 nm) was found to be significantly higher than values in the literature. Our results are supported when the sublimation rates of TNT and PETN are compared using various types of samples with different local densities.

Author Contributions: Conceptualization, Y.L. and B.L.W.; methodology, Y.L.; formal analysis, Y.L.; investigation, Y.L.; resources, B.L.W.; data curation, Y.L.; writing—original draft preparation, Y.L.; writing—review and editing, B.L.W.; supervision, B.L.W.; project administration, B.L.W.; funding acquisition, B.L.W.

Funding: This research received no external funding.

Conflicts of Interest: The authors declare no conflict of interest.

References

1. Steinfeld, J.I.; Wormhoudt, J. Explosives Detection: A Challenge for Physical Chemistry. *Annu. Rev. Phys. Chem.* **1998**, *49*, 203–232. [CrossRef]
2. Moore, D.S. Recent Advances in Trace Explosives Detection Instrumentation. *Sens. Imag.* **2007**, *8*, 9–38. [CrossRef]
3. Caygill, J.S.; Davis, F.; Higson, S.P.J. Current Trends in Explosive Detection Techniques. *Talanta* **2012**, *88*, 14–29. [CrossRef]
4. Gee, R.H.; Wu, C.; Maiti, A. Coarse-Grained Model for a Molecular Crystal. *Appl. Phys. Lett.* **2006**, *89*, 2004–2007. [CrossRef]
5. Hikal, W.M.; Weeks, B.L. Sublimation Kinetics and Diffusion Coefficients of TNT, PETN, and RDX in Air by Thermogravimetry. *Talanta* **2014**, *125*, 24–28. [CrossRef] [PubMed]

6. Burnham, A.K.; Qiu, S.R.; Pitchimani, R.; Weeks, B.L. Comparison of Kinetic and Thermodynamic Parameters of Single Crystal Pentaerythritol Tetranitrate using Atomic Force Microscopy and Thermogravimetric Analysis: Implications on Coarsening Mechanisms. *J. Appl. Phys.* **2009**, *105*, 1–7. [CrossRef]
7. Mu, R.; Ueda, A.; Liu, Y.C.; Wu, M.; Henderson, D.O.; Lareau, R.T.; Chamberlain, R.T. Effects of Interfacial Interaction Potential on the Sublimation Rates of TNT Films on a Silica Surface Examined by QCM and AFM Techniques. *Surf. Sci.* **2003**, *530*, 293–296. [CrossRef]
8. Gershanik, A.P.; Zeiri, Y. Sublimation Rate of TNT Microcrystals in Air. *J. Phys. Chem. A* **2010**, *114*, 12403–12410. [CrossRef] [PubMed]
9. Gershanik, A.P.; Zeiri, Y. Sublimation Rate of Energetic Materials in Air: RDX and PETN. *Propellants Explos. Pyrotech.* **2012**, *37*, 207–214. [CrossRef]
10. Hikal, W.M.; Weeks, B.L. Determination of Sublimation Rate of 2,4,6-Trinitrotoluene (TNT) Nano Thin Films Using UV-Absorbance Spectroscopy. *J. Therm. Anal. Calorim.* **2012**, *110*, 955–960. [CrossRef]
11. Hikal, W.M.; Paden, J.T.; Weeks, B.L. Thermo-Optical Determination of Vapor Pressures of TNT and RDX Nanofilms. *Talanta* **2011**, *87*, 290–294. [CrossRef]
12. Hikal, W.M.; Paden, J.T.; Weeks, B.L. Simple Method for Determining the Vapor Pressure of Materials Using UV-Absorbance Spectroscopy. *J. Phys. Chem. B* **2011**, *115*, 13287–13291. [CrossRef] [PubMed]
13. Pitchimani, R.; Burnham, A.K.; Weeks, B.L. Quantitative Thermodynamic Analysis of Sublimation Rates Using an Atomic Force Microscope. *J. Phys. Chem. B* **2007**, *111*, 9182–9185. [CrossRef]
14. Östmark, H.; Wallin, S.; Ang, H.G. Vapor Pressure of Explosives: a Critical Review. *Propellants Explos. Pyrotech.* **2012**, *37*, 12–23. [CrossRef]
15. Lenchitz, C.; Velicky, R.W. Vapor Pressure and Heat of Sublimation of Three Nitrotoluenes. *J. Chem. Eng. Data* **1970**, *15*, 401–403. [CrossRef]
16. Leggett, D.C. Vapor Pressure of 2,4,6-Trinitrotoluene by a Gas Chromatographic Headspace Technique. *J. Chromatogr. A* **1977**, *133*, 83–90. [CrossRef]
17. Cundall, R.B.; Palmer, T.F.; Wood, C.E.C. Vapour Pressure Measurements on Some Organic High Explosives. *J. Chem. Soc. Faraday Trans.* **1978**, *74*, 1339–1345. [CrossRef]
18. Pella, P.A. Measurement of the Vapor Pressures of TNT, 2,4-DNT, 2,6-DNT, and EGDN. *J. Chem. Thermodyn.* **1977**, *9*, 301–305. [CrossRef]
19. Oxley, J.C.; Smith, J.L.; Shinde, K.; Moran, J. Determination of the Vapor Density of Triacetone Triperoxide (TATP) Using a Gas Chromatography Headspace Technique. *Propellants Explos. Pyrotech.* **2005**, *30*, 127–130. [CrossRef]
20. McLean, J.G.; Krishnamachari, B.; Peale, D.R.; Chason, E.; Sethna, J.P.; Cooper, B.H. Decay of Isolated Surface Features Driven by the Gibbs-Thomson Effect in an Analytic Model and a Simulation. *Phys. Rev. B* **1997**, *55*, 1811–1823. [CrossRef]
21. Kodambaka, S.; Petrova, V.; Vailionis, A.; Petrov, I.; Greene, J.E. In Situ High-Temperature Scanning Tunneling Microscopy Studies of Two-Dimensional TiN Island Coarsening Kinetics on TiN(001). *Surf. Sci.* **2003**, *526*, 85–96. [CrossRef]
22. Parmeter, J.E.; Eiceman, G.A.; Preston, D.A.; Tiano, G.S. Sandia Report SAND-96-2016C. In Proceedings of the Conference-960767-21; Sandia National Laboratories: Albuquerque, NM, USA, 9607.
23. Eiceman, G.A.; Peterson, D.; Tiano, G.; Rodrigues, J. Quantitative Calibration of Vapor Levels of TNT, RDX, and PETN Using a Diffusion Generator with Gravimetry and Ion Mobility Spectrometry. *Talanta* **1997**, *45*, 57–74. [CrossRef]
24. Horcas, I.; Fernandez, R.; Gomez-Rodriguez, J.M.; Colchero, J.; Gomez-Herrero, J.; Baro, A.M. WSXM: A Software for Scanning Probe Microscopy and a Tool for Nanotechnology. *Rev. Sci. Instrum.* **2007**, *78*, 013705. [CrossRef]

Sample Availability: Samples of the compounds are available from the authors.

© 2019 by the authors. Licensee MDPI, Basel, Switzerland. This article is an open access article distributed under the terms and conditions of the Creative Commons Attribution (CC BY) license (http://creativecommons.org/licenses/by/4.0/).

Article

Kinetic Analysis of Digestate Slow Pyrolysis with the Application of the Master-Plots Method and Independent Parallel Reactions Scheme

Pietro Bartocci [1,*], Roman Tschentscher [2], Ruth Elisabeth Stensrød [2], Marco Barbanera [3] and Francesco Fantozzi [1]

1. Department of Engineering, University of Perugia, Via G. Duranti 67, 06125 Perugia, Italy; francesco.fantozzi@unipg.it
2. SINTEF Industry AS, Forskningsveien 1, 0373 Oslo, Norway; Roman.Tschentscher@sintef.no (R.T.); RuthElisabeth.Stensrod@sintef.no (R.E.S.)
3. Department of Economics, Engineering, Society and Business Organization, University of Tuscia, 01100 Viterbo, Italy; m.barbanera@unitus.it
* Correspondence: bartocci@crbnet.it; Tel.: +0039-0755853773

Academic Editor: Sergey Vyazovkin
Received: 7 March 2019; Accepted: 25 April 2019; Published: 27 April 2019

Abstract: The solid fraction obtained by mechanical separation of digestate from anaerobic digestion plants is an attractive feedstock for the pyrolysis process. Especially in the case of digestate obtained from biogas plants fed with energy crops, this can be considered a lignin rich residue. The aim of this study is to investigate the pyrolytic kinetic characteristics of solid digestate. The Starink model-free method has been used for the kinetic analysis of the pyrolysis process. The average Activation Energy value is about 204.1 kJ/mol, with a standard deviation of 25 kJ/mol, which corresponds to the 12% of the average value. The activation energy decreased along with the conversion degree. The variation range of the activation energy is about 99 kJ/mol, this means that the average value cannot be used to statistically represent the whole reaction. The Master-plots method was used for the determination of the kinetic model, obtaining that n-order was the most probable one. On the other hand, the process cannot be modeled with a single-step reaction. For this reason it has been used an independent parallel reactions scheme to model the complete process.

Keywords: anaerobic digestion; kinetic model; lignin rich; activation energy; thermogravimetric analysis; pre-exponential factor

1. Introduction

The Importance of Digestate Slow-Pyrolysis Process

Coupling of anaerobic digestion and pyrolysis in integrated processes has become more and more interesting [1]. Anaerobic digestion is a very promising technology to be adopted for biomasses with important moisture content (at least more than 50%). The residue of the anaerobic digestion process is called digestate and contains ashes and components that cannot be decomposed efficiently by the microbia, which are present in the anaerobic digestor (mainly belonging to the following species: Clostridium, Peptococcus, Bifidobacterium, Desulfovibrio, Corynebacterium, Lactobacillus, Actinomyces, Staphylococcus, Streptococcus, Micrococcus, Bacillus, Pseudomonas, Selemonas, Veillonella, Sarcina, Desulfobacter, Desulfomonas, and Escherichia coli) [2]. Thus, digestate is a lignin rich substrate, which is obtained as a coproduct of anaerobic digestion and can be used as a fertilizer or it can be composted. Digestate can also be used to produce energy through the subsequent steps of solid liquid separation and pyrolysis. The University of Perugia has designed and operated a prototypal pyrolysis plant: The Integrated Pyrolysis Regenerated Plant (IPRP) [3]. To understand how the digestate would react in slow pyrolysis conditions some thermogravimetric tests have been performed at SINTEF Norway laboratories (Oslo site) during the project: "Optimization of catalytic pyrolysis of digestate and sewage sludge" funded by the European Commission through the Brisk2 project. Digestate pyrolysis has already been performed in other plant concepts, like the thermocatalytic pyrolysis plant developed at Fraunhofer Institut, Sulzbach-Rosenberg, Germany [4]. In that case TGA tests were performed with 25 mg of dried digestate in argon atmosphere with a heating rate of 20 °C/min. Major weight loss happens before 400 °C. The peak of weight loss is reported at 320–330 °C. The final charcoal mass at 1000 °C was about 35.3 wt% and no kinetic analysis was performed. In the work of Gomez et al. [5] TGA is used to perform an analysis of the thermal stability of digestate, but also in this case kinetic analysis is not performed. To the authors knowledge there are only two works which perform kinetic analysis of digestate: the work of Otero et al. 2011 [6] and the work of Zhang et al. 2017 [7]. In the work of Otero et al. [6], cattle manure is used as a feedstock in laboratory tests aiming at the characterization of its Biochemical Methane Potential (BMP). The obtained digestate is used for the TGA analysis. In our case the digestate is produced from a real anaerobic digestion plant, which is fed with a mixture of animal manure, energy crops (mainly corn an sorghum), and olive pomace. Thus, it is clear that the pyrolysis behavior is deeply influenced by the nature and composition of the digestate. It has also to noticed that, in the work of Otero et al. [6], two models are used for kinetic analysis: OFW [8–10] and Vyazovkin [11]. These two isoconversional models are used to mainly obtain the activation energy (E). No pre-exponential factor is derived. This makes this kind of analysis of biomass kinetics quite limited.

In the work of Zhang et al. [7], corn stover digestate is analyzed to obtain the Activation Energy and then the pyrolysis process is simulated using a distributed Activation Energy Model (DAEM). In this case, the digestate is produced from a starting feedstock (corn) which is quite similar to the one which is also analysed in this study. The approach of this work is different because the final aim is to answer the question: Is it possible to calculate also the pre-exponential factor of the digestate pyrolysis reaction?

In the kinetic study of biomass pyrolysis in fact two problems have to be considered with particular care:

1. First of all, there has been much discussion recently on how to determine correctly the pre-exponential factor in biomass pyrolysis and nowadays there are several approaches that can be used, see also the ASTM norm E698-16 on "Standard Test Method for kinetic parameters for thermally unstable materials using Differential Scanning Calorimetry and the Flynn/Wall/Ozawa Method", the ASTM norm E1641-16 on "Standard test method for decomposition kinetics by thermogravimetry using the Ozawa/Flynn/Wal Method" and also in References [12,13]. Interesting comments are also reported in Reference [14] on the correct use of the aforementioned norms.

2. Another aspect that should be carefully considered is the thermodynamic calculations, which are often performed using a set of equations based on Eyring's theory of the activated complex [15]. This approach can be hardly adopted for complex processes like pyrolysis of biomass, which involves many reaction steps, the production of intermediates, and complex mass and heat transfer phenomena.

Dealing with the application of the Master Plots method to the analysis of biomass kinetics one of the most interesting contribution is represented by the work of Sanchez-Jiménez et al. [16], also coauthored by Criado, who was one of the first to apply the Master Plots to the kinetic analysis of non-isothermal data [17]. In Reference [16], the Master Plots method is used with the key goal of identifying clearly the kinetic model ($f(\alpha)$) of cellulose pyrolysis. Usually isoconversional kinetic models are coupled with the Master Plots method because they basically can be used to find if the requirements for the Master Plots method are met. The main assumption to use the different methods presented in the ICTAC recommendation on kinetic computations [12] to calculate the pre-exponential factor is that of "single-step kinetics". This assumption can be easily checked with an isoconversional method. In particular, in the study in Reference [16], the random scission kinetic model was found to govern cellulose pyrolysis reaction. In this case the activation energy of cellulose was found to be constant and was estimated to be 191 kJ/mol. On the other hand, in the work of de Carvahlo et al. [18], which deals with the kinetic decomposition of energy cane, applying the Master Plots method, it was found that it was not possible to find a unique kinetic model to describe the experimental data. The most consistent model was found to be F7 (7th order reaction model) for conversion lower than 0.5 and F4 (for $0.5 \leq \alpha \leq 0.67$), F3 (for $0.67 \leq \alpha \leq 0.75$) and F2 (for conversion higher than 0.9). Reaction order models are generally based on the fact that the driving force depends on the remained concentration of the reactants. The approach used in de Carvahlo et al. [18] maybe is more appropriate for the specific case of digestate, compared to the approach of Sanchez-Jiménez et al. [16], which is more focused on cellulose pyrolysis. Digestate in fact is a lignin rich subproduct, where lignin concentration prevails on the concentration of cellulose and hemicellulose. Dealing with lignin pyrolysis kinetics, an interesting work is done by Jiang et al. [19]. In this work, a review on previous studies on kinetics of different lignin types is presented (e.g., Kraft lignin, Klason lignin, organosolv lignin, Alcell lignin, etc.). From the results, we infer that there is no agreement at the moment on unique values of activation energy and pre-exponential factor for lignin. Based on the results of other literature works, we can infer two important points:

- If the reaction mechanism is not single-step, we cannot use the Master Plots method to calculate the pre-exponential factor of digestate pyroysis (in that case, a multi-step model based on independent parallel reactions can be used);
- We can use the Master Plots method to have an idea of what is the most probable kinetic model for digestate pyrolysis.

Taking inspiration from References [12,13] and what has been said above, the authors decided to apply the Master-plots method to study the kinetics of digestate and understand how to model it better. This approach has not yet been adopted on digestate pyrolysis.

2. Results

2.1. TG-DTG Curves

Thermochemical decomposition of solid digestate during pyrolysis has been analyzed using thermogravimetric curves, TG, and DTG. Figure 1a shows the weight loss curves obtained during the pyrolysis of solid digestate at different heating rates under inert nitrogen atmosphere. Being a lignocellulosic material, the thermal degradation profile of solid digestate can be divided into three stages, influenced by its chemical and physical composition in terms of hemicellulose, cellulose, and lignin. The first stage started from room temperature and ended at about 180 °C, the mass loss is due to the removal of moisture and the hydrolysis of some extractives [20]. The second stage was the main decomposition region, involving degradation of hemicellulose, cellulose, and a small amount of lignin at a temperature range comprised between 180 °C to 392 °C. The characteristic temperatures of the different stages are shown in Table 1, with the relative standard deviations. It should be considered that to define rigorously T_i, T_f, and T_m the following assumptions have been made:

- T_i represents the temperature at which a conversion of about 5% of the initial mass is obtained;
- T_m represented the temperature at which the maximum conversion is obtained. The average is about 330 °C, which is in agreement with what is reported in Reference [4];
- T_f represents the temperature at which about 80% of the conversion happened.

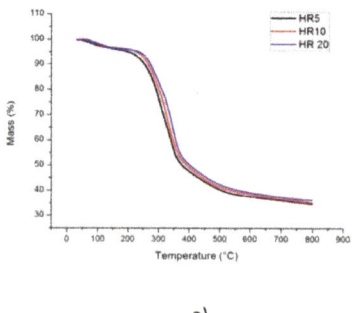 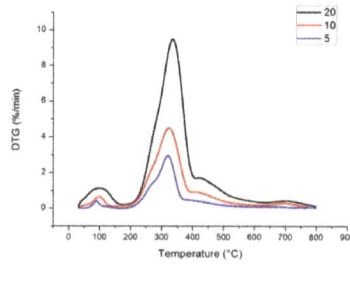

Figure 1. Thermogravimetric (**a**) and Differential thermogravimetric (**b**) curves generated during solid digestate pyrolysis at different heating rates.

Table 1. Thermal degradation characteristics of solid digestate at different heating rates.

Heating Rate (°C/min)	Temperature *			DTGmax *
	T_i (°C)	T_f (°C)	T_m (°C)	
5	184 (1)	377 (3)	319 (1)	2.9 (0.5)
10	188 (1)	382 (2)	329 (1)	4.5 (0.7)
20	190 (1)	392 (3)	346 (2)	9.5 (0.9)

* SD values are indicated in brackets.

It is well known that decomposition of hemicellulose, cellulose, and lignin occurs at the temperature range of 160–360 °C, 240–390 °C, and 180–900 °C, respectively [21,22]. Moreover, each DTG curve (Figure 1b) is characterized by a lower temperature shoulder at around 290 °C, corresponding to the decomposition of hemicellulose and a higher temperature peak that can be attributed to cellulose devolatilization. In particular, White et al. [23] pointed out that the cellulose decomposition happens into two ways: (1) Depolymerization with the formation of CO, CO_2, and carbonaceous residues at low temperature; and (2) integration of bonds at high temperature with the formation of liquid product containing a wide range of organic compounds. After 400 °C, the third stage of pyrolysis began where the slow decomposition of lignin causes the typical long tail of TG-curves. Biochar yield at 800 °C was in the range of 35.08%–36.45%, which was higher than the char yield of other lignocellulosic biomass, such as rice straw (23.68%) and rice bran (25.17%) at 700 °C [24], and camel grass (30.46%) at 550 °C [25], while it was comparable with the char yield of empty fruit bunch (35.14%) at 500 °C [26], reflecting that the lignin content of biomass plays a significant role in biochar formation.

Moreover, as shown in Figure 1a, the shape of the mass loss curve of the solid digestate is not influenced by the heating rate, there is only a little shift to the right in the temperature range from 250 °C to 450 °C, passing from 5 °C/min to 20 °C/min. This result confirms that the degradation chemistry is quite independent from the heating rate and suggests that lower heating rates could be employed in order to optimize the pyrolysis conversion of solid digestate. However, as seen in Figure 1b, DTG curves show an increase in maximum mass loss rates and a slight shift of the major peak to higher temperatures with higher heating rates, mainly due to the combined effects of the heat transfer process at different heating rates and of the kinetics of the thermal volatilization, which result in delayed degradation [27].

2.2. Determination of Activation Energy

The knowledge of kinetic parameters is essential for effective modeling and design of thermochemical processes because biomass pyrolysis is a heterogeneous reaction that is strongly affected by kinetic parameters, such as: activation energy, pre-exponential factor and kinetic model (also known as kinetic triplet) [16]. Decomposition kinetics during the solid digestate pyrolysis process was calculated using Starink model at degrees of conversion (α) ranging from 0.05 to 0.95 with a step of 0.05 according to the ICTAC recommendations [28]. According to Starink model, the activation energy can be calculated from $\ln(\beta/T^{1.92})$. The plots used for the determination of activation energy at different conversion rates are shown in Figure 2. In particular in the linear plot of $\ln(\beta/T^{1.92})$ versus 1/T the slopes obtained at different conversion rates are equal to -1.0008E/R. Figure 3 presents the values of E and the standard deviation, calculated using the data retrieved from 3 repetitions of the same experiment. The average value of the Activation Energy is about 204.1 kJ/mol, with a standard deviation of 25 kJ/mol, which is about 12% of the average value. The variation range of the Activation Energy is about 99 kJ/mol, which is a high value. This means that the average E value cannot be used to statistically represent the activation energy variation.

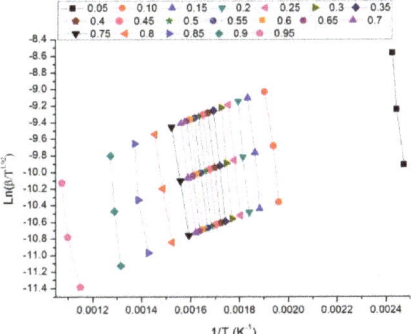

Figure 2. Linear plots in the 0.05–0.95 conversion range for determining activation energy of solid digestate, calculated according to the Starink method.

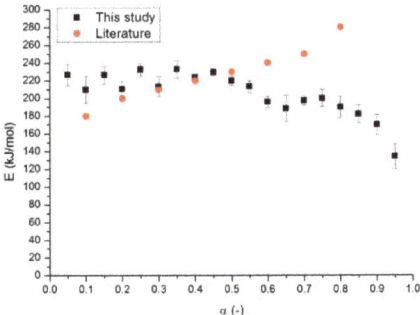

Figure 3. Activation energy distribution (with standard deviations) for solid digestate pyrolysis.

Correlation coefficients (R^2) are shown in Figure 4. As recommended by the ICTAC Committee [28], since this work was performed with three heating rates, the number of degrees of freedom (calculated as n-2) is only 1, so "in statistical terms, such a plot can be accepted as linear with 95% confidence only when its respective correlation coefficient, R is more than 0.997 (equal to R^2 of 0.994)". In our case (see Figure 3), the first two points have a correlation coefficient that is lower than that 0.994. This happened also in the publication of de Carvalho et al. [18] and denotes high uncertainty of the measure activation energies (at least for conversion of 0.05 and 0.1).

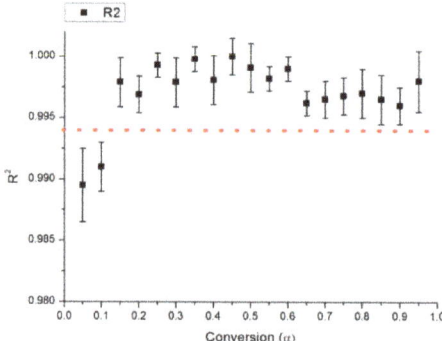

Figure 4. R^2 coefficient for the calculation of activation energy of solid digestate.

The kinetic results show that activation energy is quite dependent on the conversion rate, which means that the pyrolysis of solid digestate is characterized by a complex degradation mechanism that involves different types of reactions, so it cannot be completely considered as a single-step process. The relationship of the activation energy with the conversion rate suggests that the activation energy is almost constant within the conversion range of 0.05–0.55, and then, in the conversion zone of 0.55–0.95, a decline is observed. It can be inferred that there are at least two kinetic models working in sequence. This is mainly due to the fact that digestate is composed by at least four pseudo-components: Cellulose, hemicellulose, lignin, and extractives. Vamvuka et al. [27] reported that the activation energy values for hemicellulose, cellulose, and lignin are in the range of 145–285, 90–125, and 30–39 kJ/mol, respectively. These values are quite accepted in literature at least for cellulose and hemicellulose. For lignin, higher values have also been reported, see Reference [22].

In Figure 3 we see the comparison between the Activation Energy values obtained in this study and the values reported in Literature [7]. The two data sets are not always comparable, especially for higher conversion values. Anyway the decreasing trend of the activation Energy values seems to be more reasonable, also considering other publications on biomass thermal behavior [18] and the composition of the raw material.

2.3. Identification of the Reaction Model

The reaction model and the pre-exponential factor are not evaluated directly by the isoconversional methods. Once the activation energy has been calculated the reaction model has to be identified. The identification of a reaction model without a previous verification with the Master-plots model is not recommended, since the solid-state reaction rate can be influenced by diffusion, solid geometry, and reagent concentrations models [18]. Thus in this case, the average value of activation energy (204.1 kJ/mol) is used in the Master-plots method, in order to predict the reaction mechanism of solid digestate. This is an approximation and the method is more accurate when the Activation Energy is constant, but this is done only to have some more hints on the uniqueness of the kinetic model as performed also in [18].

Using Equation (9) (see the materials and methods section), the temperature integral, p(x), can be calculated as a function of α by employing the average value of E. Figure 5a,b show the theoretical plots of g(α)/g(0.5) as a function of α, and the experimental plots p(u)/p(u0.5), against α, also the experimental data obtained at β = 10 °C/min, respectively, for α ≤ 0.5 and α ≥ 0.5, are reported. Since the experimental master plots are practically overlapped it was chosen to use in this screening analysis only one heating rate (i.e., 10 °C/min). It can be noted that the Fn model is the most reliable, because the experimental data have the same trends as F4, F5, F6, and F7 models. It is also clear that it is not possible to define any unique function that describes the entire kinetic process for the pyrolysis of solid digestate. The Fn model can be written as:

$$g(\alpha) = \frac{(1-\alpha)^{1-n} - 1}{n - 1} \tag{1}$$

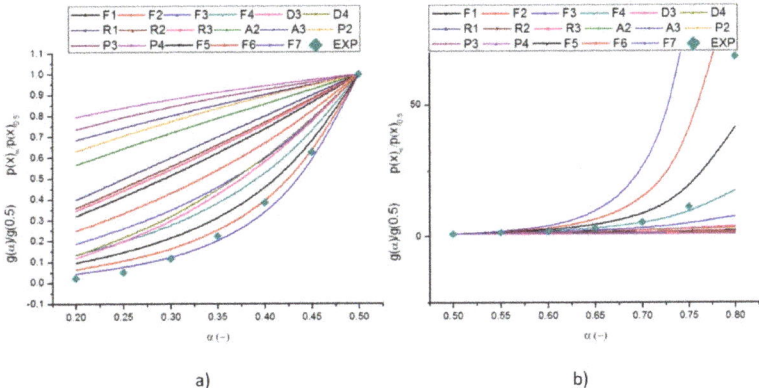

Figure 5. Theoretical and experimental master plots functions from (**a**) 0.2–0.5 and (**b**) 0.5–0.8 of conversion.

It can be noted that the experimental data lie between the theoretical master-plots F6 and F7 for $0.20 \leq \alpha \leq 0.50$, F4, and F5 for $0.50 \leq \alpha \leq 0.75$, and F5 and F6 for $0.75 \leq \alpha \leq 0.80$. It was chosen to refer to the conversion interval 0.2–0.8 because it was thought to be the more stable by the point of view of the pyrolysis reaction. We can conclude that despite the interval takes into account the phases in which the pyrolysis reaction should be more stable, we could not identify a unique theoretical master-plot, which approximates the experimental data perfectly. For this reason, the authors decided to model pyrolysis reaction with an independent parallel reaction scheme. This approach and these types of conclusions are also reported in the work of de Carvahlo et al. [18]. This is the reason why, in the literature, this approach is also gaining more and more interest, see Reference [29].

2.4. Independent Parallel Reactions Scheme

The results of the peak deconvolution calculations are shown in Figure 6. The heating rate of 5 °C/min is taken as an example, but tests have been performed on all the three heating rates and also repeated three times the final results have been averaged and the standard deviation has been calculated. The fit correlation coefficient (R^2) between the experimental data and the multi peak fitting result is higher than 0.992 for each of the three considered heating rates.

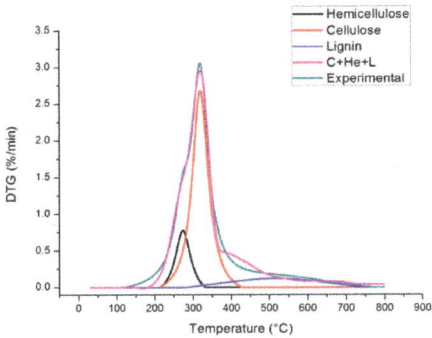

Figure 6. Peak deconvolution result (Heating Rate 5 °C/min).

In Table 2 Activation Energy, Pre-exponential factor and reaction order are presented for each digestate pseudo-component.

Table 2. Thermal degradation characteristics of solid digestate at different heating rates.

Pseudo-Component	Activation Energy		Pre-Exponential Factor		Reaction Order	
	Value	SD	Value	SD	Value	SD
Cellulose	189 kJ/mol	15 kJ/mol	4.7×10^{17} min^{-1}	1.5×10^{16} min^{-1}	1.0	0.1
Hemicellulose	151 kJ/mol	21 kJ/mol	4.4×10^{14} min^{-1}	5.0×10^{12} min^{-1}	1.1	0.2
Lignin	64 k/mol	7 kJ/mol	6.3×10^{3} min^{-1}	1.2×10^{3} min^{-1}	1.6	1.1

Dealing with cellulose Activation Energy, values reported in literature (see [22]) are quite variable and they range from 175 to 235 kJ/mol, our value falls in this range. The pre-exponential factor for cellulose pyrolysis is usually comprised between 1.2×10^{10} and 2.2×10^{19} min^{-1} [30]. In particular the work of Conesa et al. [30] reports a value of 3.0×10^{17} min^{-1}, which is quite similar to the one obtained in this study. Dealing with the order of reaction usually a first order reaction is assumed by Antal and Varheghyi [31], and also confirmed in Reference [32].

The values reported in the work of [22] on Activation Energy of hemicellulose pyrolysis are comprised between 149 kJ/mol and 174 kJ/mol. The pre-exponential factor ranges from 10.6 to 15.0 logA/s^{-1}. Both values are in agreement with this study. The reaction order is also in agreement with Reference [32].

Dealing with lignin a review of kinetic parameters is reported in the study of Jiang et al. [19]. The reported values for Activation Energy range from 25.2 kJ/mol to 361 kJ/mol, so there is a huge variation. Nevertheless many studies report low values of activation energy and pre-exponential factor for lignin, confirming the results obtained in this study. Thus, if the low values of activation energy and pre-exponential factor can be easily explained, the high standard deviation of the pre-exponential factor indicates that this value in particular has a high level of uncertainty (the same consideration applies to the reaction order).

In Figure 7 the comparison between the experimental DTG data and the combined kinetics of the three-parallel-reaction model is shown. This is obtained by integrating the Equation (13) (see Material and Methods section) for each pseudo-component and adding the results to obtain the multi peak trend. To check the quality of the fitting between experimental data and model data Equation (2) is used to calculate the variance (as reported in Reference [28]):

$$S(\%) = 100 \times \sqrt{\frac{\sum_{j=1}^{N}\left(x_j^c - x_j^e\right)^2}{(N_d - N_p)}} \qquad (2)$$

where j denotes the j-th experimental point; N_d denotes the total number of experimental points; N_p denotes the total number of unknown parameters (3 in this case); x_j^e and x_j^c denote the values of experimental and calculated x, respectively.

The calculated value of S is equal to 0.79%. The correlation coefficient (R^2) between calculated and experimental data is about 0.990. These values are higher with respect to those shown in the paper of Wang et al. [29]. However, it has to be considered that, compared to the methods used in literature, the one used in this work performs two fitting stages: The first one during peaks deconvolution and the second one when each peak is fitted to a curve finding optimal values of activation energies, pre-exponential factor and reaction order for each pseudo-component. For this reason the method followed in this work is more easy and quick to implement, but probably less accurate. This can also be seen from the correspondence of the blue line shown in Figure 6 with the purple line, there is still space for improving the correlation coefficient and also improve peak deconvolution.

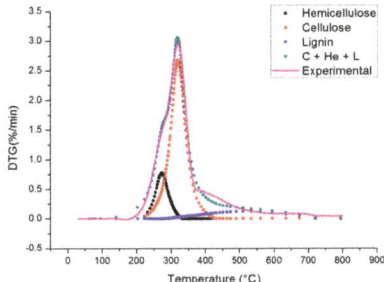

Figure 7. Comparison between experimental DTG data and the combined kinetics of the three-parallel-reaction model (Heating Rate 5 °C/min).

3. Materials and Methods

3.1. Sample Preparation

The solid digestate used in this study was collected from an on-farm biogas plant of the capacity of 1 MWel located in Central Italy (Umbria Region, province of Perugia), which is fed with a substrate consisting of pig slurry (15 m^3/d), olive pomace (19 t/d), maize silage (19.6 t/d), sorghum silage (36.4 t/d), and onion scraps (1 t/d). The feedstock is mainly constituted by lignocellulosic biomasses. The set-up of the biogas plant consisted of two anaerobic digesters in parallel (operating at a temperature of 43–44 °C), followed by a post fermenter (operating at a temperature of 37 °C). The solid fraction was obtained by mechanical separation with a screw press separator fed with raw digestate. The sample has been air-dried for 24 h and then oven-dried in a muffle furnace at 105 °C for 8 h. The dried digestate has been ground using an ultracentrifugal mill (mod. ZM200, Retsch) and sieved to obtain a particle size lower than 500 μm. This was done in particular to ensure a heat transfer rate within the kinetic regime of decomposition. The chemical composition of the solid digestate is reported in Table 3.

Table 3. Characterization of the digestate sample [33].

	Solid Digestate
Proximate analysis (wt.%, dry basis)	
Ash	12.38
Volatile Matter	67.07
Fixed Carbon	20.55
VM/FC	3.29
Ultimate analysis (wt.%, dry basis)	
C	42.52
H	5.94
N	1.79
O	49.75
Compositional analysis (wt.%, dry basis)	
Cellulose	21.64
Hemicellulose	15.08
Lignin	40.88
Extractives	10.02
Calorific value (MJ/kg, dry basis)	
Higher Heating Value	19.74

3.2. Experimental Setup

Pyrolysis tests have been carried out to evaluate the rate of mass loss of solid digestate versus temperature, using a thermogravimetric analyzer (NETZSCH STA 449F1, Selma Cloth, Germany). The sample with mass of 20 mg was inserted directly into a small alumina crucible and temperature was ramped from 30 to 800 °C in nitrogen atmosphere with a flow rate of 50 mL min^{-1}. According to ICTAC recommendations [28], kinetic experiments should be performed using three to five different heating rates (less than 20 °C min^{-1}); therefore, solid digestate was tested at three heating rates of 5, 10, and 20 °C min^{-1}. Thermogravimetric (TG) and differential thermogravimetric (DTG) curves were obtained as a function of temperature during each test. Blank tests have been carried out without sample for TG baseline correction in order to avoid any buoyancy effects. All the thermal analyses were repeated for three times to decrease the test error, and the reproducibility was good. The standard deviation on the TG residues is always lower than 0.5 wt%. The experimental thermogravimetric analysis data have been treated, as recommended in Reference [34].

3.3. Kinetic Analysis through Iso-Conversional Methods

Biomass pyrolysis is a complex process consisting of several reactions due to the different chemical composition of biomass material. Biomass in general exhibits a three stage pyrolytic reduction with the formation of chars, volatiles and gases [35]. In this study an isoconversional model has been employed for the calculation of the Activation Energy and the Master Plots method is employed for the determination of the reaction model, f(α). The decomposition rate is given by Equation (3) [13]:

$$\frac{d\alpha}{dt} = k(T)f(\alpha) \tag{3}$$

where t is time, T is the absolute temperature, f(α) is the differential form of the reaction model, k(T) is the temperature dependence of the rate constant and α is the conversion degree, expressed as:

$$\alpha = \frac{m_i - m_t}{m_i - m_f} \tag{4}$$

where m_i is the initial mass of the sample, m_t is the mass of the sample at temperature T, m_f is the final mass of the sample.

The rate constant k(T) is defined by the Arrhenius equation, Equation (5):

$$k(T) = A\exp\left(-\frac{E}{RT}\right) \tag{5}$$

where E (kJ mol^{-1}) is the activation energy, A (s^{-1}) is the pre-exponential coefficient and R (J mol^{-1} K^{-1}) is the universal gas constant. Then, under non-isothermal conditions at a constant heating rate, β = dT/dt, Equation (3) is transformed into Equation (6):

$$\frac{d\alpha}{dT} = \left(\frac{A}{\beta}\right)\exp\left(-\frac{E}{RT}\right)f(\alpha) \tag{6}$$

The integration of Equation (6) gives Equation (7):

$$\int_0^\alpha \frac{d\alpha}{f(\alpha)} = g(\alpha) = \frac{A}{\beta}\int_{T_0}^T \exp\left(-\frac{E}{RT}\right)dT = \frac{AE}{\beta R}p(x) \tag{7}$$

where p(x), with x = E/RT, and g(α) are the temperature integral and the integral form of the reaction model, respectively.

This equation does not present an analytical solution and $p(x)$ can be obtained by some approximations, depending on the applied kinetic model. In this study, the activation energy was determined employing two kinetic models based on the isoconversional method: Starink [36]. The approximated linear equations of the model are given in Equation (8).

$$\text{Starink}: \ln\left(\frac{\beta}{T^{1.92}}\right) = -1.0008\frac{E}{RT} + \text{constant} \tag{8}$$

The activation energy can be determined by the slope of the regression lines in the graph realized by plotting $\ln(\beta/T^{1.92})$ vs. $1/T$ for the Starink method.

3.3.1. Master-Plots Method

The Master-plots method was employed for the determination of the reaction mechanism. Moreover, when the value of activation energy is obtained from the isoconversional methods, the Master-plots method allows identifying the reaction model.

The temperature integral, p(x), can be expressed by an approximation. The master-plots method employs Doyle's approximation [7] to solve the value of p(x):

$$p(x) = 0.00484 e^{-1.0516x} \tag{9}$$

In Equation (6), the determination of the pre-exponential factor is affected by the reaction model $g(\alpha)$; therefore, adopting a conversion reference point ($\alpha = 0.5$) Equation (6) becomes as follows:

$$g(0.5) = \frac{AE}{\beta R} p(x_{0.5}) \tag{10}$$

where $x_{0.5} = E/RT_{0.5}$, $T_{0.5}$ is the temperature at $\alpha=0.5$ and $g(0.5)$ is the integral form of the reaction model at $\alpha = 0.5$.

The integral master-plots equation can be obtained by dividing Equation (7) by Equation (10).

$$\frac{g(\alpha)}{g(0.5)} = \frac{p(x)}{p(x_{0.5})} \tag{11}$$

In order to determine the reaction model, which better describes the thermal decomposition reaction, the theoretical, $g(\alpha)/g(0.5)$, and experimental, $(p(x)/p(x_{0.5}))$, master plots are plotted as a function of the conversion rate. In particular, for a single step decomposition process with a constant $g(\alpha)$ expression, the master-plots method allows to obtain the proper kinetic model with a high degree of certainty [37]. Table 4 shows the most common kinetic functions $f(\alpha)$ and their integral forms $g(\alpha)$.

Table 4. Most frequently used mechanism functions and their integral forms [38].

Mechanism	Symbol	$f(\alpha)$	$g(\alpha)$ *
Order of reaction			
First-order	F_1	$1 - \alpha$	$-\ln(1 - \alpha)$
Second-order	F_2	$(1 - \alpha)^2$	$(1 - \alpha)^{-1} - 1$
Third-order	F_3	$(1 - \alpha)^3$	$[(1 - \alpha)^{-2} - 1]/2$
Diffusion			
One-way transport	D_1	0.5α	α^2
Two-way transport	D_2	$[-\ln(1 - \alpha)]^{-1}$	$(1 - \alpha)\ln(1 - \alpha) + \alpha$
Three-way transport	D_3	$1.5(1 - \alpha)^{2/3}[1 - (1 - \alpha)^{1/3}]^{-1}$	$[1 - (1 - \alpha)^{1/3}]^2$
Ginstling-Brounshtein equation	D_4	$1.5[(1 - \alpha)^{-1/3}]^{-1}$	$(1 - 2\alpha/3) - (1 - \alpha)^{2/3}$
Limiting surface reaction between both phases			
One dimension	R_1	1	α
Two dimensions	R_2	$2(1 - \alpha)^{1/2}$	$1 - (1 - \alpha)^{1/2}$
Three dimensions	R_3	$3(1 - \alpha)^{2/3}$	$1 - (1 - \alpha)^{1/3}$
Random nucleation and nuclei growth			
Two-dimensional	A_2	$2(1 - \alpha)[-\ln(1 - \alpha)]^{1/2}$	$[-\ln(1 - \alpha)]^{1/2}$
Three-dimensional	A_3	$3(1 - x)[-\ln(1 - x)]^{2/3}$	$[-\ln(1 - x)]^{1/3}$
Exponential nucleation			
Power law, n =1/2	P_2	$2\alpha^{1/2}$	$\alpha^{1/2}$
Power law, n = 1/3	P_3	$3\alpha^{2/3}$	$\alpha^{1/3}$
Power law, n = 1/4	P_4	$4\alpha^{3/4}$	$\alpha^{1/4}$

* $g(\alpha)$ is the integral form of $f(\alpha)$.

3.3.2. Independent Parallel Reactions Scheme

To develop an independent parallel reactions scheme for digestate pyrolysis we have based our methodology on the following assumptions:

- The DTG diagram can be decomposed in three peaks representing, respectively: Hemicellulose, cellulose, and lignin. Extractives are considered together with cellulose, because in the differencial thermogram, their presence cannot be easily distinguished;
- To deconvolute the DTG diagram peaks two approaches can be used: Gaussian and Lorentzian. Both approaches can be implemented in Matlab (Mathworks, Natick, MA, USA) and Origin (OriginLab Corporation Northampton, Massachusetts, USA) software. In this case, the second one was used because it gave better results. In fact once that the DTG diagram is decomposed in three curves (see Figure 6), each one is identified by the three parameters that identify the Lorentzian fit curve equation:

$$y = a \frac{1}{1 + \left(\frac{x-x_0}{dx}\right)} \tag{12}$$

where a is the amplitude; dx is half width at half maximum (HWHM); x_0 is the maximum position.

In this case it was checked that the area of each peak obtained from the deconvolution operation was proportional to the concentration in weight of the pseudo-components (cellulose, hemicellulose, and lignin) inside biomass.

Once the peaks had undergone the deconvolution process, the single peaks (each one corresponding to one pseudo-component) were fitted with the following equation:

$$\frac{d\alpha_{theor}}{dt} = A * \exp\left[-\frac{E}{RT}\right](1-\alpha)^n \tag{13}$$

Three variables were calculated for each pseudo-component: Activation Energy, pre-exponential factor and reaction order. The process was repeated for the three heating rates and also for the three repetitions of the experimental tests. The fitting procedure was based on the Matlab patter search tool, which minimized the difference between the deconvoluted values and the calculated values.

$$LSF = \left\{\sum_{\beta=1}^{3}[(d\alpha/dt)_{deconv} - (d\alpha/dt)_{theor}]^2\right\} \tag{14}$$

Concluding, with this method two fitting steps were performed: the first to deconvolute the peaks (based on Lorentz fitting function) and the second to find the optimal activation energy, pre-exponential factor and reaction order for each of the three analysed pseudo-components. The advantage of this method was to avoid fitting the sum of the three Equations (13), corresponding to each pseudo-component, focusing the attention only on one deconvoluted differential curve at a time.

4. Conclusions

In this study, the solid fraction of biogas digestate was studied as a potential feedstock for pyrolysis, by analyzing its decomposition kinetics. Pyrolysis of solid digestate comprises three stages. In the first stage the moisture is removed, in the second stage, where the main pyrolysis process happens, where the decomposition of hemicellulose, cellulose, and small amount of lignin occurs. In the third stage, the solid residual is slowly decomposed with the formation of char. Compared to biomass digestate is a lignin rich residue in which the more recalcitrant fractions of cellulose and hemicellulose probably remained, for this reason, a quite high activation energy corresponding to the pyrolysis process has to be noted. The average activation energy determined through the Starink method is about 204.1 kJ/mol, with a standard deviation of 25 kJ/mol, which is about 12% of the average value. The variation range of the Activation Energy is about 99 kJ/mol, which is a high value. This means that the average E value cannot be used to statistically represent the activation energy of the whole reaction. For this reason the application of the Master plots is not fully appropriate in this case and for sure will lead to the obtainment of a value of the pre-exponential factor which is not reliable. So in this case the Master-Plots method was used to have some hints only on the kinetic model. Our study indicated that the most probable thermal degradation mechanism function was the nth order reaction model f(α) = (1 − α)n, with a variable reaction order along with the conversion degree. For this reason, we decided to apply an independent parallel reaction scheme to describe the pyrolysis process of digestate. We chose to apply first a deconvolution process to identify three peaks corresponding to the degradation of the three main pseudo-components in biomass (cellulose, hemicellulos and lignin) and then to apply fitting to the identified peaks. In this way we obtained a less precise estimation of the experimental data, but we sped up the implementation of the model. Some limits of the proposed approach are the following:

- High uncertainty of the activation energies measure (at least in the first conversion values), more tests at different heating rates are required;
- The method has still an important error, especially in the deconvolution phase;
- Master plots method should be used only with single-step reactions;
- In the master plots method all the interval of conversion should be considered (from 0.05 to 0.95);
- In the development of the three independent parallel reactions scheme extractives are considered lumped with cellulose;
- Interactions among the digestate pseudo-components are neglected;
- the pseudo components still have some differences from the real lignin, cellulose and hemicellulose. Especially the kinetic data found for lignin are still uncertain.

Author Contributions: P.B. was Principal Investigator of the project (he designed the experimental campaign and collected data), R.T. was supervisor in SINTEF (he coordinated the experiments), R.E.S. performed the TGA analysis with Netzsch STA 449F1), M.B. implemented the model, F.F. was supervisor of the project for University of Perugia.

Funding: This research was funded by European Commission, grant number 731101 under the Brisk 2 project.

Acknowledgments: The authors would like to thank the Brisk 2 European project, grant n. 731101, for funding the project. Project acronym was B2PB-SIN2-1001, project title: "Optimization of catalystic pyrolysis of digestate and sewage sludge". The authors would like to acknowledge i-REXFO LIFE (LIFE16ENV/IT/000547), a project funded by the EU under the LIFE 2016 program. The authors would like to acknowledge the COST Action CA17128 "Establishment of a Pan-European Network on the Sustainable Valorisation of Lignin (LignoCOST)"for useful discussion of lignin valorization. The authors would like to acknowledge the help provided by the reviewers and academic editor in improving the quality of this paper. The authors also want to acknowledge the help of dr.ess Federica Barontini of the Department of Civil and Industrial Engineering, University of Pisa, Italy for useful discussions on DSC analysis.

Conflicts of Interest: The authors declare no conflict of interest.

References

1. Feng, Q.; Lin, Y. Integrated processes of anaerobic digestion and pyrolysis for higher bioenergy recovery from lignocellulosic biomass: A brief review. *Renew. Sustain. Energy Rev.* **2017**, *77*, 1272–1287. [CrossRef]
2. Kosaric, N.; Blaszczyk, R. Industrial effluent processing. In *Encyclopedia of Microbiology*; Lederberg, J., Ed.; Academic Press Inc.: New York, NY, USA, 1992; Volume 2, pp. 473–491.
3. D'Alessandro, B.; D'Amico, M.; Desideri, U.; Fantozzi, F. The IPRP (Integrated Pyrolysis Regenerated Plant) technology: From concept to demonstration. *Appl. Energy* **2013**, *101*, 423–431. [CrossRef]
4. Neumann, J.; Binder, S.; Apfelbacher, A.; Gasson, J.R.; Ramirez Garcia, P.; Hornung, A. Production and characterization of a new quality pyrolysis oil, char and syngas from digestate—Introducing the thermo-catalytic reforming process. *J. Anal. Appl. Pyrolysis* **2015**, *113*, 137–142. [CrossRef]
5. Gomez, X.; Cuetos, M.J.; Garcia, A.I.; Moran, A. An evaluation of stability by thermogravimetric analysis of digestate obtained from different biowastes. *J. Hazard. Mater.* **2007**, *149*, 97–105. [CrossRef] [PubMed]
6. Otero, M.; Lobato, A.; Cuetos, M.J.; Sanchez, M.E.; Gomez, X. Digestion of cattle manure: Thermogravimetric kinetic analysis for the evaluation of organic matter conversion. *Bioresour. Technol.* **2011**, *102*, 3404–3410. [CrossRef]

7. Zhang, D.; Wang, F.; Yi, W.; Li, Z.; Shen, X.; Niu, W. Comparison Study on Pyrolysis Characteristics and Kinetics of Corn Stover and Its Digestate by TG-FTIR. *BioResources* **2017**, *12*, 8240–8254.
8. Doyle, C.D. Estimating isothermal life from thermogravimetric data. *J. Appl. Polym. Sci.* **1962**, *6*, 639–642. [CrossRef]
9. Flynn, J.H.; Wall, L.A. A quick, direct method for the determination of activation energy from thermogravimetric data. *Polym. Lett.* **1966**, *4*, 323–328. [CrossRef]
10. Ozawa, T. A new method of analyzing thermogravimetric data. *Bull. Chem. Soc. Jpn.* **1965**, *38*, 1881–1886. [CrossRef]
11. Vyazovkin, S. Evaluation of activation energy of thermally stimulated solid state reactions under arbitrary variation of temperature. *J. Comput. Chem.* **1997**, *18*, 393–402. [CrossRef]
12. Vyazovkin, S.; Burnham, A.K.; Criado, J.M.; Pérez-Maqueda, L.A.; Popescu, C.; Sbirrazzuoli, N. ICTAC Kinetics Committee recommendations for performing kinetic computations on thermal analysis data. *Thermochim. Acta* **2011**, *520*, 1–19.
13. Vyazovkin, S. *Isoconversional Kinetics of Thermally Stimulated Processes*; Springer: Basel, Switzerland, 2015; pp. 1–239.
14. Imtiaz, A. Comments on "Evaluating the bioenergy potential of Chinese Liquor-industry waste through pyrolysis, thermogravimetric, kinetics and evolved gas analyses" by Ye et al. [Energy Convers. Manage. 163 (2018) 13-21]. *Energy Convers. Manag.* **2018**, *165*, 869–870.
15. Eyring, H. The activated complex and the absolute rate of chemical reactions. *Chem. Rev.* **1935**, *17*, 65–77. [CrossRef]
16. Sánchez-Jiménez, P.E.; Pérez-Maqueda, L.A.; Perejón, A.; Criado, J.M. Generalized master plots as a straightforward approach for determining the kinetic model: The case of cellulose pyrolysis. *Thermochim. Acta* **2013**, *552*, 54–59. [CrossRef]
17. Criado, J.M.; Málek, J.; Ortega, A. Applicability of the master plots in kinetic analysis of non-isothermal data. *Thermochim. Acta* **1989**, *147*, 377–385. [CrossRef]
18. de Carvalho, V.S.; Tannous, K. Thermal decomposition kinetics modeling of energy cane Saccharum robustum. *Thermochim. Acta* **2017**, *657*, 56–65. [CrossRef]
19. Jiang, G.; Nowakowski, D.J.; Bridgwater, A.V. A systematic study of the kinetics of lignin pyrolysis. *Thermochim. Acta* **2010**, *498*, 61–66. [CrossRef]
20. Özsin, G.; Pütün, A.E. Kinetics and evolved gas analysis for pyrolysis of food processing wastes using TGA/MS/FT-IR. *Waste Manag.* **2017**, *64*, 315–326. [CrossRef]
21. Varhegyi, G.; Antal, M.J.; Szekely, T.; Szabo, P. Kinetics of the thermal decomposition of cellulose, hemicellulose, and sugarcane bagasse. *Energy Fuel* **1989**, *3*, 329–335. [CrossRef]
22. Wang, S.; Dai, G.; Yang, H.; Luo, Z. Lignocellulosic biomass pyrolysis mechanism: A state-of-the-art review. *Prog. Energy Combust.* **2017**, *62*, 33–86. [CrossRef]
23. White, J.E.; Catallo, W.J.; Legendre, B.L. Biomass pyrolysis kinetics: A comparative critical review with relevant agricultural residue case studies. *J. Anal. Appl. Pyrolysis* **2011**, *9*, 11–33.
24. Xu, Y.; Chen, B. Investigation of thermodynamic parameters in the pyrolysis conversion of biomass and manure to biochars using thermogravimetric analysis. *Bioresour. Technol.* **2013**, *146*, 485–493. [CrossRef]
25. Mehmood, M.A.; Ye, G.; Luo, H.; Liu, C.; Malik, S.; Afzal, I.; Xu, J.; Ahmad, M.S. Pyrolysis and kinetic analyses of Camel grass (Cymbopogon schoenanthus) for bioenergy. *Bioresour. Technol.* **2017**, *228*, 18–24. [CrossRef] [PubMed]
26. Lee, X.J.; Lee, L.Y.; Gan, S.; Thangalazy-Gopakumar, S.; Ng, H.K. Biochar potential evaluation of palm oil wastes through slow pyrolysis: Thermochemical characterization and pyrolytic kinetic studies. *Bioresour. Technol.* **2017**, *236*, 155–163. [CrossRef]
27. Vamvuka, D.; Kakaras, E.; Kastanaki, E.; Grammelis, P. Pyrolysis characteristics and kinetics of biomass residuals mixtures with lignite. *Fuel* **2003**, *82*, 1949–1960. [CrossRef]
28. Vyazovkin, S.; Chrissafis, K.; Di Lorenzo, M.L.; Koga, N.; Pijolat, M.; Roduit, B.; Sbirrazzuoli, N.; Sunol, J.J. ICTAC Kinetics Committee recommendations for collecting experimental thermal analysis data for kinetic computations. *Thermochim. Acta* **2014**, *590*, 1–23. [CrossRef]
29. Wang, X.; Hu, M.; Hu, W.; Chen, Z.; Liu, S.; Hu, Z.; Xiao, B. Thermogravimetric kinetic study of agricultural residue biomass pyrolysis based on combined kinetics. *Bioresour. Technol.* **2016**, *219*, 510–520.

30. Conesa, J.A.; Caballero, J.A.; Marcilla, A.; Font, R. Analysis of different kinetic models in the dynamic pyrolysis of cellulose. *Thermochim. Acta* **1995**, *254*, 175–192. [CrossRef]
31. Antal, M.J.; Varhegyi, G. Cellulose Pyrolysis Kinetics: The Current State of Knowledge. *Ind. Eng. Chem. Res.* **1995**, *34*, 703–717. [CrossRef]
32. Bartocci, P.; Anca-Couce, A.; Slopiecka, K.; Nefkens, S.; Evic, N.; Retschitzegger, S.; Barbanera, M.; Buratti, C.; Cotana, F.; Bidini, G.; et al. Pyrolysis of pellets made with biomass and glycerol: Kinetic analysis and evolved gas analysis. *Biomass Bioenergy* **2017**, *97*, 11–19. [CrossRef]
33. Barbanera, M.; Cotana, F.; Di Matteo, U. Co-combustion performance and kinetic study of solid digestate with gasification biochar. *Renew. Energy* **2018**, *121*, 597–605. [CrossRef]
34. Cai, J.; Xu, D.; Dong, Z.; Yu, X.; Yang, Y.; Banks, S.W.; Bridgwater, A.V. Processing thermogravimetric analysis data for isoconversional kinetic analysis of lignocellulosic biomass pyrolysis: Case study of corn stalk. *Renew. Sustain. Energy Rev.* **2018**, *82*, 2705–2715. [CrossRef]
35. Anca-Couce, A. Reaction mechanisms and multi-scale modelling of lignocellulosic biomass pyrolysis. *Prog. Energy Combust. Sci.* **2016**, *53*, 41–79. [CrossRef]
36. Starink, M. The determination of activation energy from linear heating rate experiments: A comparison of the accuracy of isoconversion methods. *Thermochim. Acta* **2003**, *404*, 163–176. [CrossRef]
37. Aslan, D.I.; Parthasarathy, P.; Goldfarb, J.L.; Ceylan, S. Pyrolysis reaction models of waste tires: Application of Master-Plots method for energy conversion via devolatilization. *Waste Manag.* **2017**, *68*, 405–411. [CrossRef]
38. Ceylan, S. Kinetic analysis on the non-isothermal degradation of plum stone waste by thermogravimetric analysis and integral Master-Plots method. *Waste Manag. Res.* **2015**, *33*, 345–352. [CrossRef] [PubMed]

© 2019 by the authors. Licensee MDPI, Basel, Switzerland. This article is an open access article distributed under the terms and conditions of the Creative Commons Attribution (CC BY) license (http://creativecommons.org/licenses/by/4.0/).

Article

Effects of Lipid Solid Mass Fraction and Non-Lipid Solids on Crystallization Behaviors of Model Fats under High Pressure

Musfirah Zulkurnain [1], V.M. Balasubramaniam [2,3,*] and Farnaz Maleky [2,*]

1. Food Technology Division, School of Industrial Technology, Universiti Sains Malaysia, Minden 11800, Penang, Malaysia
2. Department of Food Science and Technology, The Ohio State University, 2015 Fyffe Court, Columbus, OH 43210, USA
3. Department of Food Agricultural and Biological Engineering, The Ohio State University, 2015 Fyffe Court, Columbus, OH 43210, USA
* Correspondence: balasubramaniam.1@osu.edu (V.M.B.); maleky.1@osu.edu (F.M.); Tel.: +1-614-292-1732 (V.M.B.); +1-614-688-1491 (F.M.)

Academic Editor: Sergey Vyazovkin
Received: 8 July 2019; Accepted: 1 August 2019; Published: 6 August 2019

Abstract: Different fractions of fully hydrogenated soybean oil (FHSBO) in soybean oil (10–30% *w/w*) and the addition of 1% salt (sodium chloride) were used to investigate the effect of high-pressure treatments (HP) on the crystallization behaviors and physical properties of the binary mixtures. Sample microstructure, solid fat content (SFC), thermal and rheological properties were analyzed and compared against a control sample (crystallized under atmospheric condition). The crystallization temperature (T_s) of all model fats under isobaric conditions increased quadratically with pressure until reaching a pressure threshold. As a result of this change, the sample induction time of crystallization (t_c) shifted from a range of 2.74–0.82 min to 0.72–0.43 min when sample crystallized above the pressure threshold under adiabatic conditions. At the high solid mass fraction, the addition of salt reduced the pressure threshold to induce crystallization during adiabatic compression. An increase in pressure significantly reduced mean cluster diameter in relation to the reduction of t_c regardless of the solid mass fraction. In contrast, the sample macrostructural properties (SFC, storage modulus) were influenced more significantly by solid mass fractions rather than pressure levels. The creation of lipid gel was observed in the HP samples at 10% FHSBO. The changes in crystallization behaviors indicated that high-pressure treatments were more likely to influence crystallization mechanisms at low solid mass fraction.

Keywords: lipid gel; fully hydrogenated soybean oil; microstructure; additive effects; high pressure; heat compression

1. Introduction

In food products, fats and oils contribute to the structural and sensory characteristics of the product that are intrinsically linked to consumer acceptability. The crystallization kinetics of lipids greatly influences the final development of the crystalline structure, polymorphic forms, and their rheological properties [1]. Hence, the crystallization and modifications of fats aim to obtain lipid bases with acceptable characteristics specially when there are zero trans fats, lower saturated fats, and higher mono and polyunsaturated fatty acids content. Recently, studies indicated the use of hard fats, in particular fully hydrogenated oils (FHO) fats, as potential crystallization modifiers of liquid oils and other fats [2–4]. FHO consists of homogeneous saturated triacylglycerols (TAGs) of a high melting point (>40 °C), such as tristearin, and acts as a structuring element that accelerates crystallization in systems composed of low and medium melting TAGs [5]. When diluted with oil, the high melting point TAGs form a three-dimensional network of fat crystals inside liquid oils that have the ability to extend the plastic range of food products [6,7].

However, the naturally occurring crystallization on these lipids usually favor formation of the most thermodynamically stable polymorphic form β on dilution. This may result in high melting points and large spherulitic crystals with poor macroscopic properties [3,8]. Hence, it is recommended that the appropriate processing conditions, such as cooling rate, agitation (shear), pressurization and processing temperature, are applied to affect the crystallization behavior of TAG molecules [9–12]. High pressure treatments of lipids have been investigated to gain a fundamental understanding of phase transition [13–15] and its applications on fatty acids [16,17] and fats [18–21]. Interestingly, studies have also documented that minor components which are either present indigenously or added in fats (such as emulsifiers, mono- and diacylglycerols, and non-fat ingredients) can influence the system's crystallization and final properties [9,10]. Despite molecular dissimilarity to fats, foreign crystals such as inorganic (magnesium silicate, carbon nanotube, graphite) and organic (terephthalic acid) materials have been shown to promote fat crystallization [22,23]. It has also been shown that the effect of processing conditions is more pronounced at low solid mass fraction where the capacity of the crystalline fraction of the TAGs to form a solid network in the liquid oil plays crucial roles in the structuring ability [22,24].

Recently our group investigated the crystallization behavior of different binary mixtures of fully hydrogenated soybean oil and soybean oil under high pressure treatments, consisting of adiabatic compression to a targeted pressure followed by isobaric cooling [25]. The findings revealed distinct crystallization properties at nano-, micro- and macroscale when the crystallization started during adiabatic compressions and continued during isobaric cooling. Regardless of pressure levels and maximum temperatures achieved under pressure, the rapid onset of crystallization during adiabatic compression resulted in smaller homogeneous crystal structures [25]. This finding agreed with other studies on pressurization of lipids that showed that forced convection during compression prior to the onset of phase transition can elucidate the enhancement of structural modification via an improvement in the kinetic mobility of TAGs during the initial stage of lipid crystallization (nucleation) [26].

To further investigate the effects of high pressure treatments on lipids, here we studied the crystallization behavior of the same lipid systems (a mixture of fully hydrogenated soybean oil in soybean oil) in the presence of additives (as a non-lipid solid particle). Paying attention to the effects of solid mass fraction and salt particles as a non-lipid solid particle, we monitored the crystallization behaviors during the compression step to identify and quantify structure characteristics responsible for differences in macroscopic properties formed under high-pressure treatments.

2. Results and Discussion

2.1. Crystallization Temperature (T_s) and Pressure Threshold

The effects of pressure and salt on crystallization temperature (T_s) of three model fats made of 10, 20, 30% fully hydrogenated soybean oil (FHSBO) in soybean oil are shown in Figure 1. The T_s of the model fats increased in a quadratic manner with an increase in pressure and was well fitted into a quadratic function with a regression coefficient of $R^2 > 0.99$ and the sum of squared errors of prediction (SSE) < 6.94, as shown in Table 1. A quadratic relation of the melting temperature with pressure has been reported by Ferstl et al (2010) as up to 450 MPa for triolein [18]. Several studies have shown that high pressure treatments promote crystallization of lipids by increasing the melting temperature between 10 to 24 °C/100 MPa [13,14,16,18]. The phase boundary coefficient, dT_m/dP of homogeneous fatty acids have been reported at 12.7 °C/100 MPa and 23.9 °C/100 MPa for α and β polymorphic forms, respectively [27]. In the absence of melting point under pressure, a phase transition diagram based on the crystallization temperature can be a close substitute to understand the fundamental response of lipid materials during high pressure processing [15,19]. As shown from the values of coefficient a_1 in Table 1, an increase in the solid mass fraction from 10% to 30% FHSBO increased the phase boundary coefficients based on the T_s from 18.4 to 20.3 and 21.2 °C/100 MPa for the model fats at 10%, 20%, and 30% FHSBO, respectively. The table also shows that the addition of salt did not significantly affect these values.

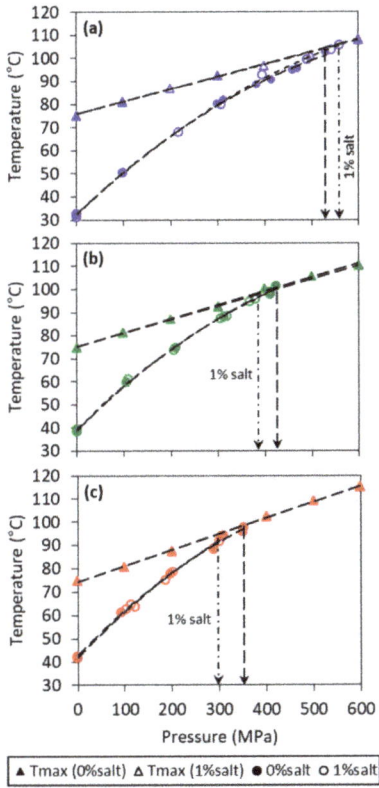

Figure 1. Maximum temperature under pressure (T_{max}) (triangle symbol) and phase diagram (circle) of model fats at (**a**) 10%, (**b**) 20%, and (**c**) 30% of fully hydrogenated soybean oil (FHSBO) without (filled symbol) and with 1% salt (empty symbol). Vertical arrows show pressure threshold where crystallization temperature (T_s) crosses T_{max}.

At an initial temperature (T_i) of 75 °C (a temperature higher than the fat models melting point at atmospheric condition), the maximum temperature under pressure (T_{max}) achieved during compression step increased linearly with increase in pressure levels from 100 MPa to 600 MPa, (shown in Figure 1). The linear relation of T_{max} with pressure is confirmed by showing a high regression coefficient for all mass fraction ($R^2 > 0.99$). The linear coefficient (b_0) of the slope of T_{max} shows small increments with the increase in solid mass fraction from 5.5 to 6.9 °C/100 MPa for 10% and 30% FHSBO, respectively (Table 1), which was not affected by the addition of salt. As indicated by vertical arrows in Figure 1, the phase boundaries of all model fats were truncated when the T_{max} crossed T_s, as the onset of crystallization shifted from the isobaric condition (during the pressure holding step) to adiabatic condition (during the rapid compression step). Instantaneous phase transition was evident from a strong increase in temperature during adiabatic compression when the crystallization happened before reaching T_{max}. This set of data are shown in Figure S1 (Supplementary Materials). The occurrence of the sudden increase in temperature during the compression of lipid materials has been reported in other studies and is linked to the emission of the latent heat of crystallization and the existence of small crystalline bodies [17,28].

Table 1. Quadratic coefficient relating crystallization temperature (T_s) with pressure and linear coefficient of maximum temperature (T_{max}) with pressure for different model fats at 10%, 20%, and 30% fully hydrogenated soybean oil (FHSBO) without and with 1% salt. SSE is sum of squared of errors of prediction.

Coefficients	10% FHSBO		20% FHSBO		30% FHSBO	
	0% Salt	1% Salt	0% Salt	1% Salt	0% Salt	1% Salt
a_0 (°C/(100 MPa)^2)	−1.17	−1.06	−1.36	−1.34	−1.57	−1.65
a_1 (°C/100 MPa)	18.4	19.0	20.3	19.9	21.2	21.1
a_2 (°C)	32.72	32.45	38.94	39.57	41.86	42.70
R^2	0.999	0.999	0.999	0.999	0.999	0.998
SSE	6.94	6.49	3.63	4.54	5.3	8.63
b_0 (°C/100 MPa)	5.5	5.4	6.0	5.8	6.9	6.9
R^2 (T_{max})	0.999	0.996	0.997	0.999	0.998	0.998
Pressure threshold (MPa)	531.7	560.1	423.6	351.0	351.0	300.5

Table 1 also reports the pressure threshold of 531.7, 423.6, and 351.0 MPa for model fats without salt at 10%, 20%, and 30% FHSBO, respectively. The increase in solid mass fraction reduced the pressure threshold required to induce crystallization because lower unsaturated fatty acid content makes less free volume available for compression [29]. Interestingly, the addition of 1% salt increased the pressure threshold for 10% FHSBO sample from 531.7 MPa to 560.1 MPa but reduced the pressure threshold from 423.6 to 379.6 MPa for 20% FHSBO and from 351.0 to 300.5 MPa for 30% FHSBO samples. When the supersaturation was low at low solid mass fraction (10% FHSBO), longer pressurization may impart forced convection in the liquid state before phase transition take place, which was observed by Tefelski et al. (2011) during the pressurization of oleic acid [26]. Whether the addition of salt leads to convection effects that affect nuclei formation is not clear.

2.2. Induction Time of Crystallization (t_c)

Further, the effects of salt particle during crystallization was evaluated from the induction time of crystallization shown in Table 2. For atmospheric crystallization, the increase in the solid mass fraction from 10% to 30% FHSBO significantly reduced the induction time from 6.5 to 3.6 min, respectively. High pressure crystallization resulted in significantly lower induction time compared to the atmospheric crystallization. An increase in pressure significantly decreased the induction time, as reported previously by Ferstl et al. (2011) [29]. At 100 MPa, the induction time during the isobaric condition (between 2.33 to 2.74 min) was not affected by the increase in solid mass fraction or salt addition. At 300 MPa, the addition of salt significantly reduced the induction time of crystallization for the 30% FHSBO model fat from 0.92 min during isobaric condition to 0.46 min during adiabatic condition, as the pressure threshold had been reached. At 600 MPa, when crystallization happened under an adiabatic condition, the increase in solid mass fraction (from 10% to 30% FHSBO) significantly reduced the induction time of crystallization (from 0.66 to 0.46 min). This is directly related to the reduction of the pressure threshold with the increase in the solid mass fraction (Table 1).

Table 2. Induction time of crystallization for fat models with 10%, 20%, and 30% fully hydrogenated soybean oil (FHSBO) without and with 1% salt.

Pressure (MPa)	Induction time of crystallization [1] (min)					
	10% FHSBO		20% FHSBO		30% FHSBO	
	0% Salt	1% Salt	0% Salt	1% Salt	0% Salt	1% Salt
0.1	6.18 ± 0.2 [Ai]	6.47 ± 0.0 [ai]	4.53 ± 0.3 [Aii]	4.20 ± 0.2 [aii]	3.69 ± 0.1 [Aiii]	3.68 ± 0.3 [aiii]
100	2.69 ± 0.3 [Bi]	2.74 ± 0.1 [bi]	2.72 ± 0.1 [Bi]	2.61 ± 0.0 [bi]	2.49 ± 0.3 [Bi]	2.33 ± 0.1 [bi]
300	1.39 ± 0.1 [Ci]	1.44 ± 0.1 [ci]	0.85 ± 0.0 [Cii]	0.82 ± 0.0 [cii]	0.92 ± 0.2 [Cii]	0.49 ± 0.1 [ciii]*
600	0.66 ± 0.1 [Di]	0.72 ± 0.0 [di]	0.49 ± 0.0 [Dii]	0.44 ± 0.0 [dii]	0.46 ± 0.1 [Dii]	0.43 ± 0.0 [cii]

[1] Means in the same column followed by different upper-case letters are significantly different ($p < 0.05$) between pressure levels. Means in the same row followed by different Roman letters are significantly different ($p < 0.05$) between solid mass fractions. Star symbol (*) denotes significant difference between means in the same row for samples without and with addition of 1% salt.

2.3. Melting Properties of the Crystallizaed Model Fats

The fatty acid composition of all model fats, without and with salt, is reported in Table 3. The addition of salt did not affect the fatty acid composition of the model fats. To explore the effects of the pressure level on the samples, their melting profile was studied immediately after the processing, when the crystalline solid-state structure remained unaffected. Figure 2 shows a representative of the melting thermograms of samples crystallized under atmospheric conditions compared to the high pressure treatment at 600 MPa. The melting point (T_m), onset of melting (T_o), and specific enthalpy of melting (ΔH_m) of samples crystallized at 0.1, 100, 300, and 600 MPa are shown in Table 4. The melting curves of the atmospheric crystallized samples at 30% FHSBO and 20% FHSBO (with 1% salt) show a shoulder (indicated by arrows in Figure 2a) in addition to a broad prominent peak (β phase), which suggests the presence of a β' polymorph phase in these samples [8,25]. The result of the 30% FHSBO is in agreement with the Ribeiro et al. who reported the existence of a β' phase in the mixture of FHSBO and soybean oil for samples containing more that 30% FHSBO [8]. In accordance with Ostwald's step rule, studies have shown that the crystallization of most fully hydrogenated oils favors the formation of a metastable α phase which is kinetically favored at the beginning of phase transition [30,31]. It is also shown that blending FHO with high amount of oil (>80%) allows the adoption of a more thermodynamically stable polymorphic form because a reduction in the viscosity of the mixtures enhances TAG diffusion [8,30,32]. This kinetic of crystallization may vary when other non-fat particles are added to the system, when the model fat at 20% FHSBO (with 1% salt) still shows formation of the β' phase. The presence of foreign particles may increase the rate of crystallization as it kinetically favors the formation of the less stable polymorphic phase when TAG molecules are incorporated into the crystalline surface very quickly, and therefore imperfectly.

Table 3. Fatty acid composition of fully hydrogenated soybean oil (FHSBO) and model fats at different percentages of FHSBO, without and with 1% salt.

Fatty Acid (%)	FHSBO	FHSBO:SBO (w/w%)					
		10:90		20:80		30:70	
		0% Salt	1% Salt	0% Salt	1% Salt	0% Salt	1% Salt
Linoleic acid (C18:2n6)	-	49.6	50.6	44.0	44.6	39.4	39.0
Stearic acid (C18:0)	91.1	11.8	11.4	20.4	20.2	29.0	28.8
Oleic acid (C18:1n9)	-	20.6	21.9	18.4	19.1	15.9	16.3
Palmitic acid (C16:0)	8.9	11.1	10.8	11.2	10.9	11.1	10.6
γ-Linolenic acid (C18:3n6)	-	4.8	4.9	4.2	4.8	4.5	4.9
Linolenic acid (C18:3n3)	-	0.2	0.2	0.2	0.2	0.2	0.3

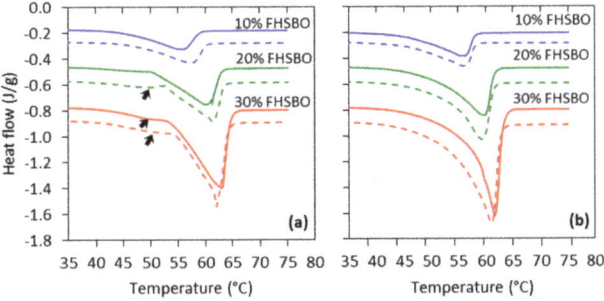

Figure 2. Melting thermograms of model fats at 10%, 20%, and 30% fully hydrogenated soybean oil (FHSBO) without (-) and with 1% salt (---) crystallized under (**a**) atmospheric pressure and (**b**) 600 MPa.

Despite the atmospheric crystallized samples, all of the high pressure crystallized samples (without and with 1% salt) showed a single peak with a sharp and narrow melting exotherm that suggest the existence of only the most stable β polymorphic phase in the high pressure crystallized samples. This is in agreement with the previous study of 30% FHSBO without the addition of salt [25]. The reduction in free space under high pressure increases the degree of supersaturation and thus, increases the thermodynamic driving force for polymorphic transformation [33]. Moreover, the high pressure treatments significantly reduced T_m and increased ΔH_m of the samples compared to the atmospheric crystallization (Table 4). Increase in pressure levels from 100 to 600 MPa further reduced T_m of the samples except for 20% FHSBO. The reduction of the pressured sample melting points can be explained by the depression of their freezing points (colligative effect) due to the higher solubility of TAGs in samples crystallized at high pressure levels [30]. These effects of TAGs solubility were not observed on the T_m of pressured fat with 30% FHSBO due to differences in the sample's initial temperature (T_i) of the pressurization [23].

Table 4. Melting point (T_m), onset of melting (T_o), and specific enthalpy of melting (ΔH_m) of 10%, 20%, and 30% fully hydrogenated soybean oil (FHSBO), without and with 1% salt, crystallized at different pressure levels.

Pressure (MPa)	Melting Point [1] (T_m, °C)					
	10% FHSBO		20% FHSBO		30% FHSBO	
	0% Salt	1% Salt	0% Salt	1% Salt	0% Salt	1% Salt
0.1	57.10 ± 0.2 A	57.64 ± 0.6 a	60.61 ± 0.6 I	61.23 ± 0.2 i	62.74 ± 0.3 A	62.11 ± 0.4 a
100	56.31 ± 0.1 B	56.89 ± 0.1 b	59.29 ± 0.5 II	59.34 ± 0.4 ii	62.11 ± 0.3 B	61.59 ± 0.3 ab
300	56.37 ± 0.2 B	56.30 ± 0.1 c	59.48 ± 0.1 II	59.58 ± 0.2 ii	61.57 ± 0.1 C	61.40 ± 0.1 b
600	55.90 ± 0.2 C	55.84 ± 0.1 d	59.38 ± 0.2 II	59.32 ± 0.3 ii	61.83 ± 0.2 BC	61.08 ± 0.2 b*

Pressure (MPa)	Onset of Melting [1] (T_o, °C)					
	10% FHSBO		20% FHSBO		30% FHSBO	
	0% Salt	1% Salt	0% Salt	1% Salt	0% Salt	1% Salt
0.1	46.02 ± 0.5 A	45.81 ± 0.4 a	52.72 ± 2.0 A	53.25 ± 2.3 a	53.41 ± 0.8 A	54.11 ± 0.9 ab
100	46.37 ± 0.7 AB	48.20 ± 0.1 d*	52.50 ± 0.9 A	52.14 ± 2.1 a	56.12 ± 0.7 B	53.98 ± 0.8 ab*
300	47.23 ± 0.9 AB	47.85 ± 0.2 c	52.50 ± 1.6 A	52.39 ± 2.0 a	57.07 ± 1.6 B	54.00 ± 0.2 a*
600	47.72 ± 1.0 B	46.87 ± 0.4 b	51.76 ± 1.1 A	53.58 ± 1.5 a*	56.96 ± 1.0 B	55.21 ± 0.6 b*

Pressure (MPa)	Specific Enthalpy of Melting [1] (ΔH_m, J/g)					
	10% FHSBO		20% FHSBO		30% FHSBO	
	0% Salt	1% Salt	0% Salt	1% Salt	0% Salt	1% Salt
0.1	17.67 ± 0.1 A	16.86 ± 0.6 a*	34.64 ± 1.1 A	35.16 ± 0.4 a	58.45 ± 2.1 A	52.58 ± 2.3 a*
100	18.86 ± 0.6 B	16.62 ± 0.6 a*	37.93 ± 1.0 B	39.92 ± 1.6 b*	60.88 ± 2.7 AB	61.66 ± 2.7 b
300	18.87 ± 0.5 B	18.79 ± 0.2 b	38.64 ± 2.0 B	40.49 ± 2.3 b	60.66 ± 1.6 AB	67.51 ± 1.2 b*
600	18.80 ± 0.3 B	19.11 ± 0.2 b	37.89 ± 1.0 B	40.39 ± 1.0 b*	63.34 ± 0.3 B	70.61 ± 3.2 b*

[1] Means in the same column followed by different upper-case letters are significantly different ($p < 0.05$). Star symbol (*) denotes significant differences between means in the same row for samples, without and with addition of 1% salt.

Further analysis of Figure 2 (dotted thermogram in Figure 2b) shows that salt addition resulted in the formation of a broader exothermic peak (with higher ΔH_m) for samples containing 20% and 30% FHSBO. The comparison of the samples ΔH_m is shown with star symbols in Table 4. The increase in ΔH_m may infer the formation of a higher amount of crystals in the samples that is in agreement with the significant increment of these sample's solid fat content (SFC) shown in Table 5 [25]. This finding may translate into differences in crystallization behavior and the resulting network properties of the samples.

2.4. Samples Solid Fat Content (SFC), Shear Storage Modulus, and Microstructure

The effects of atmospheric (0.1 MPa) and high pressure treatments (100, 300, and 600 MPa) on the solid fat content (SFC) of the model fats, without and with 1% salt, are shown in Table 5. Although there were no significant differences observed between the SFC values of the 10% FHSBO samples, sample homogeneity was different in this set of samples (data are not shown here). For 10% FHSBO, the atmospheric crystallized samples showed sedimentation of large fat crystals at the bottom of the glass tube upon storage for 24 hours at 20 °C, while all high pressure crystallized samples maintained their gel texture. This observation indicates a clear phase separation of the atmospheric sample's solid crystals and liquid oil, in which the samples can be classified as having low plasticity and low resistance to oiling-out [34–36]. This could be confirmed by the low G' values (15.1 Pa) measured for this set of samples, as shown in Table 6. Although the addition of salt increased the sample G' to 21.4 Pa, sample storage modulus remained lower than their loss modulus ($G'' = 44.1 \pm 3.7$ Pa). This indicates that the dispersed low concentration of solid crystallites was not able to bind the liquid oil and oil remains as the continuous phase to induce liquid-like properties to the system [35]. In contrast, Table 6 shows significantly higher G' values (from 2.9×10^4 to 9.4×10^4 Pa) for the pressurized samples at 10% FHSBO. An increase in pressure levels significantly increased their solid-like behavior with a significant increase in the G' values. They formed a translucent lipid gel with a homogeneous distribution of high density small stable β crystals that helps them hold their shape and solid texture.

Table 5. Solid fat content (SFC) of model fats at 10%, 20%, and 30% fully hydrogenated soybean oil (FHSBO), without and with 1% salt, crystallized at different pressure levels.

Pressure (MPa)	Solid Fat Content [1] (SFC, %)					
	10% FHSBO		20% FHSBO		30% FHSBO	
	0% Salt	1% Salt	0% Salt	1% Salt	0% Salt	1% Salt
0.1	10.92 ± 0.12 [A]	10.97 ± 0.11 [a]	19.31 ± 0.24 [II]	19.58 ± 0.31 [ii]	28.78 ± 0.15 [B]	20.07 ± 0.46 [b]
100	11.07 ± 0.04 [A]	11.15 ± 0.12 [a]	20.54 ± 0.05 [I]	21.21 ± 0.04 [i*]	29.83 ± 0.37 [A]	31.28 ± 0.48 [a*]
300	11.12 ± 0.03 [A]	11.18 ± 0.17 [a]	20.58 ± 0.05 [I]	21.23 ± 0.06 [i*]	29.66 ± 0.08 [A]	32.38 ± 0.20 [a*]
600	11.17 ± 0.05 [A]	11.19 ± 0.17 [a]	20.44 ± 0.04 [I]	20.33 ± 0.07 [i*]	29.78 ± 0.07 [A]	31.63 ± 0.27 [a*]

[1] Means in the same column followed by different upper-case, lower-case, and Roman letters are significantly different ($p < 0.05$) between pressure levels. Star symbol (*) denotes significant difference between means in the same row for samples without and with addition of 1% salt.

Table 6. Shear storage modulus (G') of model fats at 10%, 20%, and 30% fully hydrogenated soybean oil (FHSBO), without and with 1% salt, crystallized at different pressure levels.

Pressure (MPa)	Shear Storage Modulus [1] (G', Pa)					
	10% FHSBO		20% FHSBO		30% FHSBO	
	0% Salt	1% Salt	0% Salt	1% Salt	0% Salt	1% Salt
0.1	15.1 ± 3.2 [C]	21.4 ± 2.0 [d]	$4.1 \times 10^5 \pm 1.2 \times 10^5$ [I]	$4.5 \times 10^5 \pm 9.9 \times 10^4$ [ii]	$2.4 \times 10^6 \pm 2.0 \times 10^5$ [A]	$2.3 \times 10^6 \pm 3.2 \times 10^5$ [ab]
100	$2.9 \times 10^4 \pm 4.3 \times 10^3$ [B]	$3.3 \times 10^4 \pm 1.4 \times 10^3$ [c]	$4.7 \times 10^5 \pm 5.4 \times 10^4$ [I]	$3.7 \times 10^5 \pm 3.7 \times 10^4$ [ii*]	$2.1 \times 10^6 \pm 7.1 \times 10^4$ [B]	$2.2 \times 10^6 \pm 4.0 \times 10^4$ [b]
300	$5.3 \times 10^4 \pm 1.8 \times 10^3$ [A]	$5.7 \times 10^4 \pm 2.7 \times 10^3$ [b]	$4.6 \times 10^5 \pm 1.2 \times 10^4$ [I]	$5.0 \times 10^5 \pm 2.9 \times 10^4$ [i,ii]	$2.3 \times 10^6 \pm 9.4 \times 10^4$ [AB]	$2.5 \times 10^6 \pm 1.6 \times 10^5$ [ab]
600	$6.1 \times 10^4 \pm 2.4 \times 10^3$ [A]	$9.4 \times 10^4 \pm 1.3 \times 10^4$ [a]	$5.9 \times 10^5 \pm 2.9 \times 10^4$ [I]	$6.4 \times 10^5 \pm 1.4 \times 10^4$ [i*]	$2.4 \times 10^6 \pm 1.2 \times 10^5$ [A]	$2.7 \times 10^6 \pm 1.1 \times 10^5$ [a*]

[1] Means in the same column followed by different upper-case, lower-case, and Roman letters are significantly different ($p < 0.05$) between pressure levels. Star symbol (*) denotes significant difference between means in the same row for samples without and with addition of 1% salt.

Figure 3 represents and compares the microstructures of the samples, without and with 1% salt, crystallized under atmospheric and high pressure treatment at 600 MPa. The mean cluster diameter of all the crystallized samples is summarized in Figure 4. At 10% FHSBO, the microstructure of the atmospheric crystallized samples is spherulitic in nature with well-defined dendritic aggregations of large crystallites. They have largest mean cluster diameter with dimension of 61.8 ± 1.3 μm. The growth of the spherulites was not radial, resulting in characteristic Maltese-cross shapes, parallel with observations reported by Omonov et al. (2010) [35]. Bouzidi et al. (2013) and Omonov et al. (2010) reported that a high portion of liquid oil allows high melting TAGs (tristearin) to continuously grow into larger spherulites that branch out into dendrites until the concentration of high melting TAGs diminishes [35,37]. Under a polarized light microscope, the crystallites were seen segregated from the oil with the largest open dendrites and looser network, as reported by Omonov et al. (2010) [35]. This allows the liquid phase to leave the solid network more easily compared to the denser fibrillar network at higher solid mass fractions. Under an atmospheric condition, the addition of salt to 10% FHSBO resulted in the formation of denser and smaller branched dendrites that reduced the size of the spherulites to a smaller dimension at 49.6 ± 11.5 µm (Figure 4).

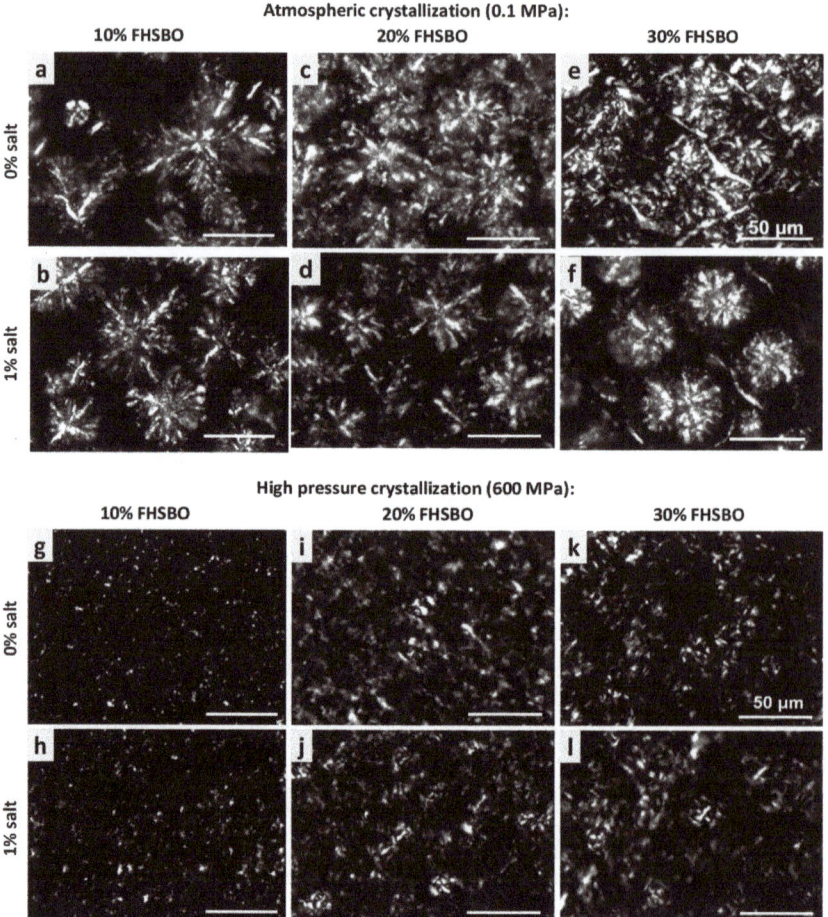

Figure 3. Polarized light microscopy (PLM) micrographs (500× magnification) of model fats of 10%, 20%, and 30% FHSBO, without salt and with 1% salt, crystallized at atmospheric condition (**a**–**f**) and under high pressure treatment at 600 MPa (**g**–**l**).

In contrast, PLM of 10% FHSBO samples crystallized under high pressure at 600 MPa (Figure 3g–h) shows unique microstructures of non-spherulitic behavior. The primary microstructural elements did not show characteristics of crystalline fibers like dendritic or needle. Rather, the microstructural elements, which originate from different individual nucleus, are small and uniformly distributed and responsible for their gel-like behaviors (formed during the rapid compression step at 600 MPa).

A similar creation of β-fat gel made of binary mixtures of high-melting and low-melting fats has been reported by Higaki et al. (2004) via melt-mediated transformation [24]. Occasionally, a small crystal cluster may form with the average diameter of 5.8 ± 1.0 μm (Figure 4). As shown in Figure 4, the mean cluster diameter increased with an increase in pressure levels by about 5 times to 30.1 ± 1.3 μm at 100 MPa. This might be explained by the slower nucleation and its subsequent growth under isobaric conditions (at lower pressure levels) at the expense of the smaller surrounding crystals. Interestingly, the addition of salt shows a significant reduction in the mean cluster diameter under this condition at 100 and 300 MPa.

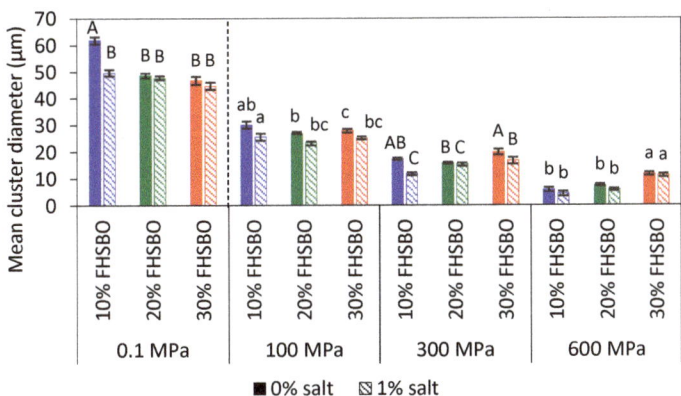

Figure 4. Mean cluster diameter of 10%, 20%, and 30% fully hydrogenated soybean oil (FHSBO), without and with 1% salt, crystallized at different pressure levels. Different upper-case or lower-case show significant difference ($p < 0.05$) between samples at different percentage of FHSBO.

At 20% FHSBO, the high pressure crystallization increased the SFC from 19.3% to the range of 20.4–20.6% for non-salted samples. The addition of 1% salt further increased the SFC values of the high pressure crystallized samples from 19.6% to the range of 21.2–21.3%. This is supported by the differences in the fat crystal structure formed under atmospheric (Figure 3c–d) versus high pressure treatments (Figure 3i–j). Although both samples have the physical state of a semi-solid fat, the atmospheric crystallization resulted in highly contrasted PLM micrographs that represent a high molecular order in the crystal structure [35]. The mean cluster diameter of these 20% FHSBO samples are between 44.6 to 48.6 μm (Figure 4) and are not affected by the addition of 1% salt. The increase in solid mass fraction from 10% to 20% FHSBO resulted in more densely packed crystal clusters, thus limiting radial growth of the spherulites by the adjacent agglomerates when formed slowly under atmospheric crystallization [38].

On the other hand, the PLM of high pressure crystallized samples at 20% FHSBO (Figure 3i–j) showed circular clusters of poorly defined boundaries embedded in high number of small primary microstructural elements, due to higher nucleation rates, and a faster initial overall crystal growth. The mean cluster diameters of these samples were significantly smaller compared to the atmospheric crystalized samples ranging from 26.9 ± 0.4 μm at 100 MPa to 5.8 ± 0.5 μm at 600 MPa (Figure 4). Similar to samples at 10% FHSBO, an increase in pressure significantly reduced the mean cluster diameter of the samples with the greatest impact seen at 600 MPa and the addition of salt reduced the diameters at low pressure levels. Interestingly, the 20% FHSBO high pressure crystallized samples showed comparable G' with the atmospheric crystallized samples ranged between 4.2×10^5 (0.1 MPa) to 5.9×10^5 Pa (600 MPa). The addition of salt significantly increased the G' of the high pressure crystallized sample at 600 MPa to 6.4×10^5 Pa when the crystallization took place during an adiabatic condition. This was also supported by the increment of the samples ΔH_m showing higher crystal formation.

At 30% FHSBO, pressurization with the addition of salt significantly increased the SFC of the samples at all pressure levels to the range between 31.3–32.4% (denoted with star symbol in Table 5). The sample's G' are comparable between the atmospheric crystallization (2.4 × 10^6 Pa) and the high pressure crystallization ranged between 2.1 × 10^6 Pa (100 MPa) to 2.4 × 10^6 Pa (600 MPa), as shown in Table 6. However, the atmospheric crystallized samples were brittle in nature, as given by a smaller linear viscoelastic region (LVR) reported earlier [12]. The LVR is defined as the region where a fat material is able to hold its texture before the bond between fat crystals network is broken. All of the high pressure crystallized samples have longer LVR than the atmospheric crystallized samples, suggesting a stronger bond between their fat crystal networks. Solid networks based on a small crystal network have increased specific surface areas that increase contact with the oil and tend to be firmer [39]. The addition of salt significantly increased the storage modulus of the sample at 600 MPa at 2.7 × 10^6 Pa. Like SFC, the effect of salt on the G' of the high pressure crystallized samples was significant when the crystallization happened during adiabatic compression at 600 MPa.

Although similar in textural properties at macroscale, the microstructures of the atmospheric crystallized samples were significantly larger compared to the high pressured crystallized samples, as shown in Figure 4. The mean cluster diameter of the atmospheric crystals was between 44.7 ± 1.3 to 46.8 ± 1.5 μm and were at least twice larger than the high-pressure crystallized samples between 27.9 ± 0.7 μm at 100 MPa to 11.1 ± 0.7 μm at 600 MPa. As shown in Figure 4, the atmospheric crystallized samples show a crystal network of discrete spherulites of large structures that contributes to their brittle texture. However, the PLM micrograph of samples crystallized with the addition of 1% salt shows a lower contrast which is well correlated with the reduction of ΔH_m of these samples (Table 4). As discussed earlier, the presence of foreign particles may suggest the formation of less perfect crystals when TAG molecules are incorporated into the crystalline surface quickly.

For the high-pressure crystallized samples, increase in pressure significantly reduced the mean cluster diameter regardless of the addition of salt. This shows that the effect of pressure levels is more prominent than solid mass fraction at the microstructure level, in contrast to the macrostructural properties. Interestingly, the influence of solid mass fraction was significant at 600 MPa when the mean cluster diameters were reduced with the reduction in FHSBO fraction from 30% to 10% in contrary to the increment of their induction time of crystallization (Table 2). High pressure crystallization at 600 MPa resulted in unique microstructures that constituted of large numbers of very small size and uniform distribution of microstructural elements between 0.1 to 5 μm, as reported in earlier work of binary mixtures at 30% FHSBO [25]. In static crystallization, an increase in SFC is expected to increase the viscosity of the melt that may limit mass transfer and crystal growth [40]. However, the opposite results may suggest a different nucleation mechanism and crystal growth during the dynamic compression step.

3. Materials and Methods

3.1. Sample Preparation

Model fats of different solid mass fractions were prepared by blending 10%, 20%, and 30% w/w fully hydrogenated soybean oil (FHSBO) with soybean oil (SBO). Iodized salt (Morton Salt, Chicago, IL, USA) was powdered and passed through a 630-mesh sieve. The lipid mixtures were heated at 90 °C for 15 min under constant stirring to erase their crystal memory. An amount of 1% w/w salt was added to the melted mixture at 90 °C under constant vigorous stirring. Both the mixtures, with and without salt, were dispensed carefully while under heating and constant stirring, to ensure the samples homogeneity, in a 5-mL polypropylene syringe (Becton Dickinson and Co., Franklin Lakes, NJ, USA) with a movable plunger and rubber stopper. This sample carrier was insulated with two layers of tape (CVS Pharmacy Inc., Woonsocket, RI, USA).

3.2. Fatty Acid Composition

The fatty acid composition of the model fats was determined according to AOCS Ce 1-62 standard method [41].

3.3. High-Pressure Crystallization Experiments

The high pressure experiments were conducted using a high-pressure kinetics tester (PT-1, Avure Technologies, Kent, WA, USA) with a 54-mL high-pressure chamber (0.02 m internal diameter). A high-pressure intensifier (M-340A, Flow International, Kent, WA, USA) was used to achieve a pressure level up to 700 MPa at the rate of 12 MPa/s. The decompression time was at 4 s regardless of holding pressure. Propylene glycol (Brenntag North America Inc., Reading, PA, USA) was used as the pressure-transmitting fluid. The pressure and temperature were monitored and recorded every 1 second using DasyLab software (Version 7.00.04, National Instruments Corp., Austin, TX, USA).

Prior to high-pressure treatment, samples were thermally equilibrated in a water bath at 90 °C to erase crystal memory. A K-type thermocouple (KMTSS-062G-12, Omega Engineering, Stamford, CT, USA) was positioned in the middle of the sample carrier. The direct contact of the thermocouple junction with the sample enabled temperature measurement in the sample during pressure treatment. Samples were pressurized over a range of pressures between 100 MPa and 600 MPa from a fixed initial temperature (T_i) of 75 °C, and the maximum temperature under pressure (T_{max}) was recorded. The T_i was set higher than the melting temperatures of the model fats to ensure the samples were in melt state and to avoid any chance of nucleation in the sample prior to pressurization. Then, the samples were held under the targeted pressure and allowed to cool for 10 min by maintaining the pressure's chamber at 30 °C for model fats of 20% and 30% FHSBO and 25 °C for model fat of 10% FHSBO. Finally, the samples were depressurized to the atmospheric pressure in 4 seconds and taken out for analysis. All high-pressure crystallization experiments were conducted in duplicate, and the crystallized samples were analyzed within one hour after the treatment. The solid crystallized samples were carefully removed from the sample carrier and sliced into three parts. Physical and structural analysis of the samples were made using the sample taken from the center slice. The samples were stored at 20 °C for 24 hours for analysis that demand further observations.

3.4. Atmospheric Crystallization Experiments (Control Sample)

Samples were equilibrated in a water bath at 75 °C and crystallized at atmospheric pressure (0.1 MPa). The experiments were conducted by immersing the samples in a temperature-controlled bath for 30 min at 30 °C for fat mixtures of 20% and 30% FHSBO and 25 °C for fat mixtures of 10% FHSBO, with and without salt. The setup conditions were aimed to keep the degree of supercooling (ΔT_s) which is the temperature difference between crystallization temperature (at atmospheric condition, T_s) and temperature of the pressure medium comparatively close between samples at mean value of 11.4 ± 1.7 °C. As shown in Table 1, the difference between the crystallization temperature of samples at 10% and 30% FHSBO was higher (~10 °C) compared to samples at 20% and 30% FHSBO (~3 °C). Sample temperature was recorded every 1 second using DasyLab software (Version 7.00.04, National Instruments Corp., Austin, TX, USA). The control experiments were conducted in duplicate and the crystallized samples were analyzed within one hour after the treatment. Except for the sample at 10% FHSBO (liquid state), the solid crystallized samples were carefully removed from the sample carrier and sliced into three parts. Sampling for physical and structural analysis of the samples were conducted using the center part of the slice.

3.5. Determination of Crystallization Temperature and Induction Time

Using the principles of thermometry analysis [42], the first derivative of the sample's temperature history (dT/dt) was plotted against pressure and temperature to identify the induction time of crystallization (t_c), and crystallization temperature (T_s), and pressure. The crystallization temperature

(T_s) is defined as the peak maxima of the first derivative of the sample's temperature history for samples crystallized during adiabatic compression (>300 MPa), as shown in Figure S1 (Supporting Information). For samples crystallized at atmospheric pressure and during isobaric cooling, the T_s is corresponded to the peak minima of the first derivative of the sample's temperature history. The induction time was defined as the time taken to reach T_s from the start of pressurization. Samples T_s and the corresponding pressure (*P*) can be related using a simple polynomial expression as follow:

$$T_s = a_0 P^2 + a_1 P + a_2 \tag{1}$$

The coefficients of regression of Equation (1), a_0, a_1, and a_2, was estimated by regression analysis using MINITAB version 16 statistical software (Minitab, Inc., State College, PA, USA). On the other hand, the relationship between the maximum temperature under pressure (T_{max}) and pressure was expressed as follow:

$$T_{max} = b_0 P + b_1 \tag{2}$$

where the regression coefficient of Equation (2), b_0 and b_1, was estimated by regression analysis using MINITAB version 16 statistical software (Minitab, Inc., State College, PA, USA).

3.6. Solid Fat Content (SFC)

About 3 g of the crystallized sample was placed into glass vials (Sigma-Aldrich, St. Louis, MO, USA) and the sample's SFC was measured by pulse nuclear magnetic resonance (p-NMR) spectrometer (Bruker Minispec mq20, Bruker Corporation, Billerica, MA, USA) at room temperature. The reported data corresponds to the average of five individual measurements.

3.7. Thermal Analysis

The thermal behaviors of the model fats at different solid mass fraction of FHSBO (10%, 20%, 30% *w/w*), with and without salt, was analyzed by differential scanning calorimetry (DSC) equipped with a refrigerated cooling system (Q2000, TA Instrument, New Castle, DE, USA) [23]. About 10 mg of each model fat was hermetically sealed in an aluminum pan, and an empty pan was used as a reference.

The thermal behavior of the high-pressure and atmospheric crystallized samples was studied by melting each sample (10 mg) in a sealed hermetic aluminum pan. The sample pan was heated from 20 to 80 °C at 5 °C/min. Sample melting temperatures (T_m), the onset of melting (T_o) and the enthalpy of melting (ΔH_m) were determined using TA Universal Analysis 2000 (Advantage Software v5.5.24/Universal Analysis Software, TA Instrument, New Castle, DE, USA).

3.8. Small Amplitude Oscillatory Rheological Measurement

Dynamic oscillatory measurements were conducted using a strain control rheometer (MCR 302, Anton Paar, Graz, Austria). Plate-plate geometry (20 mm) with sandblasted fixtures were used with normal force set at 3 ± 0.5 N for all measurements. The loading protocol consisted of applying an increasing normal force control (1–10 N), allowing the sample to relax the axial force to a constant value over a zero strain relaxation test for 15 min. Amplitude sweeps were performed imposing a constant frequency (ω) of 6.28 rad/s. The storage modulus (G') and loss modulus (G'') were determined within the linear viscoelastic region (LVR).

3.9. Microstructural Analysis

A polarized light microscope (Axio Imager.M2m, Carl Zeiss Microscopy GmbH, Jena, Germany) with high-resolution CCD camera was used to analyze the microstructure of the crystallized samples. To obtain satisfactory reproducibility of the microscope slides, a definite amount (~1 mg) of the crystallized sample was placed on each microscope slide, spread in all directions and gently secured with cover slips. Images were acquired using 20× and 50× objective lens and AxioVision software (Carl Zeiss Microscopy GmbH, Jena, Germany).

The threshold of the microstructure images at 20x magnification were manually set, converted to grayscale and analyzed for particle size distribution using Image J software (Version 1.50b, National Institutes of Health, Bethesda, MD, USA). The particle size distributions were fitted using the Gaussian function in PeakFit software (Seasolve, Framingham, MA, USA) to determine mean particle equivalent diameter and standard deviation from full width at half maximum of the Gaussian peak. The crystal cluster size was measured using the Circle feature of AxioVision software on images at 50× magnification. The program assumes a circular geometry and obtains the square root of the quotient of the area to π. Since the crystal lattice of sodium chloride is cubic and highly symmetric, it is not birefringence in nature. Thus, under polarized light microscopy, the salt particles appeared as black particles embedded in the crystalline matrix [38]. They can be clearly distinguished from the fat matrix by applying partial polarization.

3.10. Statistical Analysis

Reported values correspond to means with standard deviations of the measurements. Statistical analysis between means was carried out by one-way ANOVA ($p < 0.05$) with Tukey's multiple comparisons as a post-test ($p < 0.05$) using MINITAB version 16 statistical software (Minitab, Inc., State College, PA, USA).

4. Conclusions

The crystallization of binary mixtures of FHSBO and soybean oil under pressure were evaluated at different solid mass fraction and the addition of 1% salt. The investigated model fats showed a significant nonlinear increase of the crystallization temperature with pressure up to the pressure threshold when the onset of crystallization shifted from isobaric to adiabatic condition as the crystallization temperature reached the maximum temperature under pressure. The crystallization behaviors of different mixtures observed showed the domination effects of pressure levels on the induction time and size of the microstructure. At high-pressure levels, when crystallization was induced during an adiabatic condition, the short induction time resulted in the formation of a high number of small and homogenously distributed fat crystal structures. However, macrostructural properties were strongly influenced by the solid mass fraction of the mixtures. The enhancement effects of the high-pressure treatments on the SFC and shear storage modulus were observed at low solid mass fraction (10% FHSBO) with the creation of lipid gel. The promotional effects of salt were detected on the SFC, thermal, and rheological properties at high solid mass fraction especially beyond the pressure threshold. The results indicate that high pressure treatment could be used as a novel physical process to modify the structural properties of FHSBO/soybean oil blends to produce healthier lipid sources such as trans-free and low-saturated fat, tailoring to their specific functional properties and applications.

Supplementary Materials: The following are available online at http://www.mdpi.com/1420-3049/24/15/2853/s1, Figure S1: First derivative of temperature as a function of pressure during adiabatic compression at 600 MPa. Arrow indicates crystallization temperature (T_c) at local maximum of the first derivative of temperature curve.

Author Contributions: Conceptualization, M.Z., V.M.B., and F.M.; methodology, M.Z., F.M. and V.M.B.; data analysis, M.Z.; investigation, M.Z.; resources, V.M.B. and F.M.; data curation, M.Z. and F.M.; writing—original draft preparation, M.Z.; writing—review and editing, F.M. and V.M.B.; visualization, M.Z.; supervision, V.M.B. and F.M.; project administration, V.M.B.; funding acquisition, V.M.B.

Acknowledgments: The research was a joint contribution between The Food Safety Engineering and Lipid Analysis laboratories at The Ohio State University. M.Z. gratefully acknowledges the financial support from Ministry of Higher Education, Malaysian Government. Support in part provided by USDA National Institute for Food and Agriculture HATCH projects and the food industry is also gratefully acknowledged.

Conflicts of Interest: The authors declare no conflict of interest.

References

1. Sato, K. *Crystallization of Lipids: Fundamentals and Applications in Food Cosmetics and Pharmaceuticals*; Sato, K., Ed.; John Wiley & Sons Ltd.: Hoboken, NJ, USA, 2018; ISBN 9781118593929.
2. Lopes, J.D.; Grosso, C.R.F.; De Andrade Calligaris, G.; Cardoso, L.P.; Basso, R.C.; Ribeiro, A.P.B.; Efraim, P. Solid lipid microparticles of hardfats produced by spray cooling as promising crystallization modifiers in lipid systems. *Eur. J. Lipid Sci. Technol.* **2015**, *117*, 1733–1744. [CrossRef]
3. Stahl, M.A.; Buscato, M.H.M.; Grimaldi, R.; Cardoso, L.P.; Ribeiro, A.P.B. Structuration of lipid bases with fully hydrogenated crambe oil and sorbitan monostearate for obtaining zero-trans/low sat fats. *Food Res. Int.* **2018**, *107*, 61–72. [CrossRef] [PubMed]
4. Ribeiro, A.P.B.; Masuchi, M.H.; Miyasaki, E.K.; Domingues, M.A.F.; Stroppa, V.L.Z.; de Oliveira, G.M.; Kieckbusch, T.G. Crystallization modifiers in lipid systems. *J. Food Sci. Technol.* **2015**, *52*, 3925–3946. [CrossRef] [PubMed]
5. De Oliveira, G.M.; Ribeiro, A.P.B.; Kieckbusch, T.G. Hard fats improve technological properties of palm oil for applications in fat-based products. *LWT–Food Sci. Technol.* **2015**, *63*, 1155–1162. [CrossRef]
6. Pernetti, M.; van Malssen, K.F.; Flöter, E.; Bot, A. Structuring of edible oils by alternatives to crystalline fat. *Curr. Opin. Colloid Interface Sci.* **2007**, *12*, 221–231. [CrossRef]
7. Rogers, M.A. Novel structuring strategies for unsaturated fats–Meeting the zero-trans, zero-saturated fat challenge: A review. *Food Res. Int.* **2009**, *42*, 747–753. [CrossRef]
8. Ribeiro, A.P.B.; Grimaldi, R.; Gioielli, L.A.; Dos Santos, A.O.; Cardoso, L.P.; Gonçalves, L.A.G. Thermal behavior, microstructure, polymorphism, and crystallization properties of zero trans fats from soybean oil and fully hydrogenated soybean oil. *Food Biophys.* **2009**, *4*, 106–118. [CrossRef]
9. Svanberg, L.; Ahrné, L.; Lorén, N.; Windhab, E. Effect of sugar, cocoa particles and lecithin on cocoa butter crystallisation in seeded and non-seeded chocolate model systems. *J. Food Eng.* **2011**, *104*, 70–80. [CrossRef]
10. Dhonsi, D.; Stapley, A.G.F. The effect of shear rate, temperature, sugar and emulsifier on the tempering of cocoa butter. *J. Food Eng.* **2006**, *77*, 936–942. [CrossRef]
11. Maleky, F.; Smith, A.K.; Marangoni, A. Laminar Shear Effects on Crystalline Alignments and Nanostructure of a Triacylglycerol Crystal Network. *Cryst. Growth Des.* **2011**, *11*, 2335–2345. [CrossRef]
12. Zulkurnain, M.; Maleky, F.; Balasubramaniam, V.M. High pressure crystallization of binary fat blend: A feasibility study. *Innov. Food Sci. Emerg. Technol.* **2016**, *38*, 302–311. [CrossRef]
13. Delgado, A.; Kulisiewicz, L.; Rauh, C.; Benning, R. Basic aspects of phase changes under high pressure. *Ann. N. Y. Acad. Sci.* **2010**, *1189*, 16–23. [CrossRef] [PubMed]
14. Heremans, K. High pressure effects on proteins and other biomolecules. *Annu. Rev. Biophys. Bioeng.* **1982**, *11*, 1–12. [CrossRef] [PubMed]
15. LeBail, A.; Boillereaux, L.; Davenel, A.; Hayert, M.; Lucas, T.; Monteau, J.Y. Phase transition in foods: Effect of pressure and methods to assess or control phase transition. *Innov. Food Sci. Emerg. Technol.* **2003**, *4*, 15–24. [CrossRef]
16. Hiramatsu, N.; Inoue, T.; Suzuki, M.; Sato, K. Pressure study on thermal transitions of oleic acid polymorphs by high-pressure differential thermal analysis. *Chem. Phys. Lipids* **1989**, *51*, 47–53. [CrossRef]
17. Kościesza, R.; Kulisiewicz, L.; Delgado, A. Observations of a high-pressure phase creation in oleic acid. *High Press. Res.* **2010**, *30*, 118–123. [CrossRef]
18. Ferstl, P.; Gillig, S.; Kaufmann, C.; Dürr, C.; Eder, C.; Wierschem, A.; Russ, W. Pressure-induced phase transitions in triacylglycerides. *Ann. NY Acad. Sci.* **2010**, *1189*, 62–67. [CrossRef]
19. Yasuda, A.; Mochizuki, K. The behavior of triglycerides under high pressure: The high pressure can stably crystallize cocoa butter in chocolate. *High Press. Biotechnol.* **1992**, *224*, 255–259.
20. Greiner, M.; Reilly, A.M.; Briesen, H. Temperature- and pressure-dependent densities, self-diffusion coefficients, and phase behavior of monoacid saturated triacylglycerides: Toward molecular-level insights into processing. *J. Agric. Food Chem.* **2012**, *60*, 5243–5249. [CrossRef]
21. Roßbach, A.; Bahr, L.A.; Gäbel, S.; Braeuer, A.S.; Wierschem, A. Growth Rate of Pressure-Induced Triolein Crystals. *JAOCS J. Am. Oil Chem. Soc.* **2019**, *96*, 25–33. [CrossRef]
22. Yoshikawa, S.; Kida, H.; Sato, K. Promotional effects of new types of additives on fat crystallization. *J. Oleo Sci.* **2014**, *63*, 333–345. [CrossRef] [PubMed]

23. Yoshikawa, S.; Kida, H.; Matsumura, Y.; Sato, K. Adding talc particles improves physical properties of palm oil-based shortening. *Eur. J. Lipid Sci. Technol.* **2016**, *118*, 1007–1017. [CrossRef]
24. Higaki, K.; Koyano, T.; Hachiya, I.; Sato, K.; Suzuki, K. Rheological properties of β-fat gel made of binary mixtures of high-melting and low-melting fats. *Food Res. Int.* **2004**, *37*, 799–804. [CrossRef]
25. Zulkurnain, M.; Balasubramaniam, V.M.; Maleky, F. Thermal Effects on Lipids Crystallization Kinetics under High Pressure. *Cryst. Growth Des.* **2017**, *17*, 4835–4843. [CrossRef]
26. Tefelski, D.B.; Kulisiewicz, L.; Wierschem, A.; Delgado, A.; Rostocki, A.J.; Siegoczyński, R.M. The particle image velocimetry method in the study of the dynamics of phase transitions induced by high pressures in triolein and oleic acid. *High Press. Res.* **2011**, *31*, 178–185. [CrossRef]
27. Hiramatsu, N.; Inoue, T.; Sato, T.; Suzuki, M.; Sato, K. Pressure effect on transformation of cis-unsaturated fatty acid polymorphs. 3. Erucic acid (cis-ω9-docosenoic acid) and asclepic acid (cis-ω7-octadecenoic acid). *Chem. Phys. Lipids* **1992**, *61*, 283–291. [CrossRef]
28. Siegoczyński, R.; Kościesza, R.; Tefelski, D.B.; Kos, A. Molecular collapse–modification of the liquid structure induced by pressure in oleic acid. *High Press. Res.* **2009**, *29*, 61–66. [CrossRef]
29. Ferstl, P.; Eder, C.; Ruß, W.; Wierschem, A. Pressure-induced crystallization of triacylglycerides. *High Press. Res.* **2011**, *31*, 339–349. [CrossRef]
30. Maleky, F.; Acevedo, N.C.; Marangoni, A.G. Cooling rate and dilution affect the nanostructure and microstructure differently in model fats. *Eur. J. Lipid Sci. Technol.* **2012**, *114*, 748–759. [CrossRef]
31. DeMan, L.; DeMan, J.M.; Blackman, B. Polymorphic behavior of some fully hydrogenated oils and their mixtures with liquid oil. *J. Am. Oil Chem. Soc.* **1989**, *66*, 1777–1780. [CrossRef]
32. Rousseau, D.; Hodge, S.M.; Nickerson, M.T.; Paulson, A.T. Regulating the β′→β polymorphic transition in food fats. *J. Am. Oil Chem. Soc.* **2005**, *82*, 7–12. [CrossRef]
33. Sevdin, S.; Yücel, U.; Alpas, H. Effect of High Hydrostatic Pressure (HHP) on Crystal Structure of Palm Stearin Emulsions. *Innov. Food Sci. Emerg. Technol.* **2017**, *42*, 42–48. [CrossRef]
34. Lida, H.M.D.N.; Ali, A.R.M. Physicochemical characteristics of palm-based oil blends for the production of reduced fat spreads. *JAOCS J. Am. Oil Chem. Soc.* **1998**, *75*, 1625–1631. [CrossRef]
35. Omonov, T.S.; Bouzidi, L.; Narine, S.S. Quantification of oil binding capacity of structuring fats: A novel method and its application. *Chem. Phys. Lipids* **2010**, *163*, 728–740. [CrossRef] [PubMed]
36. Wassell, P.; Young, N.W.G. Food applications of trans fatty acid substitutes. *Int. J. Food Sci. Technol.* **2007**, *42*, 503–517. [CrossRef]
37. Bouzidi, L.; Omonov, T.S.; Garti, N.; Narine, S.S. Relationships between molecular structure and kinetic and thermodynamic controls in lipid systems. Part I: Propensity for oil loss of saturated triacylglycerols. *Food Funct.* **2013**, *4*, 130–143. [CrossRef] [PubMed]
38. Murphy, D.B. *Fundamentals of Light Microscopy and Electronic Imaging*; John Wiley & Sons Ltd.: Hoboken, NJ, USA, 2002.
39. Bot, A.; Veldhuizen, Y.S.J.; den Adel, R.; Roijers, E.C. Non-TAG structuring of edible oils and emulsions. *Food Hydrocoll.* **2009**, *23*, 1184–1189. [CrossRef]
40. Himawan, C.; Starov, V.M.; Stapley, A.G.F. Thermodynamic and kinetic aspects of fat crystallization. *Adv. Colloid Interface Sci.* **2006**, *122*, 3–33. [CrossRef]
41. American Oil Chemists' Society. Fatty acid composition by packed column gas chromatography. In *Official Methods and Recommended Practices of the AOCS*; AOCS Press: Champaign, IL, USA, 2011.
42. Wunderlich, B. *Thermal Analysis*, 1st ed.; Academic Press: New York, NY, USA, 1990.

Sample Availability: Samples of the binary model fats are available from the authors.

© 2019 by the authors. Licensee MDPI, Basel, Switzerland. This article is an open access article distributed under the terms and conditions of the Creative Commons Attribution (CC BY) license (http://creativecommons.org/licenses/by/4.0/).

MDPI
St. Alban-Anlage 66
4052 Basel
Switzerland
Tel. +41 61 683 77 34
Fax +41 61 302 89 18
www.mdpi.com

Molecules Editorial Office
E-mail: molecules@mdpi.com
www.mdpi.com/journal/molecules